The Trichomycetes

Two thalli of *Asellaria ligiae*, a commensalistic fungus growing in the hindgut of an isopod, *Ligia italica*, on the island of Rab, Yugoslavia. Collected and prepared by Dr. Jan J. Kohlmeyer, who kindly provided the photomicrograph. Scale bar = 10 μm.

Robert W. Lichtwardt

The
Trichomycetes
Fungal Associates of Arthropods

With 85 Figures

Springer-Verlag
New York Berlin Heidelberg Tokyo

Robert W. Lichtwardt
Department of Botany, University of Kansas, Lawrence, KS 66045-2106, U.S.A.

On the front cover: Fig. 7.10 E. SEM micrograph of the holdfast of *Enterobryus* sp. from *Californiobolus uncigerus* (p. 56). (Kindly furnished by R.E. Reichle)

Library of Congress Cataloging in Publication Data
Lichtwardt, Robert W.
  The trichomycetes, fungal associates of arthropods.
  Bibliography: p.
  Includes index.
  1. Trichomycetes. 2. Arthropoda—Parasites.
I. Title.
QK621.A1L53   1986   589.2′58   85-27964

© 1986 by Springer-Verlag New York Inc.
All rights reserved. No part of this book may be translated or reproduced in any form without written permission from Springer-Verlag, 175 Fifth Avenue, New York, New York 10010, U.S.A.
The use of general descriptive names, trade names, trademarks, etc., in this publication, even if the former are not especially identified, is not to be taken as a sign that such names, as understood by the Trade Marks and Merchandise Marks Act, may accordingly be used freely by anyone.

Media conversion by David E. Seham Associates, Inc., Metuchen, New Jersey.
Printed and bound by Halliday Lithograph, West Hanover, Massachusetts.
Printed in the United States of America.

9 8 7 6 5 4 3 2 1

ISBN 0-387-96237-9 Springer-Verlag New York Berlin Heidelberg Tokyo
ISBN 3-540-96237-9 Springer-Verlag Berlin Heidelberg New York Tokyo

Preface

Associations and interactions between species of organisms are phenomena shared by all living things. What varies is the extent to which the more long-lasting interactions are beneficial or destructive to a given species and the degree of intimacy and reliance which one organism may have developed in association with another. Many of the more highly evolved relationships that have been studied involve microorganisms, either in consort with other microorganisms or with so-called higher forms of life. Mycologists are rarely surprised—but often fascinated—by the variety of kinds of living substrates and specialized organismal relationships that evolutionary processes have produced among the fungi. The present book deals in some detail with the specialized dependence of a unique group of fungi, the trichomycetes, upon certain arthropods.

There has been no comprehensive and worldwide treatment of the trichomycetes since their discovery by Joseph Leidy in 1848. The literature is scattered and in several languages, and many articles are now not only a bit old but out of date as well. As in many areas of biology, our knowledge about trichomycetes has increased somewhat exponentially in recent years. That is not to say that the time has arrived when *the* definitive monograph can be written, for there is much about the biology and evolution of these gut fungi and their arthropod hosts that cannot be known at present. One of my hopes, in truth, is that this book will encourage more biologists to search for the intriguing answers yet to be revealed. I have attempted to make the present monograph sufficiently self-contained and useful so as to be a firm and reliable foundation upon which to build new knowledge without an excessive need to go to the primary literature.

The monograph is divided into two major sections, one covering the various known biological aspects of the trichomycetes and their hosts, the other one being a systematic treatment of the fungi. I have tried to bring together in one book all pertinent published information on trichomycetes, and therefore it summarizes the research efforts of many biologists. Nevertheless, included are many original observations, opinions, data, and taxonomic information that have not previously appeared in print. Although no new taxa are to be found in this treatise, I have in some cases altered or emended description of species, or extended the geographic range and kinds of hosts infested by the fungi, on the basis of unpublished data.

One of the most time-consuming efforts on my part has been the sorting out of the good species from the invalid, illegitimate, or questionable ones. More named species have been discarded for these reasons than have been accepted. I have had the opportunity to examine living material, collected by myself or with collaborators, of more than half of the species recognized in the formal taxonomic treatment in Chapter 11. I have made only limited use of type specimens for two reasons. One is that designated types of valid species prior to 1958 are nonexistent. The other reason is that where types do exist, and are available, they usually consist of preserved slides which, in most cases, have little value in resolving particular taxonomic problems such as the kind outlined in Chapter 10. Mlle. Manier kindly made available during my visits to her laboratory a few slides or unpublished photographs of some species of trichomycetes about which I had some question. Throughout this book I have used the word **trichomycetes,** with a lower case **t,** when referring to the gut fungi as a biological group. This seems preferable in view of the possibility that the class Trichomycetes is polyphyletic and therefore would be considered by some as a convenient, but not a natural, grouping of fungi.

This monograph culminates many enjoyable years of research on trichomycetes conducted both in the field and the laboratory and made even more stimulating by the active contributions of various collaborators and graduate students. To all of them, whom I acknowledge in various parts of this book, I am most grateful. In particular, I would like to mention the following persons with whom I have had many lively discussions about trichomycetes: J.–F. Manier, S.T. Moss, H.C. Whisler, Y. Kobayasi, H. Indoh, and M.C. Williams. The fieldwork undertaken while away from my university in most cases has required access to nearby laboratory facilities for doing dissections of living specimens, microscopy, culture work, EM preparations, and the like. Persons at numerous field stations and other institutions in the United States and abroad kindly provided the necessary facilities, and are acknowledged in my separate publications. As well, I wish to call attention to the many biologists who have identified arthropods (an often tedious and sometimes impossible task with some larval forms); they, likewise, have been named and acknowledged in my publications.

The research which led to the present monograph was in large part supported over many years by the National Science Foundation (current grant no. BSR–8019724). The monograph could not have been written, and would not have been attempted, without the background of experience that NSF made possible. Also, a number of my graduate students who contributed materially to our knowledge of trichomycetes have had direct or indirect support from NSF. My sincere appreciation goes to this fine organization and its dedicated personnel and advisors. Of course, any opinions, findings, and conclusions or recommendations expressed in this publication are mine and do not necessarily reflect the views of the National Science Foundation.

November 1985   Robert W. Lichtwardt

Contents

PART I INTRODUCTION

1. Summary of Trichomycete Characteristics 3
2. Historical Résumé 6
3. Methods 11
   Collection of Hosts 11
   Maintenance and Preservation of Hosts 14
   Dissection Techniques 16
   Preparation of Specimens for Microscopy 19
   Axenic Culturing 21

PART II BIOLOGY

4. Arthropod Hosts and Habitats 29
5. Geographic Distribution 35
6. Host Specificity 38
7. Morphology, Cytology, and Fine Structure 43
   Life Cycles 43
   Thallus Types and Development 48
   Holdfasts 55
   Asexual Reproduction 68
   Sexual Reproduction 87
   Viruses 94
8. Host–Fungus Relationships 95
   Nutritional Relationships 96
   Pathogenicity 97
   Ecdysis: Effects on Fungi 98
   Survival and Distribution Mechanisms 101

x   Contents

9. Experimental Studies on Cultured Species 105
   Nutrition 106
   Growth Dynamics and Culture Conditions 107
   Sterol and Lipid Production 110
   Effects on Host Development 112
   Sporulation 116
   Spore Germination and Longevity 120
   Host Inoculations 126
   Amoebagenesis 126
   Wall Composition 127
   Serology 128
   RNA Molecular Weights 129
   Abnormalities in Cultures 130

Part III Systematics

10. Taxonomic Problems 135

11. Taxonomic Treatment 138
    Keys to Orders and Families of Trichomycetes 138
    Trichomycetes 139
      Harpellales 139
        Harpellaceae 139
        Legeriomycetaceae 152
      Asellariales 199
        Asellariaceae 200
      Eccrinales 209
        Eccrinaceae 210
        Palavasciaceae 255
        Parataeniellaceae 258
      Amoebidiales 262
        Amoebidiaceae 262
      Excluded and Doubtful Taxa 269

12. Phylogeny 274
    Relationships with Other Fungi 275
    Relationships Among the Trichomycetes 279
    Morphologically Similar but Unrelated Organisms 284

Part IV Appendices

A. List of Fungi and Their Arthropod Hosts 291
B. List of Arthropods and Their Fungal Associates 297
C. Axenic Isolates of Trichomycetes and Their Sources 310

References 313

Glossary 329

Index 333

Part I Introduction

CHAPTER 1

Summary of Trichomycete Characteristics

Fungi belonging to the class Trichomycetes live hidden within the digestive tract of several types of insects and other arthropods. On dissection of the gut, they may be seen as small unbranched or branched fungal bodies (thalli) firmly attached to the gut lining and lying within the gut lumen from which they obtain their nutrients. Many are minute and require suitable preparation to be seen with the microscope. Some are larger, so that in guts that have been cut open they may be observed, with the unaided eye or with a low–power dissecting microscope, growing as individual thalli or in dense clusters. In the aggregate, longer thalli may make the inside of the gut look fuzzy or hairy, hence the name trichomycetes ("hair fungi"). In a very few species, part of the fungus may protrude from the anus in such a way that its presence can be detected without cutting open the gut. One trichomycete *(Amoebidium parasiticum)* grows externally on the exoskeleton of a variety of aquatic arthropods, usually daphnids and insect larvae, but the thalli are small; even when they are numerous, some magnification may be required to distinguish the trichomycete from bacteria and protozoans that sometimes occupy similar sites.

Regardless of where trichomycetes are attached to their respective hosts, they have an obligate association with them. That is, the fungi are not capable of growing, metabolizing, or reproducing in the natural environment other than in association with their living hosts, at least insofar as we currently understand these microorganisms. In a few instances released propagules functioning as dissemination agents may undergo some development while separated from their arthropods. These include the amoebae and cysts of the Amoebidiales and the resistant primary infestation spores of some Eccrinales.

The existence of these gut fungi in many groups of commonly encountered arthropods is not generally realized, nor is the fact that they have been found on all continents of the world (except Antarctica) as well as on many larger and smaller islands. As presently known, their distribution ranges from north of the Arctic Circle to south of the Tropic of Capricorn and from below sea level to mountain altitudes as high as the larval stages of some insects are capable of development.

The majority of arthropods known to be associated with these gut fungi live in freshwater habitats. They include, among others, larvae of a number of families of flies (Diptera), mayflies (Ephemeroptera), and stoneflies (Plecoptera), and adult stages of some isopods, amphipods, cladocerans, and copepods (Crustacea), and even a few species of springtails (Collembola). Certain arthropods from terrestrial habitats, especially millipedes (Diplopoda), but including some beetles (Coleoptera), isopods, and amphipods, are frequently infested with trichomycetes. Trichomycetes in marine habitats are known to be associated with several kinds of true crabs (Brachyura) and shrimplike crabs (Anomura), as well as isopods and amphipods. Most of these marine animals live in the intertidal or subtidal zones; others live on higher reaches of marine shores.

Virtually all of the arthropod hosts of trichomycetes are detritivores, scavengers, or consumers of living aquatic algae, and they all have chewing mouthparts (are mandibulate). Most of the fungal species live in the hindgut region; some attach to the peritrophic membrane in the midgut of Diptera larvae; a few grow in the foregut of crustaceans. They exhibit a greater or lesser degree of host specificity, but all develop spore types to ensure successful host–to–host transmission. Further, they have adapted in special ways to integrate their development with that of their arthropod hosts, including dealing with the molting process that all arthropods periodically undergo during growth and which in turn causes the fungi to be expelled from the gut.

Although trichomycetes are obligately associated with their particular hosts, the reverse is not necessarily true. Natural populations of arthropods that might be infested can be devoid of trichomycetes, or they may have a low percentage of infested individuals. In some species of arthropods, such as certain millipedes and passalid beetles, virtually all adult individuals contain trichomycetes in their guts. Because these gut fungi do not usually affect their hosts adversely or beneficially in any obvious way, they are often called commensals. Nevertheless, there are several reports, one well substantiated, that at least one species of the fungal genus *Smittium* (Harpellales) can cause high mortality to mosquito larvae cultured in the laboratory. On the other hand, there is limited experimental evidence that mosquito larvae grown on sterol– and vitamin–deficient media develop faster and go through more instars when they are infested with another species of *Smittium*. Thus, there may be a range of symbiotic relationships between these fungi and their arthropod hosts. This important biological question has not been addressed to the extent it should be, in part because only about 6% of the recognized species of trichomycetes have been cultured axenically, and those that have been represent only two of the seven trichomycete families.

TABLE 1.1. Summary of trichomycete orders.

Harpellales	Reproduce by trichospores and zygospores; branched or unbranched thalli attached to hindgut cuticle or peritrophic membrane of immature aquatic insects
Asellariales	Reproduce by disarticulating cells (arthrospores); branched thalli attached to hindgut cuticle of freshwater, marine, or terrestrial isopods and insects (Collembola)
Eccrinales	Reproduce by sporangiospores; unbranched coenocytic thalli attached to cuticle of hindgut or foregut of terrestrial, marine, or freshwater crustaceans, diplopods, or insects
Amoebidiales	Reproduce by amoebae and cystospores, or by sporangiospores; unbranched coenocytic thalli attached to hindgut cuticle of freshwater insect larvae or to exoskeleton of freshwater arthropods

The class Trichomycetes is placed in the Zygomycotina, among the lower fungi. There are four orders according to the current classification (see Table 1.1). It has not been determined to everyone's satisfaction that all of the orders are phylogenetically related, and there is a possibility that the Trichomycetes may not be a natural class. Included in this treatise are 7 families, 39 genera, and 131 species. They are morphologically and ecologically quite distinct from all other fungi. The known trichomycete species probably represent only a fraction of the actual extant taxa.

CHAPTER 2

Historical Résumé

Contributions to our understanding of trichomycetes have come from a large number of individuals, as can be confirmed by perusing the list of references at the end of this treatise. The historical sketch that follows will be limited to the relatively few individuals who have made especially notable contributions or who have in some way influenced concepts or helped to establish trends within the field. The reader is referred to publications by Duboscq et al. (1948), Manier (1950, 1969b), and Moss (1972) for additional comments of a historical nature, particularly concerning the earlier developmental years. The writer is indebted to Mlle. Manier for some of the biographical information in this chapter. Manier (1969b) offered the opinion that trichomycete studies can be conveniently divided into four periods of development, an opinion that this writer shares and follows in his presentation. These periods are defined not so much by the span of contributions by individual investigators as they are on new discoveries and emphases within each time frame.

The period 1848 to 1904 was initiated by Joseph Leidy's discovery of several species of Eccrinales within the hindguts of millipedes and a beetle and publication of their descriptions under the generic name *Enterobryus* (and *Eccrina,* now a synonym). Leidy, a noted American naturalist with a talent for investigating new organisms ranging from bacteria and nematodes to fossil mammals, thought these eccrinids were colorless algae related to the Confervaceae. Thus began a conflict on the concepts of relationships of such gut inhabitants, now known as the Trichomycetes, which is still not fully resolved (see Chapter 12). Charles Robin (1853), in his classic book on "plants" that parasitize man and animals, discovered in France another species of *Enterobryus,* but thought Leidy's organisms and his might be related to the fungal order Saprolegniales. Shortly thereafter, in the 1850s, a different kind of arthropod associate, which lived on the exoskeleton of a variety of aquatic arthropods, was discovered in Europe by Lieberkühn and, independently, by Schenk; it was named *Amoebidium parasiticum* by Cienkowski a few years later. This organism was variously ascribed to the algae, lower fungi, or protozoans by these and other biologists of that period. In 1895, Hauptfleish described a second genus of Eccrinales, *Astreptonema*, from the hindgut of an amphipod, and suggested his new species belonged to the Saprolegniaceae. This ended the first period, covering more than half a century, wherein two different groups but very few species of these unusual organisms were found, studied, and debated.

The period 1905 to 1928 was marked by the discovery of numerous other Eccrinales, primarily by French protozoologists, most of whom were interconnected through their professional interests and careers. The most notable collaborators of this period were L. Léger and O. Duboscq; the latter had been Léger's student at Grenoble. In 1905 the two initiated a series of morphological and taxonomic studies on eccrinids from marine crustaceans, millipedes, and hydrophilid beetles. These and subsequent studies by Léger and Duboscq eventually culminated in their 1948 monograph, published posthumously with Odette Tuzet. In 1906(a), Edouard Chatton began the most thorough investigations of the time on the biology of *Amoebidium* species, including a few experimental studies on host specificity and the effects of environmental parameters on morphological development. In 1927, Raymond Poisson started publishing on new genera and species of eccrinids from amphipods and isopods, and his later studies through the mid-1930s were expanded to include other groups of trichomycetes as well. The major contributions from 1905 to 1928 by these French biologists were the discovery and study of species of Eccrinales and Amoebidiales, the establishment of some early classification systems for the known species, and the consensual belief that they were some kind of fungi. The only mycologist to study any of the gut fungi during this period was Roland Thaxter, who described *Enterobryus compressus* from the passalid beetle.

The period 1929 to 1959 began with the discovery of the first species of Harpellales, *Harpella melusinae*, by Léger and Duboscq. That same year (1929) they also named the only endocommensal genus of Amoebidiales, *Paramoebidium*, and Poisson described the interesting eccrinid genus *Parataeniella* (Parataeniellaceae). Two years later Poisson discovered the genus *Asellaria*, which is now placed in the last of the four orders of trichomycetes, the Asellariales. Léger began a series of publications in 1932 with Marcelle Gauthier, who had been his assistant at Grenoble, describing many new genera of branched Harpellales (Legeriomycetaceae), and Gauthier by herself published on several new taxa of this order. By 1947, when Tuzet and Manier discovered the eccrinid genus *Palavascia* (Palavasciaceae), representatives of all presently recognized families had been described.

The years 1929 to 1959 were primarily a period of description and enlargement of the number of known species and host types, as well as studies on the morphology and ecology of these fungi. There was a growing realization that the trichomycetes had a worldwide distribution, thanks to the curiosity of many scattered investigators who became aware of their existence. This awareness was aided by the publication of the first comprehensive treatment of the eccrinids (considered then to be the Eccrinales and Amoebidiales) by Duboscq, Léger, and Tuzet in 1948. They introduced the term Trichomycetes and provided a classification of all genera recognized at the time, but the body of the monograph excluded what are now called the Harpellales and Asellariales. Odette Tuzet had served as Duboscq's assistant at the Mediterranean laboratory of the University of Paris at Banyuls, and some years later she became head of the invertebrate zoology laboratory at the Université de Montpellier (later to become part of the Université des Sciences et Techniques du Languedoc) where Duboscq had been located from 1903 to 1926. Duboscq died in 1943, and using in part his extensive notes, Tuzet and Léger completed the 1948 monograph. Léger passed away in Grenoble a few months before it came off the press.

Jehanne–Françoise Manier in 1950 (1951) published the thesis of her Docteur d'État, which she earned from the University of Paris under Mlle. Tuzet. It was a major publication on trichomycetes, for it covered all groups of these fungi. Mlle. Manier worked in Tuzet's laboratory in Montpellier, in time becoming a Maître de Recherches of the Centre National de Recherches Scientifiques. Tuzet and Manier started publishing a series of jointly authored papers in 1947, and this continued for two decades. After 1950 Manier, by herself and with her students and other collaborators, published extensively on many aspects of the biology of trichomycetes, and she was undoubtedly the most influential investigator of these fungi in Europe on her retirement in 1981.

The period 1960 to 1984 represents a time of considerable change and maturation in trichomycete studies. In 1960 Howard C. Whisler, then a graduate student at the University of California at Berkeley, reported the successful isolation of the first trichomycete, the ectocommensal *Amoebidium parasiticum*. These axenic isolates permitted Whisler to conduct the first *in vitro* experiments on the nutrition of *Amoebidium* and to study under controlled conditions the factors that lead to amoebagenesis. Three years later the entomologists Clark, Kellen, and Lindegren in the Fresno mosquito laboratory in California isolated the first two species of the endocommensal genus *Smittium* from mosquito hindguts. This was followed by Lichtwardt's isolations of many strains of several species of *Smittium* from various kinds of dipteran larvae and, by Peterson, of species of two monotypic genera, *Genistelloides* and *Capniomyces,* from stonefly nymphs. The *Smittium* isolates have served as the basis for a variety of experimental studies, reviewed in Chapter 9.

Lichtwardt's interest in the trichomycetes began with the Eccrinales when he was a graduate student at the University of Illinois, and his investigations expanded to all of the families of trichomycetes after he joined the faculty of the University of Kansas. There he was fortunate to direct the studies of a number of graduate students who contributed materially to our understanding of the biology of trichomycetes, primarily through the use of experimental methods, but also by investigating the fine structure of cultured and uncultured species. These students, listed chronologically, were David F. Farr, Mary E. Chapman, Vijay K. Sangar, Marvin C. Williams, A.M. El–Buni, Thaddeus R. Preisner, Anne M. Starr, Si–nan Dang Mayfield, Bruce W. Horn, and Stephen W. Peterson.

Early in the 1960s, before axenic cultures of the gut fungi were generally available, Manier, Tuzet, and collaborators began laboratory studies on cross–infestation and host specificity of trichomycetes (Chapter 6). In the mid 1960s electron microscopic studies began with Farr's work on cultured *Smittium culisetae* and electron micrographs of *Enterobryus oxidi* by Tuzet, Manier, and Oustau. Such studies were extended very soon to many other species by investigators in France, the United States, and England. Stephen T. Moss was a student at the University of Reading in the early 1970s when he completed his dissertation on the fine structure, ecology, and general biology of trichomycetes, and he has since provided some of the most elegant electron micrographs and interpretative studies of some of these fungi. Moss spent 18 months in the mycology laboratory at the University of Kansas after earning his degree, and this led to several collaborative studies with Lichtwardt that continued after Moss's return to England.

By 1960 the taxonomy of trichomycetes had become confusing, if not rather chaotic, for the literature contained many illegitimate type species and ill–defined or synonymous genera. Manier and Lichtwardt attempted to rectify this problem by publishing a manuscript in 1968 (1969) that dealt with the trichomycetes to the generic level. In 1969(b) [1970(b)] Manier published a paper on the Trichomycetes of France, and it has remained the major monograph on the systematics of these fungi prior to the present worldwide treatise. Aspects of the biology of the trichomycetes were reviewed by Lichtwardt in 1976 and by Moss in 1979. Another publication of note in the 1970s was Jolly Hibbits' (1978) publication on marine Eccrinales, based on her Master's thesis done under the direction of Whisler at the University of Washington.

The consequences of the studies from 1960 to date include a better appreciation of the complexities of the relationships between the trichomycetes and their arthropod hosts (Chapter 8). Sweeney's (1981a) discovery and axenic isolation of *Smittium morbosum* substantiated earlier reports from Italy and Russia that some species were highly lethal to mosquito larvae in laboratory cultures, and at the same time demonstrated that within the wide range of trichomycete species not all were necessarily benign. The biological impact of these associations, both deleterious and beneficial, are being elucidated. Some plausible concepts regarding the evolution of the trichomycetes from other fungi are beginning to emerge (Chapter 12), but the phylogenetic relations within the group remain controversial. Today it is probable that many—even a majority—of trichomycete species remain undescribed. Some of those that have been recorded are poorly understood. The physiological and structural adaptations that have allowed these gut fungi to associate successfully with such a wide range of host types are just beginning to be revealed, as are those features that are unique and those that have a commonality with other organisms. The history of trichomycete investigations is hardly at an end.

CHAPTER 3

Methods

Collection of Hosts

Except for the few species available in axenic culture, trichomycetes must be studied from material obtained in arthropod hosts. The wide range of host types, size, and habitats requires many different collecting methods. In this section some of the techniques we have found to be useful will be described. Living hosts are more easily dissected than preserved ones and, more importantly, provide superior material for study. Consequently, not only must the arthropods be found and collected, but they need to be kept in a living condition until dissected in the laboratory. Freshly dead specimens may be used. However, as decomposition progresses the thalli degenerate rapidly and sporulating parts may produce artifacts that can lead to misinterpretations. Molts of hosts should also be collected, because some stages of the fungi such as zygospores and resistant spores may develop only during ecdysis. A list of trichomycete species arranged by host types can be found in Appendix B.

Hosts of trichomycetes can be handled safely without special precautions, with two exceptions. One obvious group consists of those crustaceans with large chelae such as the crabs. The other group is the millipedes, which, when disturbed, may secrete a noxious liquid from glands located laterally along the body. This substance contains cyanide in some species and can stain the skin and has an unpleasant odor, but is not harmful provided it does not contact membranes of the mouth or eyes.

Freshwater hosts are more common in flowing waters than in quiet ponds and pools. They consist mostly of immature insects attached to stones and sticks or to immersed living or dead vegetation, and include the larvae of blackflies and midges and nymphs of mayflies and stoneflies, among others. Their trichomycetes are species of Harpellales and Amoebidiales (*Paramoebidium*). Clean, small streams are generally preferable to larger rivers. The insects often will be located in riffles and other parts of the stream where water flows rapidly and provides good aeration. Infested larvae of Chironomidae, Dixidae, etc. may be found along waterfalls where they are bathed by running or trickling water. The Dixidae also occur on aquatic vegetation where the water interfaces with air. Winter–emerging stonefly nymphs (Capniidae) are probably in diapause and hyporheic (deep in the substratum) most of the year; they can be collected in the winter and early spring in leaf packs or on sticks and stones in flowing streams when they come up from the substrate to complete their development. Amphipods and insect larvae may live in aquatic vegetation or decomposing plant material near the border of streams, or in pools, and isopods are found on rocks and submerged plants and decomposing vegetation. Some of these crustaceans are hosts to species of Eccrinales.

A pair of fine–pointed (but not sharp) forceps is useful to pick up arthropods from stones and sticks lifted from the water. Mayfly and stonefly nymphs on the underside of rocks often drop into the water when disturbed; consequently, stones should be turned over quickly and, preferably, held over a pan to catch those that do drop off. A net positioned downstream can also be used to catch nymphs that release themselves. A white–enamelled pan is useful to examine the emptied contents of nets or strainers that have been swept through aquatic vegetation, or to hold plucked vegetation until blackfly and chironomid larvae are transferred to jars. Stones may have chironomid larvae visible on the surface or concealed in tubes. Others are found in the algae that coat some rocks, or in decaying wood. Blackfly larvae may occur in profusion on rocks or submerged leaves in streams, but care should be taken not to scrape them off in masses, because under these conditions they become entangled in the silky threads spun from their salivary glands and do not survive long.

Insects from fast–flowing streams should be placed in collecting jars without crowding and with a minimum amount of water. As soon as possible these jars should be placed on ice. Jar lids can be tightened during transport to prevent loss of specimens, but the lids should be opened occasionally to insure adequate aeration. It is generally best to separate the families of insects, and care should be taken that predaceous forms are not included. Aquatic arthropods that contain trichomycetes are not carnivorous, but under crowded conditions in time some may resort to cannibalism. On arrival at the laboratory, the specimens should be further sorted and placed in shallow layers of prechilled stream or distilled water in containers with loose lids such as petri dishes, and kept in a refrigerator. Depending on the species, the care in their handling, and transportation conditions, insects can be maintained alive in a refrigerator without additional food for several days to several weeks.

Arthropods from lakes, pools, and other quiet waters that are likely to contain Harpellales include the larvae of mosquitoes and midges. In general, ephemeral pools and ditches are less likely to have trichomycetes than more permanent bodies of water, even though the appropriate arthropods may be present. One of the most useful tools we have found to collect insects from aquatic vegetation or muddy bottoms is a round–bottomed, woven–metal food strainer, 15 cm or less in diameter, which we use with the two metal prongs (at the edge opposite the handle) bent backwards against the strainer. Contents of such a strainer, or a net, can be emptied into a pan for examination. Smaller specimens can be removed and transferred to jars using a small pipette fitted with a bulb, larger specimens with forceps. Bloodworms (*Chironomus* spp.) build tunnels in muddy bottoms of pools, ditches and even slow–running streams, and if the mud is disturbed they can be scooped from the water. They can sometimes be removed effectively from concentrated populations by slowly dragging a small wire loop (~8 cm in diameter) attached to a handle through the mud, which results in the larvae becoming looped around the wire, or by simply dredging with a net. Other midges with trichomycetes are sometimes found in small transparent tunnels, which they build on the surfaces or within the senescent tissues of hydrophytes. Cladocerans are best collected with a plankton net. Collembola (springtails) may be a challenge, but can be collected in a pan and transferred to a jar with water for transport and storage.

As the majority of known marine hosts live in or near the intertidal zone, collecting is easiest at low tide. Mud–burrowing crabs may be caught while they are walking on the surface, or removed from their burrows by inserting a trowel or spade through the mud so as to cut across the burrow and prevent their downward escape. Collecting anomurids such as *Callianassa* and *Upogebia* may require deep and considerable digging with a spade, or with a special device in Australia called a "yabbie" pump that consists of a cylinder with a plunger. Nonburrowing isopods and amphipods may seek shelter and moisture under stones, or under wood, seaweed, etc. deposited at the high tide line. Some of these animals that do not live within the water at high tide might more properly be classified as terrestrial, but they are tolerant of water with high salinity. The ubiquitous rock louse, *Ligia* spp., moves rapidly on rocky shores, seawalls, and docks, and may be difficult to catch. But specimens can often be collected from walls and pilings by placing a pail or plastic bucket under them into which they will fall when disturbed or when prodded from cracks. They are somewhat delicate and must be handled carefully. In rock piles, smaller rocks lifted quickly over a pail may allow catching several ligias at a time. Most marine arthropods can be kept for days in a cool place in a container with a shallow layer of seawater and some objects onto which they can crawl. Freshwater, rather than seawater, should be used to replenish the water as it evaporates in order to prevent excessive increases in salinity. Aerated seawater tanks may also be used, or fresh seawater can be used to replace the stale seawater once or twice a day.

Terrestrial arthropods with trichomycetes include millipedes, beetles, isopods, and amphipods. They are found predominantly in or around the moist, decomposing vegetation on which they feed. Populations may fluctuate considerably in number in areas where droughts occur. Some millipedes and beetles either normally inhabit logs or seek shelter in logs during drier periods, and can be located by tearing apart the decomposing wood. All terrestrial hosts can be placed in containers with some of their natural substrate for transporting to the laboratory for subsequent maintenance.

The time of year when trichomycetes can be obtained depends on habitat and climatic conditions. Especially important may be the life cycles of some of the insect hosts whose immature forms, but not pupal or adult stages, contain trichomycetes. The timing of collections is generally more critical in temperate climates with severe winters than it is in tropical or subtropical areas of the world where developmental stages of populations of multivoltine species may overlap considerably. Species of Harpellales are essentially restricted to immature insects. At high altitudes and latitudes, the larval stages may be found only during relatively short seasons. Nevertheless, the abundance and variety of larval insects may be so intense in many of these regions that they offer some of the best collecting of Harpellales. When infested hosts consist of adults as well as sexually immature stages of the same species (many crustaceans and millipedes), the fungi may be collected at any season when the hosts can be found. Marine habitats are usually more stable than freshwater and terrestrial ones in temperate climates, so that many marine trichomycetes can be obtained throughout the year. However, even in such cases there may be some variation in abundance and stages of development of the trichomycetes that is tied to environmental factors and molting cycles of the host (Hibbits, 1978).

Living arthropods with trichomycetes may sometimes be obtained from commercial biological supply houses or professional collectors, but such sources are usually unreliable because of unpredictable infestation among different populations. Daphnid and bloodworm cultures (the latter, for instance, available as fish food from merchants in Japan and possibly other countries) can yield excellent *Amoebidium parasiticum* infestation in some cases. Likewise, it is possible to obtain from commercial sources living hosts like the fiddler crab (*Uca* spp.) and spiroboloid millipedes containing species of *Enterobryus*. These may be satisfactory only if shipped soon after they are collected from natural sources.

Maintenance and Preservation of Hosts

Successful collecting expeditions invariably require that the living arthropods be maintained for at least a few days in the laboratory while they are being dissected and their fungi studied. Longer periods of maintenance may be desirable or necessary where collecting sites are distant from a laboratory, or when hosts must be available during experimental investigations involving them. With some species the percentage of infested individuals in laboratory populations can actually increase over time, whereas in others the infestation may decrease as the hosts molt and shed their fungi and then do not become reinfested, or if they are fed an unsuitable diet. Many arthropods molt in the laboratory, and thus provide a stage of fungal development that may not always be recoverable in nature.

Insect larvae and nymphs from streams can be kept alive in a refrigerator in shallow layers of water, as described previously. Those that remain alive may molt, pupate, or metamorphose into adults. The larvae of some dipterans cannot be identified to species easily, if at all; consequently, it may be helpful to the specialist to be given preserved specimens of pupae or adults that are known to have developed from immature forms identical to those that have been dissected.

Hosts from quiet waters include dipteran larvae and cladocerans. Most can be kept at room temperature in jars half filled with water, or in a refrigerator if arrestment of the rate of host development is desired. Fungal transmission can be fairly good among crowded mosquito larvae (with *Smittium* spp.) and cladocerans such as daphnids (with *Amoebidium parasiticum*). If sample specimens in such collections are examined and found to be uninfested, the remainder can be left a few days to a week or more for subsequent examination on the possibility that infestation undetected in a few individuals has spread to others. It is best to remove the pupae of mosquitoes as they appear in order to avoid problems later with escaping adults; with one exception (Sweeney, 1981a), only larval stages of mosquitoes are known to harbor trichomycetes. Bloodworm larvae that live in muddy bottoms will survive better in shallow layers of water, either with or without mud. Infested cultures of cladocerans, in contrast to insect larvae, can be kept indefinitely under appropriate culture conditions.

Many marine hosts, such as hermit crabs, true crabs, and isopods, can be maintained satisfactorily in the laboratory provided there are facilities with running seawater or aquaria with aeration and filtration systems. Intertidal species should be provided with stones or other objects onto which they can climb. The author has kept fiddler crabs and isopods (*Sphaeroma quadridentatum* and *Ligia exotica*) for several months to more than a year in small aquaria with slowly recirculating filtered seawater, feeding them small amounts of rolled oats. The *Sphaeroma* isopods, however, lost their infestation by *Palavascia sphaeromae* under these artificial conditions (Lichtwardt, 1961b).

Most terrestrial arthropods can be kept in terraria for prolonged periods of time. Many species of millipedes and pill bugs (*Armadillidium*, etc.) will go through one generation after another if provided with ample decomposing leaves and wood, and will maintain their trichomycete infestation. The substrate must be kept moist, but not wet. Passalid beetles can survive for more than a year on decomposing wood. When collecting these kinds of arthropods, extra supplies of their natural substrates can be gathered and kept in plastic bags, and added to the terraria as needed. Substrates obtained from gardens or urban sites are sometimes contaminated with pesticides and should be selected with caution.

All hosts of trichomycetes can be satisfactorily preserved in 70% ethanol for use as vouchers or for subsequent identification. Specimens that have been dissected may be used for identification provided essential anatomical parts are not missing. Generally, however, it is better to preserve undissected specimens from the same collection for this purpose.

The author knows of no satisfactory method to kill and preserve hosts that are to be dissected and studied critically at a later date. Fixatives such as alcohol or formalin–alcohol harden the host tissues making dissection considerably more difficult, as well as causing the thalli of trichomycetes to become brittle and subject to damage, especially if they are intermixed with undigested contents of the gut lumen. Later complete dissections are done more easily if the guts are removed from the host and fixed in 10% lactophenol (Amman's medium). Reasonably satisfactory slides can be prepared from such material stored for a few weeks.

Dissection Techniques

Several methods can be employed to dissect the guts of arthropods, depending on the investigator's individual preference and objective. The methods described in this section have been found to be satisfactory for general trichomycete studies.

The digestive tract of arthropods is divided into three principal parts, which differ considerably in their anatomy, relative length, and functions among various groups of arthropods. The hindgut and foregut arise embryologically from the ectodermal layer, whereas the midgut has an endodermal origin. The hindgut and foregut are thus an invaginated extension of the exoskeleton, and are lined with a noncellular substance composed largely of chitin. It is to this cuticular layer, produced by the underlying epithelial cells, that the thalli of trichomycetes are normally attached. As arthropods grow in size or metamorphose and shed their outer integument, they also shed these gut linings including any attached thalli.

The midgut is the primary region of digestion and assimilation in many arthropods. In some groups the midgut is lined with a chitinous peritrophic membrane. Two major types are recognized. One of these is produced more or less continuously from a ring of cells at the anterior end of the midgut, and it consists of a transparent tube that is unattached to the epithelial cells of the midgut except at its origin. This type of membrane is found in several families of dipteran larvae and is the site of growth of most of the Harpellaceae. Young thalli attach and begin to grow at the anterior end of the peritrophic membrane and are slowly shifted posteriorly as the membrane grows, with the result that mature sporulating thalli occur near the posterior end of the membrane. Peritrophic membranes in some Diptera may extend some distance into the hindgut, where the membrane disintegrates. Dissection methods to recover the peritrophic membrane will be described later.

Fungal growth in the foregut region can be found in species of two genera of Eccrinaceae, *Arundinula* and *Enteromyces*. Their hosts are crabs, anomurans (shrimplike crabs), and crayfish. Species of *Arundinula* occur in the hindgut as well. The anomurans and most crabs live in marine or brackish habitats, consequently, they should be dissected in seawater or seawater diluted up to 50% with freshwater. Care should be taken that evaporation does not result in excessive buildup of salinity during dissection procedures. The carapace can be cut dorsally with scissors along two parallel lines, one on either side of the center, beginning near the eyes and extending down the abdomen to the telson. Removal of the skeletal strip between the cuts will expose the large stomach and the hindgut. The abdomen of crabs is folded ventrally under the body. It can be lifted backward and torn off; this piece will include the entire hindgut. Stomachs and hindguts can be excised and cut open using fine iris scissors with the aid of a dissecting microscope and flushed with seawater to remove intestinal debris. Crayfish and freshwater crabs should, of course, be dissected in freshwater. Preparation of the fungal material for microscopic examination will be described in the next section of this chapter.

The remaining genera of trichomycetes (except for the external species, *Amoebidium parasiticum*) live in hindguts. Those of millipedes are easily and quickly exposed by cutting off the anterior portion of the animal about one-quarter of the distance down from the head with a sharp razor blade or scalpel, then cutting off the last couple of segments near the anus. The posterior part of the hindgut is grasped with forceps and pulled out of the body. Alternatively, the gut can be removed by making lateral incisions on both sides of the body (Reichle, 1978). The junction of the hindgut with the midgut can be recognized by the presence of a sphincter muscle and by the zone of attachment of the Malpighian tubes. Starting with the anterior end, the hindgut is cut open carefully with fine scissors so as not to penetrate too deeply, and is flushed with water. In addition to Eccrinaceae, there may be numerous filamentous bacteria, protozoans, or nematodes within the hindgut. Thalli of *Enterobryus* spp. may attach to the cuticle of nematodes in some species of millipedes. Many Eccrinaceae select particular regions of the hindgut for attachment and growth, such as the very anterior part, whereas others may be found throughout its length.

The hindgut of beetles can be cut out of the abdomen after removal of the elytra and wings. However, the author has found it possible in most instances to remove their hindguts by firmly grasping with forceps and pulling at the anal plates of the intact beetle. Coleoptera often have looped hindguts that are structurally complex and long. Most isopod hindguts can also be removed by placing the host on its back and grasping the anal region and pulling. It may be desirable first to tear off some of the pleopods in those isopods in which they are large. With amphipods and nymphs of mayflies and stoneflies, the hindgut can be obtained by grasping the posterior segment and associated appendages and pulling; the unwanted parts can then be torn away from the hindgut with fine forceps. Some of these guts are too small to be cut open with scissors, and one may wish to use a minuten insect needle mounted in a handle, or some similar instrument of small size.

Aquatic larvae of blackflies, mosquitoes, chironomids, and other Diptera are among the smallest hosts of trichomycetes. Some measure only a few millimeters in length. The Collembola can be a special challenge to the dissector. Despite the small size of such insects, with experience it is often possible to prepare slides for microscopy in less time than it takes with some of the larger hosts. Dissections should be made under the lenses of a dissecting microscope with variable magnification. Either incident light oriented onto a dark, opaque background or light coming from beneath a glass stage can be used. The most valuable tools are high–quality fine–pointed jeweler's forceps and micro dissecting needles. The head and anal segment are cut off with a sharp razor blade, and the gut is removed from the posterior end of the body. The dissection should be made in a drop or two of water in a petri dish or on a slide.

With Diptera, careful removal of the hindgut often withdraws the peritrophic membrane as well, or the membrane can be withdrawn by careful removal with the head region. The peritrophic membrane may contain masses of algae and other ingested materials. These are conveniently removed by grasping one end of the peritrophic membrane with forceps and lifting it several times from the water until cleared of all loose materials. The membrane is sufficiently transparent that it need not be cut open, and it can be mounted directly on a slide for microscopic examination. It is necessary to remove the epithelium of the hindgut, however. A simple way to accomplish this with most small dipteran guts is to grasp the hindgut at one end with forceps and pull it gently through the partially spread tips of another pair of forceps so as to strip off part of the epithelium. The remaining epithelium can be stripped off by grasping the other end and repeating the procedure. If desired, the unopened gut can be examined microscopically for the presence of thalli before final preparation. The guts can be either torn open carefully with two pairs of forceps, or a micro needle can be used while one end of the gut is held in place.

Most small thalli of trichomycetes can be seen in opened guts at magnifications of $\times 30$ or lower, but because of their lack of pigmentation they do not always stand out against the background of the gut lining. With larger guts, where the chitinous lining is not easily separated from the epithelial tissue for quick examination at higher magnifications, it is possible to overlook small thalli. Before discarding apparently uninfested material, a drop or two of lactophenol cotton blue can be placed on the inner surface of the gut and then flushed off with water after a few seconds. Trichomycete thalli, if present, will stain sufficiently to be seen against a white background.

Preparation of Specimens for Microscopy

Standard methods for fixing and staining trichomycetes can be used for light microscopic studies, of course, but the author prefers to study living material with phase–contrast microscopy for most purposes. This produces fewer artifacts and has the additional advantage of permitting limited stages of development to be observed over time in some of the unculturable species. Water mounts on microscope slides may show maturation, release, or germination of spores in some species of Harpellales, Asellariales, and Eccrinales; or amoeba release, locomotion, and encystment in the Amoebidiales. Water mounts can be kept in moist chambers 1 to several days before all development ceases or the thalli begin to deteriorate. Cysts of the Amoebidiales can be kept for longer periods of time so as to obtain cystospores and cystospore release.

Arthropod molts and small intact gut linings with the epithelial layer removed can be mounted on slides in distilled water. Trichomycete material from some marine hosts can also be mounted in distilled water often without undesirable osmotic effects, although 50% seawater may be preferable if developmental stages are sought. With some larger guts, it may be difficult during the intermolt stage to separate the entire lining from epithelial tissue. If the lining adheres tenaciously, small pieces of gut lining can be peeled off for mounting, or individual intact thalli can be removed by careful manipulation with a fine needle.

It is sometimes desirable to be able to prepare slides containing all stages of the fungus from one gut. The following technique works well with foreguts and hindguts whose linings cannot otherwise be removed easily. The gut is placed to one side in a 100–mm petri dish with sufficient water to cover it. About 20 drops of full–strength lactophenol (~1 drop/ml of water) are added to the inside of the dish opposite the gut and allowed to diffuse through the water. After a few hours, or overnight, the lining usually can be peeled from the tissues with ease. It should be rinsed in dilute (10%) lactophenol, and the excess liquid drained off by holding the lining momentarily against the side of the dish, before spreading the lining on a slide in full–strength lactophenol with cotton blue. As much proteinaceous material as possible should be removed before final mounting, because its coagulation in lactophenol can produce an inferior slide.

Linings or thalli can be mounted directly from water into lactophenol cotton blue, using normal procedures for making slides of fungi. The coverslips can be sealed with clear fingernail polish or other preferred sealant. Small guts or thalli mounted in water on a slide can be permanently preserved without disturbing the delicate thalli by allowing excess water to evaporate from the edges of the coverslip, then placing a drop of lactophenol cotton blue on one edge and allowing it to infiltrate passively under the coverslip. The three clean edges can be sealed, and after the sealant has hardened, the edge with excess lactophenol can be washed with running water, dried, and sealed. This technique is especially useful when specimens have been studied with phase–contrast microscopy and photomicrographed, because the structural components can be relocated after a permanent slide has been prepared.

The use of fixatives and stains other than lactophenol cotton blue may be necessary for special cytological studies with the light microscope. When these are employed there is a tendency for the nonsporulating parts to plasmolyze or to become twisted during fixation or dehydration, or long thalli (those of some Eccrinales measure more than 1 cm) may break during the embedding process. For these reasons, the use of lactophenol cotton blue is recommended for routine permanent slide preparation.

In recent years, electron microscopic studies have provided valuable information on the biology and phylogeny of the trichomycetes. Preparation procedures for transmission electron microscopy are similar to those generally used with other fungi, except for the dissection of hosts and selection and orientation of suitable material to be sectioned. The quality of fixation may vary depending on the trichomycete species, morphological structure, and the fungal condition at the time of fixation. However, good results have been obtained by several investigators whose studies are cited elsewhere in this book.

Morphological structures to be processed for electron microscopy should be dissected with special care and with a minimum of handling. Larger guts must be cut open to expose the fungi, and either the entire gut or pieces of the gut or its lining can be fixed. Peritrophic membranes and small guts that have had the epithelial layer removed from the lining for visibility need not be cut open, but should be cleansed of all loose detritus in the lumen. Aquatic hosts often contain minute grains of sand and diatoms in their guts that may interfere with sectioning. Removal of unwanted substances can be accomplished by lifting the lining through the surface of clean water several times. Where damage to delicate thalli may be a factor, it is preferable to handle the material as little as possible, even at the expense of not cleaning it.

Good fixation is possible with a 2% glutaraldehyde:3% acrolein mixture in 0.1 M cacodylate buffer at pH 6.8 for 2–3 hr at room temperature, post–fixing in 2% osmium tetroxide in the same buffer for 2 hr at 0–4°C. Some trichomycete material will fix well in freshly prepared 1% aqueous potassium permanganate for 1 hr in the dark at room temperature. The chitinous lining of arthropods is often hydrophobic, or may tend to float, owing to air bubbles that become entrapped in folded portions of the lining.

Consequently, one should insure that the material in fixative does not float. Sometimes it may be necessary to place it under vacuum for a brief period. Dehydration should be gradual, such as 15 min in each of 10% steps in a graduated ethanol series.

Shallow polypropylene dishes or special flat–embedment plates are useful for the final embedment in plastic. This permits microscopic examination and better selection and orientation of material to be sectioned. Selected portions can be cut out and cemented to short plastic stubs of suitable diameter to fit the microtome chuck.

Scanning electron micrography can be very informative, as with other biological subjects, and its use with trichomycetes probably will increase. Once the proper material to be studied has been located, it can be fixed in glutaraldehyde, permanganate, formalin–aceto–alcohol, etc., then dehydrated, critical–point dried, and coated in the conventional manner. The technique of freeze–fracturing has not been employed in published trichomycete studies to date.

Axenic Culturing

The availability of axenic cultures of trichomycetes has led to a variety of studies involving their physiology, morphogenesis, phylogeny, and host relationships. Unfortunately, species of only two genera, *Smittium* and *Amoebidium*, have been available for this experimental work. Recently Peterson et al. (1981) and Peterson and Lichtwardt (1983) were successful in culturing two new Legeriomycetaceae, *Genistelloides hibernus* and *Capniomyces stellatus*. A list of the species and numbers of isolates known to be currently in culture can be found in Appendix C. This section deals primarily with the methods that have been employed to obtain primary axenic isolates of trichomycetes, whereas the results of experimental studies using such isolates are covered in Chapter 9.

It is perhaps not surprising that the first species to be cultured was *Amoebidium parasiticum*, an ectocommensal trichomycete with a wide host range. This was accomplished by Whisler (1960), who retrieved sporangiospores released from thalli attached to *Cladocera* sp., washed them several times in sterile pond water, and streaked them on 0.1% tryptone agar. In 1962 Whisler published a study of the nutritional requirements of *Amoebidium* in axenic culture, and obtained maximum growth in a thiamine–enriched tryptone–glucose–salts medium. He was also able to grow the organism in a defined medium by substituting methionine for tryptone.

Whisler's success at axenic culturing was followed by that of Clark et al. (1963), who obtained from mosquito larvae the first endocommensal pure cultures. Clark et al. surface–sterilized larvae prior to removal of the hindguts, and used antibiotics in the dissection water as well as in a series of washes before plating on a blood–agar medium (SNB–9) used for culturing hemoflagellates. By this method they were able to isolate *Smittium culicis* and a *Smittium* species that we have identified as *S. culisetae*. Subcultures grew also on Difco brain–heart infusion agar with a 2% neopeptone overlayer and in NIH thioglycolate broth. However, they reported that trichospore production occurred only after 1 month of growth in the initial isolates, a remarkably slow development. [Subsequent studies (El–Buni and Lichtwardt, 1976a) have shown that the use of brain–heart infusion at the concentration recommended by the manufacturer, and certain other cultural conditions, are inhibitory to spore production in *Smittium* spp.; see Chapter 9.]

The following year Lichtwardt (1964a) reported the isolation of two new species, *S. culisetae* and *S. simulii*, from a mosquito and a blackfly larva, respectively, using dilute brain–heart infusion and potato dextrose–yeast extract agar media. Since that time he has obtained a large number of isolates of more than four species of *Smittium* from several families of dipteran hosts. Other investigators have likewise been able to isolate species of *Smittium* or *Amoebidium parasiticum*, but the techniques they used have not been published (Whisler, personal communication; Chapman, 1966; Manier, 1969b; Coste–Mathiez, 1970). The simplest methods that have been successful in the author's experience are described below.

Brain–heart infusion diluted to 1/10 the usual concentration has proved to be a good isolation and maintenance medium for *Smittium* and *Amoebidium* species:

BHI/10

Brain–heart infusion (Difco)	3.7 g
Glass–distilled water	1 liter
Agar	15 g

The following tryptone–glucose medium with salts and vitamins is a modification of Whisler's (1962) medium, and is also suitable for isolation of both genera. It has been the basic medium for various nutritional and physiological studies on species of *Smittium*.

TGv

Tryptone (Difco)	20 g
Glucose	5 g
KH_2PO_4	0.28 g
K_2HPO_4	0.35 g
$(NH_4)_2SO_4$	0.26 g
$MgCl_2 \cdot 6H_2O$	0.10 g
$CaCl_2 \cdot 2H_2O$	0.07 g
Thiamine HCl	200 µg
Biotin	50 µg
Glass–distilled water	1 liter
Agar	15 g

When these media are used with agar in petri dishes, they should be flooded with a shallow layer of sterile distilled water before inoculation. In agar slants, about 1 ml of distilled water is added to the test tube, and the tube is tipped daily so as to flood the agar surface during initial stages of growth. We also add thiamine and biotin to BHI/10 medium in most instances, but it has not been determined that these vitamins are actually required. However, supplemental thiamine has been shown to be stimulatory to growth (see Chapter 9).

The antibiotics most commonly used have been prepared as a single stock solution consisting of 40,000 units of Penicillin G and 80,000 units of Streptomycin sulfate per milliliter of distilled water. It can be filter-sterilized and kept in serum bottles in a refrigerator for 6 months or longer, and is dispensed with a sterile syringe.

For isolating *Smittium* species, the hindgut is dissected from a dipteran larva and the epithelium is removed. Microscopic examination of the unopened gut lining mounted in water with transmitted light will usually reveal whether or not *Smittium* is present. The fungus can then be processed in this condition, or the lining can be torn open carefully to release debris from the lumen. Individual thalli can also be removed and used, but greater care is necessary in handling and the results usually have not been as good as when the entire hindgut lining is used.

Amoebidium parasiticum is located on external parts of various aquatic crustaceans and insects, and hosts with mature thalli should be selected. Animals like small cladocerans or first instar mosquito larvae can be used intact. Parts with *Amoebidium* on larger arthropods (the antennae, anal papillae, etc.) should be excised and used for culturing.

Washing should be done in small containers with antibiotics. The author prefers 35 × 10 mm sterile plastic petri dishes partially filled with water to which is added 0.05 ml of Penicillin–Streptomycin stock solution per dish. When bacteria are abundant, larger amounts of stock antibiotics, as high as 0.2 ml per dish, can be used without apparent harm to the fungi. The material to be isolated is placed in this wash water for 15–60 sec and transferred serially and aseptically through two or more washings, then to the agar medium with a water overlayer. Antibiotics may be avoided at this stage, or concentrations up to 0.2 ml per dish may be added to the water. It is generally best to vary the concentration in replicate cultures. The use of 60 × 15 mm plastic petri dishes for the initial isolations allows one to examine the inoculum under a ×5 microscope objective (×50 magnification) to check on growth and contamination (Fig. 3.1). If growth occurs it is usually evident within 2–4 days. Part of the growing clump of fungus should be transferred to new medium without antibiotics. Transfer of *Smittium* can be done with a small metal loop, but a Pasteur pipette or micropipette may be desirable for spores and young thalli of *Amoebidium* that have been released onto the agar surface. Careful monitoring for contamination during the first few days may permit contaminated material to be saved, before it is overcome by bacterial growth, by rewashing

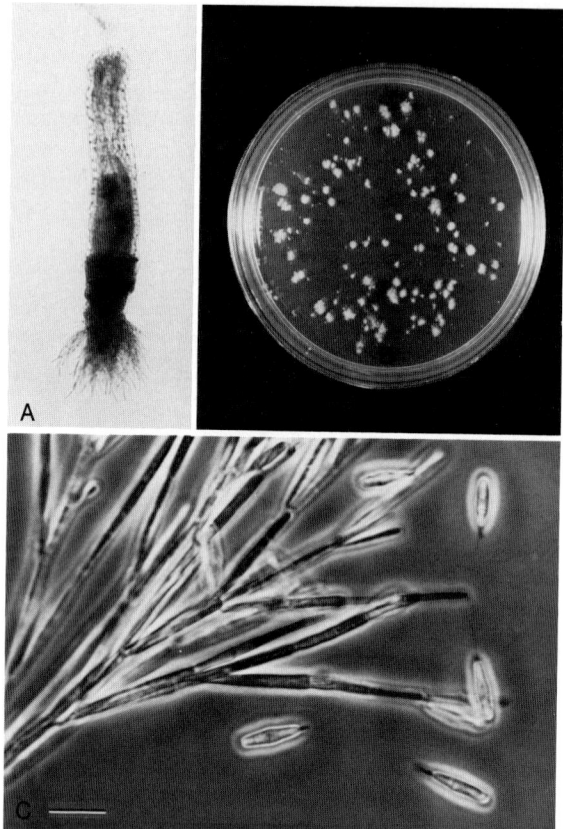

FIG. 3.1. Axenic cultivation of *Smittium simulii*. A. Fungus growing out from the posterior end of a midge hindgut 3 days after placing the gut on dilute brain–heart infusion agar with a water overlayer, as seen through the lid of the primary isolation dish using a ×5 microscope objective lens. B. Mature fungal colonies growing in a 60-mm diam petri dish. C. Water–mount preparation from a pure culture of the fungus; scale bar = 10 μm. Fig. 3.1B from Lichtwardt, 1964a, by permission of Amer. J. Bot.

in antibiotics and replating on medium with antibiotics. Contamination by yeasts and filamentous fungi is generally more difficult to control, but is less common. Such contaminants can sometimes be diluted out by serial washings of the trichomycete material.

The author has attempted to culture many other genera of trichomycetes from all four orders without success, and many other investigators have made attempts with some of the genera (Lichtwardt, 1954, 1964; Manier, 1954, 1955; Whisler, 1963; Chapman, 1966; Williams, 1971; El-Buni, 1975). Of course, most unsuccessful results are not reported in the literature. Many nutritional and physical parameters that might be significant to these gut fungi have been taken into consideration. It is obvious that the axenic culture conditions that will support growth of *Smittium* and *Amoebidium* are less demanding than for other genera. This is emphasized by the fact that the author on many occasions has attempted to isolate species of genera such as *Genistellospora, Pennella, Simuliomyces,* and *Paramoebidium* growing with *Smittium* spp. in the same *Simulium* host, and only *Smittium* has grown.

In 1971 the author was able to make about a dozen isolations of *Trichozygospora chironomidarum* from chironomid larvae collected in Abisko, Sweden. Two of these survived for as long as 18 months, during which time they were transferred repeatedly and grown under a variety of conditions without ultimate success. Likewise, he obtained very limited axenic growth of *Enterobryus* sp. from the fiddler crab, *Uca crenulata,* but only over an 8–week period (Lichtwardt, 1964). In addition to the probability that most trichomycetes have unusual and as yet undiscovered requirements for vegetative growth, there is another factor that must be considered. Extrapolating from data obtained with *Smittium* cultures and from other genera removed from hosts, trichospores and other kinds of spores do not ordinarily germinate except within the host gut (see Chapter 9). Therefore, thalli that cease to grow vegetively when sporulation commences, such as the unbranched Eccrinales and Harpellaceae, will cease to develop further when their spores mature if those spores are incapable of germination under the culture conditions used. The factor(s) that induces spore germination may therefore have to be determined and incorporated into the axenic culture conditions in order to permit such trichomycetes to complete their cycles.

Only a few of the many isolates of Harpellales produce spores that germinate *in vitro*. Consequently, when transferring these fungi to dishes or test tubes of new medium the colonies should be broken up sufficiently to provide new clones. This can be done with a small loop when working on agar media. For experiments using liquid medium, young cultures can be chopped in a Waring–type blender to obtain a homogeneous inoculum as is done with other filamentous fungi.

Cultures can be stored on slants of medium in a refrigerator after growing at room temperature for 7–10 days. It is recommended that stock cultures be transferred to new slants every few months, but some refrigerated isolates may survive for up to 1 year or more. Freeze–drying *Smittium* and *Amoebidium* cultures has not been successful. The preferred method for long term storage is in liquid nitrogen.

The value of using axenic isolates to help elucidate the biology of the trichomycetes has been considerable, despite the few taxa that have been available. Many fundamental questions raised in the succeeding chapters remain unanswered, however. Their solutions will be greatly facilitated by, or even require, the axenic cultivation of species that are at present unculturable.

Part II Biology

CHAPTER 4

Arthropod Hosts and Habitats

All known hosts of trichomycetes belong to the Subphylum Mandibulata of the Arthropoda. To the extent that these fungi have been studied, no assimilative stage occurs while they are disassociated from their hosts, although maturation of some resistant spores and the development of cysts in the Amoebidiales may take place in the shed molt or on a nonhost substrate. Three classes of arthropods are involved: Crustacea, Diplopoda, and Insecta (Figs. 4.1 and 4.2). *Asellaria scutigera (nom. nud.)* was reported by Manier in 1954 from the rectal cuticle of a centipede (Chilopoda), but the description was not sufficient to determine if it was, in fact, a trichomycete. Lists of known arthropod hosts and their habitats are given in Appendix B. It must be stressed that these lists, as well as the discussion presented in this chapter, are based on current knowledge of the range of hosts, and will undoubtedly have to be modified as new discoveries are made. When one considers that the Arthropoda may represent more species than all other groups of living organisms combined, and that the trichomycete flora has been sought in relatively few parts of the world, it becomes evident that generalizations must be to some extent tentative.

The mandibulate hosts of trichomycetes feed primarily on decaying vegetation of various kinds or living algae, or they are omnivorous. Obligately parasitic, predaceous, or carnivorous arthropods, with rare exceptions (see Chapter 6), do not appear to be suitable hosts for the fungi, nor are those that feed only on the living tissues of higher plants. We are excluding from consideration the ectocommensal trichomycete species, *Amoebidium parasiticum,* whose wide range of hosts does include the predaceous nymphs of dragonflies (Odonata). Lichtenstein (1917a) described *A. fasciculatum* [which appears to be *A. parasiticum* (Tuzet and Manier, 1951b; Manier, 1969b)], growing in the rectum of *Anax imperator* (Odonata) nymphs, but stressed that the rectal environment of this aquatic dragonfly nymph is really comparable to the outside medium and contains a variety of free-living and attached bacteria, amoebae, flagellates, vorticellans, etc. It thus appears that the gut-inhabiting trichomycetes, at least, are restricted to arthropods that feed wholly or in part on decaying

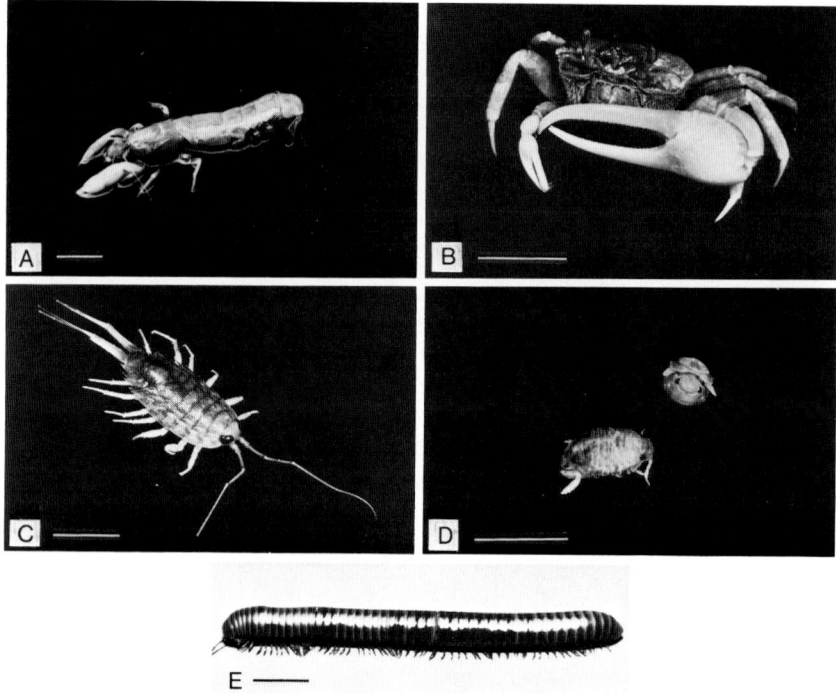

Fig. 4.1. Representative hosts of trichomycetes: Crustacea and Diplopoda. A. Red ghost shrimp, *Callianassa californiensis* (Decapoda, Anomura). B. Fiddler crab, *Uca pugilator* (Decapoda, Brachyura). C. Rock louse, *Ligia exotica* (Isopoda, Ligiidae). D. Pill bug, *Armadillidium vulgare* (Isopoda, Armadillidiidae). E. Millipede, *Narceus americanus* (Diplopoda, Spirobolidae). Scale bars = 10 mm.

plant materials or living algae or that may be only incidentally predaceous, as are some stonefly nymphs containing *Paramoebidium* sp. A case in point is that among Chironomidae larvae one finds predaceous species that feed on herbivorous species that harbor trichomycetes, yet the predaceous species have not been found infested with trichomycetes. Restricted fungal host specificity alone cannot account for the lack of infestation in these predaceous forms, because the chironomid trichomycetes in question (several species of *Smittium* and *Stachylina; Trichozygospora chironomidarum*) are not species specific, for they are known to live in species of several different genera of herbivorous Chironomidae.

As a general rule, in crustaceans and millipedes in which immature forms resemble their adult stages, feed on the same kinds of substances, and live in combined populations, both immature and mature stages can be infested by the same species of trichomycete. Immature individuals, especially young ones, may be less infested, however. This probably is due primarily to the shorter intermolt period, which may not allow time either

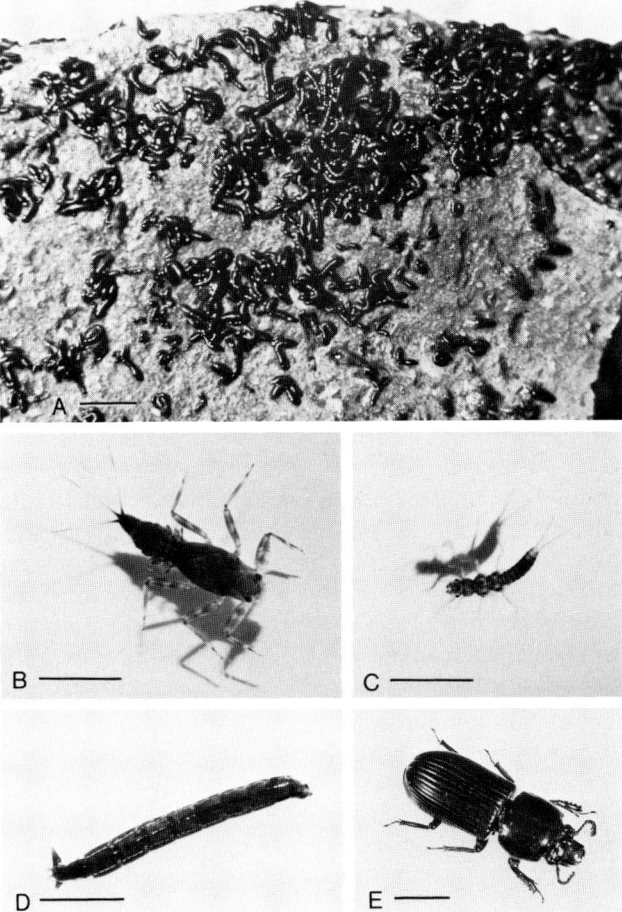

FIG. 4.2. Representative hosts of trichomycetes: Insecta. A. Blackfly larvae, *Simulium vittatum* (Diptera, Simuliidae), clinging to a rock lifted from a stream. B. Mayfly larva, *Drunella coloradensis* (Ephemeroptera, Ephemerellidae). C. Stonefly larva (Plecoptera, Chloroperlidae). D. Midge larva (bloodworm), *Chironomus* sp. (Diptera, Chironomidae). E. Beetle, *Popilius disjunctus* (Coleoptera, Passalidae). Scale bars = 10 mm.

for the chance ingestion of spores and the establishment of new thalli or, when thalli do become established, for the production of spore types (Eccrinaceae and Parataeniellaceae) that function to increase infestation endogenously. Young immature specimens in some situations may ingest abundant spores, but the thalli that develop may not sporulate in time prior to being shed with the molt (Lichtwardt, 1961b).

Virtually all of the Harpellales are restricted to the immature forms of their aquatic insect hosts, whether metamorphosis is complete (Diptera) or incomplete (Plecoptera, Ephemeroptera). In these instances the adults are not aquatic, and they feed on blood, plant juices, etc., or do not feed at all. Evidence to date indicates that adults of these insects do not contain fungal growth in their guts [except for mosquito hosts of *Smittium morbosum* (Sweeney, 1981a)], but the possibility cannot be ruled out that adults can transmit spores in some unknown way to other breeding sites (see Chapter 8).

The beetles offer examples of several patterns of infestation. *Lajasiella aphodii, Enterobryus pentodoni,* and several unnamed eccrinids (Thèodoridés, 1955) are known only from the larval stages of Scarabaeidae, whereas *Eccrinopsis leidyi (nom. dub.)* and several species of *Enterobryus* have been found only in adult Hydrophilidae. The Passalidae are widely scattered in the Americas, and most of the adults in the many populations that the author has examined contain *Enterobryus attenuatus*. Their larvae may also contain the same eccrinid, and this infestation of both developmental stages is undoubtedly related to the fact that larvae and adults live communally in tunnels excavated in logs by adults that contain considerable fecal matter, thus permitting ready transfer of inoculum.

One might expect the adaptability of the fungi to different external environments to be dependent primarily on the tolerances of the host, sheltered as the fungi are within the guts of their hosts where their ambient is at least to some extent under homeostatic control. At the same time the possibility cannot be ruled out that the absence of trichomycetes in some populations of the appropriate host species may be due to specific chemical substances such as pollutants or other factors in the environment that are deleterious to the fungi but not necessarily to the arthropods.

The degree of adaptability of arthropods and their fungi can be illustrated by several examples. Blackfly larvae of the species *Simulium vittatum* are infested with a number of trichomycetes, among them *Harpella melusinae, Genistellospora homothallica,* and *Paramoebidium curvum*. This blackfly species is widespread in North America. At one environmental extreme, larvae can be found in the clear headwaters of the northern Continental Divide in the U.S.A. that are constantly cold during the relatively brief season of larval development. At the other extreme, larvae can be found throughout the year in streams of the southern Great Plains, such as in southern Kansas, which tend to be more silted and have temperatures that fluctuate considerably from summer to winter. The author has found all three of the above trichomycete species in *S. vittatum* populations from both environmental extremes. Further, those fungi in southern Kansas where larvae can overwinter may be found throughout the year. Since the hosts are poikilothermic, the fungi must also maintain their functions over a considerable temperature range. More important, perhaps, is that the algal flora on which the larvae predominantly feed appears to be different in the two types of habitats cited.

Mosquito larvae of *Aedes detritus* have an unusual capacity for osmoregulation, being found in waters ranging from fresh to those with salinities higher than most seawater. Tuzet et al. (1961) studied 16 natural populations of this euryhaline mosquito and found that in eight of these, living in salinities equivalent to 1–14 ‰, all but one population was infested with *Smittium culicis*, whereas none of the other populations, in salinities of 15–40 ‰, was infested. They then took 76 3rd instar larvae infested with the fungus and divided them into two groups. One group (43 larvae) was placed in spring water and the remainder in lagoon water containing 38 ‰ NaCl (38 g/liter). After molting, 100% of the 4th instar larvae maintained in freshwater became reinfested, but no infestation was found in the 4th-instar larvae after they molted in saline water.

On the basis of the later experiments of El–Buni (1975), it would appear likely that the lack of growth in the molted larvae living in high salinity in the experiment of Tuzet et al. was due more to the physiological condition within the mosquito guts that adversely affected fungal development than to the effect of trichospore exposure to the saline conditions during transmission. El–Buni used three axenic isolates originally obtained from freshwater mosquito larvae: two were *S. culicis* (strains CAN–X–1 and WYO–51–11) and one was *S. culisetae* (COL–18–3). Trichospores prepared from cultures were placed for 10 hr in NaCl solutions ranging from 0 to 140 g/liter in 20-g increments. The spores were then washed in distilled water and fed to uninfested *A. aegypti* larvae. Infestation of larvae was as good using spores treated with 40 g/liter salt solution as those treated with pure water, and some infestation was obtained from trichospores exposed to as much as 120 g/liter NaCl. He also found that *in vitro* germinability of *S. culisetae* spores exposed to 40 g/liter NaCl for 10 hr and then rinsed with distilled water was only 14% less than a control set, and that 10% germinability remained when treated with 120 g/liter NaCl. Therefore, based on the studies of El–Buni and Tuzet et al. it seems evident that the trichospores of *Smittium* can be osmotolerant, but that fungal growth and development in mosquito hosts that live in high salinities may be inhibited even though the mosquitoes are biologically suitable hosts at lower salinities. *Aedes detritus* larvae kept in highly saline waters produce a fluid in the Malpighian tubes that is hyperosmotic to the hemolymph. This fluid becomes strongly concentrated in the rectum (Clements, 1963), and may account for the lack of vegetative growth of *S. culicis* under those conditions.

Isopods of the genus *Ligia* live along the coastal areas in many parts of the world. While they spend much of their time out of the water, they are capable of living submerged in seawater if provided with sufficient aeration, and are therefore considered to be marine animals. Several species of *Ligia* are hosts to *Asellaria ligiae;* the fungus has been found in populations of these isopods along shores of the Mediterranean Sea and the Atlantic and Pacific Oceans. An interesting situation was discovered by the author in 1964 on the island of Oahu, Hawaii. At a cascade known as Sacred Falls, and along the length of the river draining its waters, lives a population of an unidentified freshwater species of *Ligia*. It differs from the common *Ligia exotica* found on the coast of Oahu and may be an endemic form adapted to fresh water. This freshwater population is highly infested with *A. ligiae*. It appears in this instance that both the fungus and the isopod host became adapted to freshwater as the isopods invaded the streams and migrated away from their previous marine habitat.

The eccrinid *Enteromyces callianassae* has a wider host range than most endocommensal trichomycetes and is one of the species that grows in the stomach (foregut) of crabs and anomurids. Most of the known hosts dwell in intertidal mud flats with ample organic matter. The type specimen of *E. callianassae*, however, was found in a species of *Callianassa* taken from the tidal zone on a bottom of rather pure, fine sand. Another species of *Callianassa* containing *E. callianassae* came from a fjord, and was dredged at a depth of 34–40 m from a bottom layer consisting of fine sand and organic debris (Lichtwardt, 1961a).

Virtually no information is available on the occurrence of trichomycetes from benthic and pelagic zones. It is evident, however, that intertidal marine trichomycetes are relatively common in some groups of crustaceans, just as other trichomycetes are common in many kinds of freshwater and terrestrial mandibulate arthropods.

CHAPTER 5

Geographic Distribution

The vast majority of trichomycete species have been described from Europe and the United States, which merely reflects the location of biologists who have studied them. A systematic survey of any large land mass, and perhaps most islands with a suitable arthropod fauna, probably would reveal representatives of these fungi. Some species of trichomycetes, in fact, are known to have a much wider geographic range than the distribution of any single species of hosts which they inhabit, owing to the fact that many of the fungi are not strictly species specific and can infest other host species with different distributions (see Chapter 6). Some trichomycetes have a very limited known geographic range, but at the present time it cannot be said whether this is the actual situation or is due merely to insufficient information.

The known worldwide distribution of trichomycetes, in addition to many parts of Europe and North America, includes: Panama, Costa Rica, Puerto Rico, the West Indies, and many regions of Brazil; Israel; north and west Africa and Madagascar; India; Laos; Singapore; Taiwan; all the main islands of Japan; the Philippines; New Zealand; Australia; and the Hawaiian Islands. With few exceptions the trichomycetes described in the Western Hemisphere south of the United States, as well as in Africa and India, are members of the Eccrinales. This is probably because trichomycetes have been sought in those parts of the world in millipedes and other hosts that harbor only representatives of the Eccrinales. In the upper latitudes of the Northern Hemisphere, where eccrinid hosts are less common and aquatic insect larvae are abundant, the Harpellales appear to be more common. Limited collecting by Kobayasi (Kobayasi et al., 1969, 1971) in Alaska and Greenland has revealed several species of *Smittium* and *Stachylina,* and the author has found many genera of Harpellales, and *Paramoebidium* spp., in northern Sweden (Abisko) above the Arctic Circle.

In any given geographic region where potential arthropod hosts are varied and abundant, the trichomycetes infesting them are also likely to be varied and abundant. The author and his colleagues, Drs. Yosio Kobayasi and Hiroharu Indoh, were able to find representatives of all families of trichomycetes in short collecting trips on each of the main Japanese islands. The author has collected members of five of the seven trichomycete families on the island of Oahu, Hawaii, without attempting to survey the island thoroughly. Other parts of the world are likely to be as fruitful to future investigators.

The known distribution of individual trichomycete species is a matter of some interest, even though current knowledge is not sufficient to determine their full geographic range. A few selected examples of distribution will be cited here, but matters pertaining more strictly to host specificity will be covered in Chapter 6.

Harpella melusinae, which is restricted to the peritrophic membrane of blackfly larvae (Simuliidae), appears to be one of the most common and widespread species of Harpellales. Not all populations of blackflies are infested, but, in the author's experience, in any region where they are relatively common *H. melusinae* is likely to be found. (*Harpella leptosa,* on the other hand, has been collected only in and around Glacier National Park in Montana, U.S.A.) The current recorded distribution of *H. melusinae* has been north of the Tropic of Cancer, with one report south of the Tropic of Capricorn (Crosby, 1974). There currently are no published reports of this species, or other Harpellales, in the guts of blackfly larvae in more tropical areas, despite intensive research by the World Health Organization on the biology of *Simulium damnosum* in Upper Volta and other African countries plagued by onchocerciasis whose etiologic agent is transmitted by the adult blackfly. However, recent unpublished studies by the author have revealed that the genus *Harpella,* and several other genera of Harpellales, are virtually cosmopolitan in neotropical Costa Rica.

Species of Legeriomycetaceae in the hindguts of larval insects seem to have a somewhat more limited distribution. For example, *Legeriomyces ramosus* is relatively common in European mayfly nymphs, but less so in the United States. On the other hand, *Genistellospora homothallica* has been found in many blackfly sites in the United States, but there are, as yet, no published records from Europe. (Dr. S.T. Moss and the author did find *G. homothallica* in northern England in 1980.) *Trichozygospora chironomidarum,* in midge larvae, was known from only one collection site in the United States (Grand Teton National Park) until it was rediscovered by the author a few years later in northern Sweden and the Swiss Alps and more recently in England by Moss and the author (unpublished). This species, like some other Legeriomycetaceae, probably has a more sporadic distribution than the previous examples cited above.

Enteromyces calliannassae was first described from anomurids on the coast of Chile. Subsequently it has been found occurring in France, on the east and west coasts of the United States, and in Japan in both anomurids and true crabs. Clearly these distribution records will be expanded even further as the appropriate hosts are studied in other coastal areas of the world.

The islands of Hawaii, isolated as they are and with their highly endemic native fauna, offer an opportunity for additional studies on the distribution of trichomycetes. Since the last century, large numbers of species of plants and animals have been introduced to Hawaii, both intentionally and accidentally. In the latter category are many crustaceans, millipedes, and insects, some of which have been found to contain the same trichomycete species as their counterparts from other localities. One interesting example of accidental introduction is the mosquito, unknown in Hawaii until 1826 or shortly thereafter; *Culex quinquefasciatus* is thought to have been imported in water casks carried on ships (Hardy, 1960). Thereafter, other species were introduced so that Hawaii's fauna is now abundantly enriched with mosquitoes of at least six species. The author collected larvae from two populations of *Aedes albopictus* and one population of *A. vexans* on the island of Oahu, and found all three to contain *Smittium culisetae*, a species common in Japan and the United States but less so in Europe. The bloodworm, *Chironomus hawaiiensis*, endemic to Hawaii, is host to *Stachylina grandispora*, a species of Harpellaceae also known in the United States, Japan, and Europe. *Stachylina grandispora* is also common in *Chironomus zealandicus*, which is endemic to New Zealand (Lichtwardt, unpublished). A more thorough study of trichomycete species in the Hawaiian and other island faunas should prove interesting from the point of view of distribution, adaptation, and evolution of these fungi in island situations.

The examples given in this chapter indicate the variability in distribution patterns to be found among species of trichomycetes. Some of the restricted distributions and geographic discontinuities may be due to the limited studies to date, although some may be actual. The known geographic distribution of each species is given in the taxonomic treatment in Chapter 11.

CHAPTER 6

Host Specificity

The degree to which trichomycetes have become adapted to specific hosts appears to vary among the taxa. The lists in Appendix A show that many species of trichomycetes are reported from only one species of arthropod. Are these particular fungi species specific? Two major problems become immediately apparent before that question can be answered. First, insufficient collections and lack of studies of related arthropod species may be responsible for the presumed restricted host range. Second, there has been a tendency among some researchers to name new species of trichomycetes largely on the basis of new hosts. Unfortunately, speciation problems are not easily resolved in some instances, even when careful taxonomic judgment is employed; this is because of the variation in morphology that some species exhibit, especially in the Eccrinaceae. This problem is discussed in more detail in Chapter 10.

The ultimate determination of the degree of host specificity should employ experimental methods. Most studies of this kind, out of necessity, have used uncultured fungus inoculum and hosts that are not raised initially under axenic conditions. It is necessary to insure that these hosts are indeed devoid of trichomycetes. This can be done (1) by selecting populations that are probably uninfested, as determined by sufficiently large sample dissections; (2) by using arthropods that have molted and have been immediately separated from their exuviae and other substrates that may contain spores, in order to preclude reinfestation; or (3) by using clean or surface sterilized eggs for hatching and rearing the arthropods.

As early as 1906, Chatton (1906a) tested the host range of the ectocommensal *Amoebidium parasiticum* and found that spores developing from thalli attached to *Daphnia* would affix as well to many other arthropods, even to glass, thus confirming his contention that the spores are not particular about the substrates to which they will attach in nature. Whisler and Fuller (1968), using a pure culture of *A. parasiticum* for the purpose of studying the fine structure of holdfast formation, did not obtain adhesion of spores to chitin or cotton threads when the organism was grown in a liquid nutrient medium; however, spores did attach to cotton fibers, nylon, silk, and hair (in decreasing order of success) when thalli were transferred to a dilute salt medium. Whether or not spores of *Amoebidium* that become attached to nonarthropod substrates in the natural environment develop to maturity has not been determined.

Manier (1963a) studied the infestation of the terrestrial isopod *Armadillidium simoni* by *Asellaria armadillidii* in southern France. Ten other species of Oniscoidea collected in that general region were not infested, but she found that *Asellaria* could be transmitted to another isopod species, *Acaeroplastes melanurus*, when individuals were kept in a small container with infested *A. simoni*. She concluded from the cross–infestation tests that because of habitat preference when in their natural environment, the population of the normally uninfested isopod species did not have sufficient contact with *A. simoni* to pick up the fungus; however, once transmitted, development of *Asellaria* in the new hosts was very good. Manier's conclusion fits what we have observed in Kansas, U.S.A., where *Armadillidium vulgare* and *A. nasatum* live together and compete for the same resources. Mixed populations of the two isopod species may be infested with *Asellaria armadillidii*, others with *Parataeniella armadillidii* (Lichtwardt and Chen, 1964). Neither fungus has been found restricted to only one of the two host species.

Tuzet et al. (1961) did extensive collecting of mosquito larvae in several French Mediterranean departments, and found nine of seventeen species to be infested with *Smittium culicis*. They performed several experiments to determine whether the lack of infestation in some mosquito species in their natural habitats was due to a lack of receptivity to the fungus or to extrinsic factors. *Theobaldia longeareolata* is a mosquito species that was consistently devoid of *S. culicis* in the natural environment, even in the presence of heavily infested larvae of other species. Two tree–hole species without natural infestation, *Aedes geniculatus* and *A. berlandi,* could be readily infested experimentally with *S. culicis,* even in their native water, thus ruling out the possibility that the tree–hole water was in some way fungistatic. They theorized that the lack of natural infestation in the receptive tree–hole larvae of the two species was due primarily to the inability of the fungus inoculum to reach the tree–hole inhabitants. (However, see Chapter 8 where the author reports on several similar kinds of isolated sites with infested mosquito larvae.) Tuzet et al. also concluded that *Anopheles claviger* was not infested naturally due to the fact that it is a surface feeder and less likely to ingest spores of *Smittium*, which settle on the bottom substrate. When they kept larvae of this species in very shallow layers of water, thereby forcing the larvae to feed on the bottom substrate, all became infested in vessels containing the fungi, whereas another similar set of larvae of this species kept in water 15 cm deep did not. It seems clear from these experiments that a wide range of mosquito species, although not all, are infestable by *S. culicis*, but that environmental and behavioral factors may also determine whether the fungus becomes established in a given population.

Coluzzi (1966) in Italy was aware of the experiments by Tuzet et al. and conducted infestation tests with an undetermined species of *Smittium* (similar to *S. culisetae*) reported as being common in four local species of mosquito larvae. Using one natural host *(Culex pipiens)* and two species not normally infested *(Aedes aegypti* and *Anopheles gambiae)*, he placed laboratory colonies of 2nd-instar larvae of each species in 30-cm diameter plastic dishes filled to about 6 cm depth with well water, and added 4th-instar larval molts containing the *Smittium*. On reaching the 4th instar, 93–100% of the larvae were infested. He attributed the high infestation rate in *A. gambiae*, in contrast to the results of Tuzet et al. with *A. claviger* kept in deeper water, to the fact that *A. gambiae* is generally a bottom feeder.

Experimental infestations of a range of hosts by one fungal species were also conducted by Moss (1972) using 19 genera of midge (Chironomidae) larvae collected from an exceptionally rich 21-m stretch of a brook in England. Species of five of the tubiculous midge genera were found to be naturally infested with *Stachylina grandispora*. Moss kept 4th-instar larvae of infested and uninfested species in shallow dishes with pond water for 2 weeks, and was able to infest species of larvae of 7 genera that had not been found to be naturally infested. An interesting result was that none of the species of free-living, carnivorous genera became infested by this procedure. This study provided clear evidence of the wide host range of *Stachylina grandispora*, and that in this particular site infestation was probably determined by the feeding habits of the midge larvae.

Coste–Mathiez (1970) in Manier's laboratory attempted some more or less reciprocal infestations with two species of *Smittium*. One, *S. mucronatum*, has been found in nature only in larvae of *Psectrocladium sordidellus* (Chironomidae), and is known to produce zygospores in addition to trichospores in that host. Chironomid larvae infested with *S. mucronatum* were placed in containers with uninfested 4th instar larvae of the mosquito *Culex pipiens*. At the end of 3 days the molts of the mosquito larvae were all infested with the fungus, and 2 of the 35 molts had some zygospores as well. She found, however, that the reproductive structures were somewhat abnormal in size and shape (Manier and Mathiez, 1965). Coste–Mathiez also took uninfested larvae of *Chironomus* sp. (Chironomidae) and placed in their containers sporulating thalli from a culture of *S. culicis*, a species that grows naturally in mosquito larvae, but which is not known to infest chironomids naturally. All *Chironomus* larvae became infested with *S. culicis*, but the development was extremely slow for this species. The thalli eventually degenerated, and by the end of 2 months there was no longer infestation in the larvae that remained. Apparently reinfestation by means of trichospores produced from the original growth in the guts did not take place.

There are no records in the literature indicating that strictly carnivorous or predaceous arthropods become infested with trichomycetes. An exception was found in Australia when the author was working with Dr. A.W. Sweeney in 1983 near Innisfail, Queensland. Some specimens of the predacious mosquito larva *Culex halifaxii* were collected, placed individually in containers in the laboratory for 8 days, and, during this period, fed larvae of another, nonpredaceous species of *Culex* that were infested with *Smittium simulii* and *S. culisetae*. One of the seven specimens of *C. halifaxii* dissected proved to be highly infested with *S. simulii*. The same site also contained bloodworms (*Chironomus* sp.), most of which were infested with both *Stachylina grandispora* and *Smittium simulii*. Some predaceous midge larvae (*Pentaneura* sp.) living among them, however, were uninfested. Thus, the infestation of the predaceous mosquito larva appears to be an exception to the rule. Another interesting exception to host specificity was found in rock pools of the Georges River near Cambelltown, New South Wales, Australia. The pools contained mosquito larvae *(Aedes rupestris)*, bloodworms *(Chironomus alternans)*, and ceratopogonid larvae (*Dasyhelea* sp.), many of which were infested with *Smittium culisetae*. Among these dipteran larvae were some immature mayfly nymphs (Ephemeroptera), one of which had *S. culisetae* living in its gut, the first known instance of a nondipteran host for any *Smittium* species.

Williams and Lichtwardt (1972a) made use of a collection of axenic isolates of four species of *Smittium* obtained from mosquito, chironomid, and blackfly larvae to test their ability to infest one host species: larvae of *Aedes aegypti*. The data obtained in this experiment are presented in Chapter 9 (Table 9.3).

The natural infestation by *S. simulii* of dipteran larvae belonging to different families (most commonly Simuliidae and Chironomidae) may be due to the fact that the chironomid larvae in which this fungus has been found inhabit flowing waters, as do the larvae of simuliids. Perhaps, over time, this permitted adaptation by the fungus to both families of dipteran hosts through repeated transfer of spores. The chironomid host of *S. mucronatum,* on the other hand, is found in still waters, which is also the habitat of mosquitoes. It is interesting to note that *S. mucronatum* has a closer morphological and serological relationship to *S. culicis* than it does to *S. simulii* (Sangar et al., 1972).

The degree of specificity of many trichomycetes has been established by many data obtained in the field. For example, in a single stream one may find blackfly larvae infested with species of their particular fungal genera *(Harpella, Pennella, Genistellospora, Simuliomyces,* etc.), whereas the mayfly nymphs, even on the same rocks or sticks in the water, have their own individual flora (*Legeriomyces, Glotzia, Zygopolaris,* etc.). Species of these genera are morphologically quite distinct in the sporulating stage, and are not identified on the basis of the host (so circular logic is not a factor here). Furthermore, some of the fungal genera are restricted to mayflies of the family Baetidae, even when other families of mayfly nymphs are present on the same substrates. Refer to the genus *Zygopolaris* in Chapter 11 for further examples of specificity within families of mayflies.

The invasion of an individual arthropod by one species of trichomycete does not necessarily exclude the establishment of other trichomycetes in the same gut. The most striking example of this phenomenon is found in some populations of blackfly larvae. It is not uncommon to find larvae containing more than one trichomycete species, and frequently two or three species are found in one blackfly larva in good sites. An example of an excellent site is an inconspicuous small stream, Johnson Creek, that drains into Swan Lake in northwestern Montana, U.S.A. The author in 1975 found seven species of trichomycetes in three blackfly dissections. One of those larvae contained *Harpella leptosa, Genistellospora homothallica, Simuliomyces microsporus,* and *Paramoebidium curvum*. In that same general area of Montana, thalli of *H. melusinae* and *H. leptosa* have been found sporulating side by side in the midgut of blackflies (Moss and Lichtwardt, 1980). Also, in several populations of chironomid larvae there and in other regions of the world, the author has found combinations of two species of *Smittium* in the hindgut. We can state, therefore, that not only may one species of trichomycete infest many species of hosts, but also that in some habitats one species of host may be infested by two or more trichomycetes.

If any generalization can be made about host specificity in these gut–inhabiting fungi, it is that species of trichomycetes often are capable of infesting more than one species or one genus of hosts, but may be restricted to one arthropod family. The range of hosts can extend beyond a single family in a few trichomycetes (e.g., *Enteromyces callianassae, Taeniella carcini, Smittium simulii*). The genus of hosts, rather than the family, may delimit the range of infestation by some species of trichomycetes, especially in cases where those genera are phylogenetically distinct and when species of those genera are ecologically isolated from other related genera by virtue of their restricted habitat (e.g., the marine isopod genera *Limnoria* and *Ligia*). It is also conceivable that many trichomycetes are, in fact, species specific as the current record indicates (see Appendix A), but in all likelihood further studies will show that, in instances where closely related species of arthropods have similar feeding habits and a sympatric distribution, the fungal species will not be found restricted to a single host species.

CHAPTER 7

Morphology, Cytology, and Fine Structure

Life Cycles

HARPELLALES - The Harpellaceae and Legeriomycetaceae, commonly referred to as "harpellids," have similar asexual and sexual reproductive structures and life cycles (Fig. 7.1). In the unbranched Harpellaceae, asexual reproduction commences after the maximum but limited number of cells has been produced. Each compartment becomes a generative cell, giving rise exogenously to an appendaged trichospore. In the branched Legeriomycetaceae, only the terminal parts of branches normally become fertile, and vegetative growth and sporulation may continue as trichospores are maturing on older branches of the thallus.

Trichospores do not normally germinate immediately within the same gut that produced them, but rather are released to the outside aquatic environment where they are presumed to lie dormant until ingested by a suitable host. Germination then occurs rapidly, the inner spore body breaking through the trichospore wall so that it can attach to the peritrophic membrane (in Harpellaceae) or to the hindgut lining (in Legeriomycetaceae) during the relatively quick passage through the host gut. The attached spore body then grows and matures into a new thallus.

The sexual state in some genera of Harpellales does not develop until the host is about to molt, and in some instances trichospore development may become completely arrested as zygospores begin to form. In other (branched) genera, both zygospores and trichospores may be produced simultaneously on the same thallus. The zypospores may or may not be appendaged on release, depending on the type. Like trichospores, zygospores are thought to germinate after passage from the gut and on being subsequently ingested. Zygospores are believed to produce thalli identical to those that develop from trichospores. Details of conjugation and zygospore structure are given in a later section of this chapter.

ASELLARIALES - The Asellariales, which may be closely related to the Harpellales, produce arthrospore–like cells by fragmentation of the

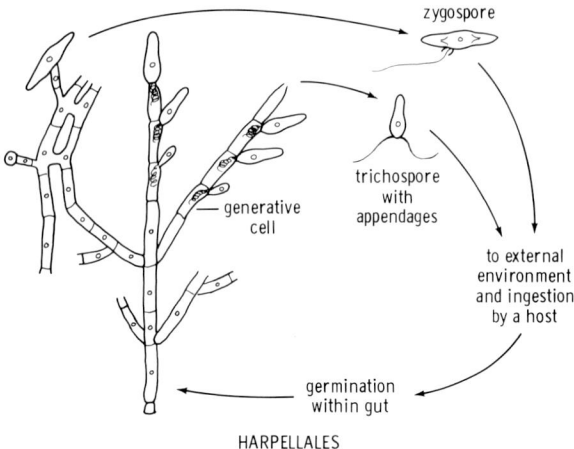

FIG. 7.1. Harpellales life cycle.

branches (Fig. 7.2). In *Asellaria ligiae* the arthrospores develop outgrowths that resemble trichospores in shape and size, and therefore arthrospores may be the equivalent of disarticulated generative cells. No further development of these structures has been directly observed; it is assumed they are ingested by their hosts and develop into new thalli. Conjugations among thalli have been seen, but zygospores have not been formed in those instances.

ECCRINALES - The Eccrinales, often called "eccrinids," have several patterns of life cycle (Figs. 7.3–7.5). In two of the three families (Eccrinaceae and Parataeniellaceae), there are two principal types of endogenous spores produced. One is multinucleate and capable of germinating immediately in the gut to produce a new thallus. The other is usually uninucleate at first, is thick walled in some species, and it serves as the propagule to

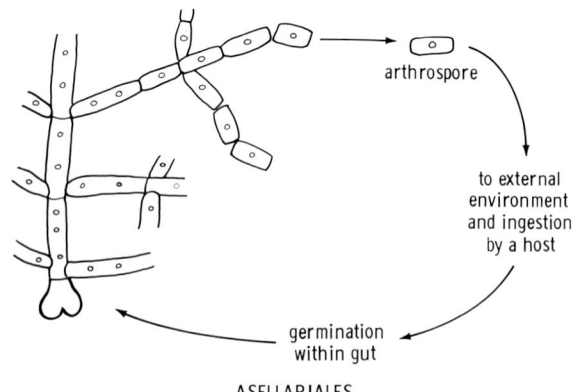

FIG. 7.2. Asellariales life cycle.

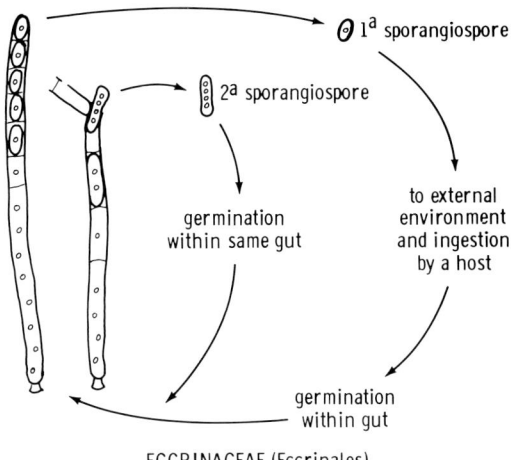

Fig. 7.3. Eccrinaceae (Eccrinales) life cycle.

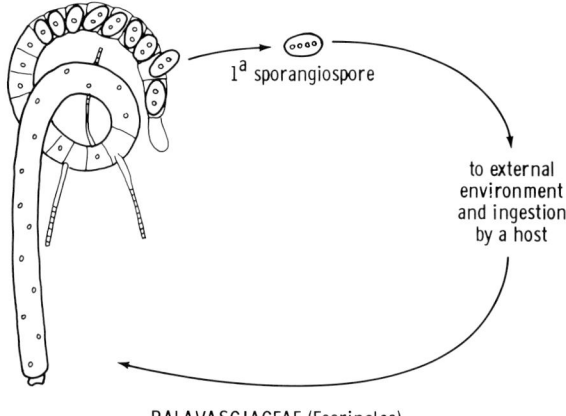

Fig. 7.4. Palavasciaceae (Eccrinales) life cycle.

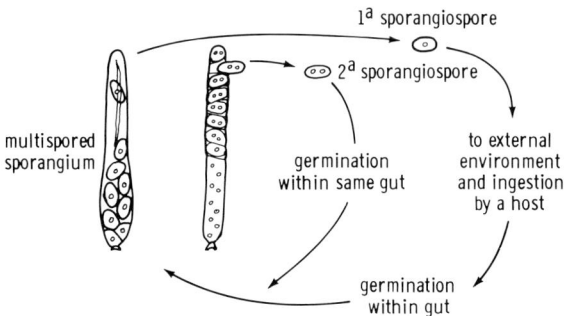

Fig. 7.5. Parataeniellaceae (Eccrinales) life cycle.

transmit the fungus to another host. Except in the Parataeniellaceae, where many uninucleate spores are produced in a common sporangium, all other spores are produced singly in separate, terminal series of sporangia. The Palavasciaceae produce but one spore type, which functions to spread infestation to other appropriate hosts. No sexual reproduction has been confirmed in the Eccrinales. In the genus *Enteropogon,* apparant fusions between cells and their nuclei have been observed (Hibbits, 1978) (Fig. 11.36), but it is not known at present if this is a normal sexual process or aberrant behavior.

AMOEBIDIALES - The ectocommensal genus *Amoebidium* has two distinct phases in its cycle (Fig. 7.6) (Manier and Raibaut, 1969). The common means of propagation is by the production of elongate sporangiospores in the mature thallus, which serves as a sporangium. When these mature they elongate and burst through the sporangial wall, normally remaining in this position until another host contacts the exposed tip of a spore. The spore adheres to its new host and is pulled from the sporangium and develops into a new thallus. When the host molts or is injured, however, the protoplasm in developing thalli cleaves into uninucleate membrane-bound portions with no cell walls. The thalli then rupture and release a number of motile, amoeboid cells. The amoebae of *Amoebidium parasiticum* crawl about the substrate for about 1–6 hr (Manier and Raibaut, 1970), then round up and encyst. [In *Paramoebidium* spp. the motility of amoebae may last only 15–30 min before encystment (Lichtwardt, unpublished)]. The cysts enlarge and eventually, sometimes several weeks later, produce one to more than a dozen elongate cystospores, each in its own chamber. These pop out of the cyst and are presumed to infest new hosts that make contact with them.

Species of the genus *Paramoebidium* live in the hindgut or rectum of their hosts, and therefore have no opportunity to transmit sporangiospores to other hosts by direct contact. Only the amoeba–cyst phase is present,

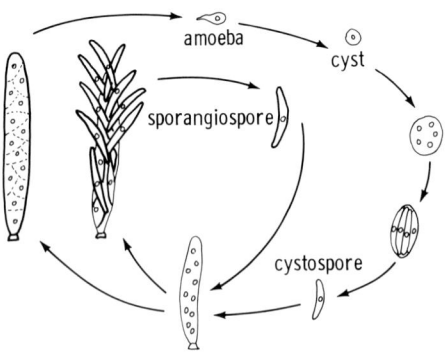

AMOEBIDIALES

FIG. 7.6. Amoebidiales life cycle.

and this part of the cycle is in all general respects identical to the development in *Amoebidium*. We must assume, in contrast to *Amoebidium*, that either the entire cysts or released cystospores are ingested before reinfestation of another host occurs. Dang and Lichtwardt (1979) have suggested that the cyst, in addition to its possible role as a resistant structure, may function as a timing device to insure discharge of the cystopores in the hindgut after ingestion. So far as is known, amoebae of *Paramoebidium* spp. do not release and emerge from the gut during the intermolt period, but rather the thalli complete their maturation and amoebae escape only after molting or injury of the host such as by dissection. Mature thalli apparently can be held in the ready stage until ecdysis, because it is sometimes possible to find amoebae being released within minutes after the exuvia is shed.

The amoebae do not appear to function as gametes. Poisson (1931b) stated that he on some 10 occasions observed amoebae fuse in pairs, but this has not been verified by other investigators who have studied this organism rather intensively; it would be interesting indeed if fusion could be observed again and nuclear behavior followed. Nor do amoebae appear to feed during their period of migration. Rather, their function may be to serve as a means of disperal and to select an appropriate substrate upon which, or within which, to encyst.

It should be noted here that Whisler (1966) reported removing three thalli of *"Paramoebidium"* from the hindgut of a stonefly nymph and that they gave rise to cultures of thalli identical to *A. parasiticum* when plated on cornmeal agar. These axenic cultures readily infested *Daphnia* as ectocommensals and produced both the sporangiospore and amoeba–cyst phases. He viewed this as evidence that the two genera may be synonymous, the morphological differences being related to different growing conditions. The author has tried unsuccessfully to culture *Paramoebidium* spp. from larval stoneflies, mayflies, and blackflies on many occasions. It would be most interesting and significant if Whisler's observations could be repeated. At the same time it should be borne in mind that nonsporulating thalli of *A. parasiticum* can be essentially indistinguishable from many young thalli of *Paramoebidium* spp. Lichtenstein (1917a) found *A. parasiticum* growing in the rectum of dragonfly nymphs, as stated in Chapter 4, and Chatton (1906b, 1920) described *A. recticola* growing attached to the hindgut lining of two species of *Daphnia*. Chatton noted that rhythmic dilations and contractions of the rectum of daphnids create a pumping action that brings water into the rectum. He demonstrated that granules of carmine could be drawn into the rectum, and surmised that in this way spores of *A. recticola* passively invaded the gut and became attached. In both Lichtenstein's and Chatton's reports, sporangiospores were produced in some of the thalli, leaving little doubt that these were species of *Amoebidium* and not *Paramoebidium*. Perhaps aquatic insects such as stonefly nymphs in rare instances can aspirate spores of *A. para-*

siticum via the anal route by a similar mechanism, resulting in growth in, or at least attachment to, the rectum.

Thallus Types and Development

The gross morphology of trichomycete thalli varies not only from simple to highly branched, but also in relative size. Width extremes range from about 2 μm in the microthalli of *Astreptonema gammari* to well over 100 μm in some robust thalli of *Arundinula haplogaster* from marine decapods. The length of mature thalli is as short as 30–55 μm in *Stachylina minuta* and 1 cm or more in length in some species of *Enterobryus* and *Arundinula*. In several genera of Eccrinaceae *(Alacrinella, Astreptonema, Enteromyces, Paramacrinella, Ramacrinella)*, a dimorphism of thalli exists. That is, there are two rather discrete size ranges into which the thalli fall. In some species of other genera (e.g., *Enterobryus borariae, Arundinula haplogaster*) the range of thallial sizes may be just as great, but the measurements intergrade to a considerable extent and are consequently not strictly dimorphic. In rare instances (e.g., *Enteropogon sexuale*) the species appear to be polymorphic, with the different thallial types occupying specific regions of the gut (Hibbits, 1978).

The walls likewise differ in their dimensions, from thin and delicate to exceptionally thick (3 μm or more) in older thalli of some species. Some *Paramoebidium* spp. become sufficiently thick walled to prevent amoebal escape, and the amoebae end up encysting within the sporangium. Electron microscopic studies reveal that the thallus wall usually consists of at least two distinct, fibrous layers in the Harpellales, the innermost being more electron transparent (Reichle and Lichtwardt, 1972; Moss, 1976; Moss and Lichtwardt, 1976), whereas in the Eccrinales there may be three or more discernible layers (Grizel, 1971; Manier and Grizel, 1972; Moss, 1975; Dang, 1979). The author has observed that outer layers of the wall occasionally slough off in older thalli of a number of species of *Enterobryus* and some other eccrinids. Also see *Arundinula galatheae* in Chapter 11.

The chemical composition of the wall has been studied in only a few species (Table 7.1). Sangar and Dugan (1973), using mechanically isolated walls from an axenic culture of *Smittium culisetae*, determined that chitin was a major component. The studies of Trotter and Whisler (1965) on an isolate of *Amoebidium parasiticum* revealed no glucose or glucosamine in their chromatographic analysis of acid hydrolysates of wall material. They concluded that the walls possibly contain hemicellulose, but no chitin or cellulose. The fact that *A. parasiticum* and thalli of *Paramoebidium* sp. are soluble in hot dilute (0.225%) KOH (Whisler, 1963) rules out the likelihood that either chitin or cellulose are significant contributors to wall structure in these organisms. The results of chemical analyses of cultured trichomycetes are given in greater detail in Chapter 9.

TABLE 7.1. Analyses of major cell wall components in trichomycetes.

Species	Chemical analysis of axenic culture		Histochemical test				Solubility in weak alkali	References
	Chitin	Cellulose	Chitin van Wisselingh	IKI–H_2SO_4	Cellulose Zn chloriodide	Mangin		
HARPELLALES								
Smittium culisetae	+							Sangar and Dugan, 1973
Legeriomyces ramosus			+	–	–		–	Whisler, 1963
Pteromaktron protrudens			+	–	–		–	Whisler, 1963
ECCRINALES								
Enterobryus halophilus				–	+	–	–	Whisler, 1963
E. elegans					+			Lichtwardt, 1954
E. iuli-terrestris					+			Robin, 1853
AMOEBIDIALES								
Amoebidium parasiticum	–	–	–	–	–	–	+	Whisler, 1961, 1963; Trotter and Whisler, 1965
A. parasiticum						+		Chatto n. 1906a
A. parasiticum				–				Schenk, 1858; Taylor, 1928
Paramoebidium sp.			–	–	–	–	+	Whisler, 1963

Histochemical tests are less reliable and precise indicators of wall composition, but are the most practical means of testing species that are not culturable. Whisler (1963) was able to get a positive van Wisselingh test for chitin with two uncultured species of Harpellales, *Legeriomyces ramosus* and *Pteromaktron protrudens,* but this same test was negative for *Enterobryus halophilus, A. parasiticum,* and *Paramoebidium* sp. These tests, plus others listed in Table 7.1, lead to the tentative conclusion that the Harpellales contain chitin in their walls, the Eccrinales may have cellulose in theirs, but the walls of Amoebidiales contain neither of these subtances. It is probable that the Asellariales, believed to be closely related to the Harpellales, will be found to have walls composed of chitin whenever a suitable chemical analysis can be run. It must be stressed that wall composition, important as it is in contributing to phylogenetic schemes (Bartnicki–Garcia, 1968, 1970), is not a settled matter in the trichomycetes.

In the Harpellales and Asellariales the thalli become septate as they develop. Each cell appears to be normally uninucleate in most species, although coenocytism is the condition in the Asellariales and some Harpellales in portions of thalli that do not become regularly septate. In contrast, the coenocytic thalli of the Eccrinales do not form crosswalls except to delimit sporangia. Farr made a significant discovery during the course of his studies on *Smittium culisetae* (Farr, 1965; Farr and Lichtwardt, 1967), namely that the septa are perforate and differ morphologically from the well known perforate septa of higher fungi such as ascomycetes and basidiomycetes. *Smittium* septa were found to be swollen around a central pore that later became occluded by an electron–opaque plug. Farr found that this type of septum also occurred at the base of trichospores. This harpellid septum (Fig. 7.7), also known to form during zygospore delimitation, has now been seen in many genera of Harpellales and in two genera of Asellariales (Manier and Coste–Mathiez, 1968; Reichle and Lichtwardt,

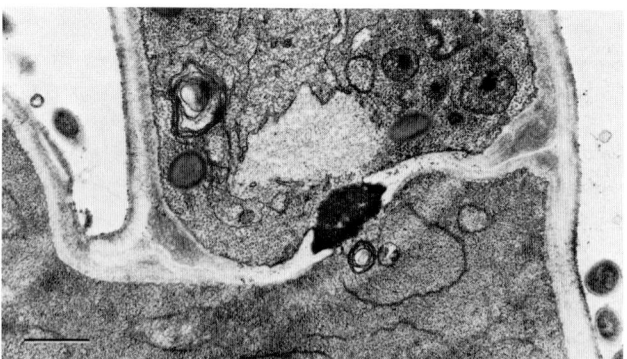

FIG. 7.7. Perforate, plugged septal apparatus of *Genistellospora homothallica.* Scale bar = 1 μm. Reprinted by permission from Mycologia, vol. 65: 14, Copyright 1973, R.W. Lichtwardt and The New York Botanical Garden.

1972; Manier, 1973a, 1973b; Moss, 1972, 1975, 1976; Moss and Lichtwardt, 1976, 1977).

In median longitudinal sections of thalli the septum is seen to have three distinct layers: two outer ones that are continuous with the inner wall of the thallus, with a more electron–transparent inner layer sandwiched between. The inner layer bifurcates at the periphery of the pore to give a flared appearance in transverse sections of the septum. The mature structure superficially resembles the bordered pit of a tracheid. In younger septa there is complete continuity of protoplasm through the 0.8– to 0.9– μm–diameter pore, and endoplasmic reticulum is commonly seen linking the two adjacent cells. The septal pore later becomes occluded by a biumbonate plug, without a bounding membrane, that effectively prevents passage of cellular organelles when fully formed. Nevertheless, intercellular continuity is maintained by the plasmalemma, which is continuous from cell to cell around the periphery of the plugged pore (Moss, 1975).

This unusual type of perforate septum was also discovered by Young (1969) in the hyphae of *Linderina pennispora* (Kickxellales; Zygomycotina), and the similarity is believed to be phylogenetically significant, as is discussed in Chapter 12. Less similar, but nevertheless having some resemblance, is the crosswall of species of Dimargaritales (Zygomycotina) with its flared, lenticular pore region enclosing a biumbonate plug attended on each side by a large globose body (see Brain et al., 1982).

Of equal interest in trichomycete morphology is the type of perforate septum constructed by species of Eccrinales during sporangium delimitation. This has been described by Manier (1979a) in *Palavascia sphaeromae*, but studied in detail only in *Astreptonema gammari* (Moss, 1972, 1975). It would be very desirable to look at septum formation and structure in other species of Eccrinales to determine if there is a common pattern. Moss described the crosswall in *A. gammari* as a single electron–transparent layer continuous with the inner layer of the three–layered lateral wall, with a central pore of 0.3–0.4 μm diameter around which the septum is dilated. The plasma membrane is continuous from one cell to the other through the pore, and cytoplasm appears to stream into the developing sporangium during this initial stage. One or more dictyosomes appear in the sporangial cell near the pore and produce secretory vesicles that migrate toward the pore and form a "pore–cap" on each side of the septal pore. This is followed by fusion of the vesicles with the plasmalemma and deposition of wall material in and around the pore so as to completely occlude the opening and isolate the sporangial plasmalemma from the rest of the thallus on completion of the process. This plugging of the pore by dictyosome–derived vesicles in the Eccrinales appears to be quite different from the formation of the septal plug in the Harpellales and Asellariales. Dictyosomes have not been seen in those two orders, but they have been reported in the Amoebidiales (Whisler and Fuller, 1968).

The cellular contents and organelles of trichomycete thalli are generally similar to those of other fungi. Lipid droplets are evident and sometimes abundant in electron micrographs. Patrick et al. (1973) found that lipid materials extracted from thalli of *Smittium culisetae* account for 9.9% of the total weight, and one can assume that lipids are a major food storage substance. Glycogen granules have also been detected and are sometimes abundant (Moss, 1972; Moss and Lichtwardt, 1977). Reichle and Lichtwardt (1972) reported finding membrane–bound microbodies containing electron–opaque crystals in *Harpella melusinae*. In the same species they also found a semicircular structure associated with the nuclear membrane and paralleled by 8–9 tubules or circular particles. The origin and function of these and other unusual cellular inclusions are yet to be determined. Virus–like particles have been found in two species of *Paramoebidium*, and are described and illustrated in the last section of this chapter. Also described in the sections that follow are special cellular features pertaining to the reproductive bodies of trichomycetes.

Very few critical cytological studies have been made of the nuclei of trichomycetes. Raabe (1911a, 1911b, 1912) described and illustrated the structure and division of *Amoebidium parasiticum* nuclei in great detail, but an interpretation of his studies, in light of our modern knowledge of karyology, is difficult. He described them as lacking a nuclear membrane, and having about 6–8 chromosomes that divide amitotically with the involvement of centrioles.

At interphase, and in some stages that appear to be dividing, the nuclei of many species of trichomycetes are clearly seen and stain well (Fig. 7.8). At other times the nuclei do not take nuclear stains or are obscured by other components of the cytoplasm. Mithramycin and acridine orange can reveal the nuclei clearly with fluorescence in such cases. Even with phase–contrast microscopy it is often difficult to see or study nuclei in many species at certain stages of development. Electron micrographs clearly show that the nuclei are bound by a typical nuclear envelope, but the behavior of the membrane during mitosis has not been studied in detail in any species. Manier (1979a) stated that in *Palavascia sphaeromae* and other (unidentified) trichomycetes the nuclear membrane persists during nuclear division and that there are no centrioles or spindles involved, but she provided no electron micrographs. The author has seen what appear to be intranuclear divisions on several occasions with the light microscope. Moss (1974) was able to describe some Feulgen–stained mitotic figures in *Stachylina grandispora* and to obtain a count of about seven chromosomes in the somatic nuclei.

It can only be assumed that the nuclei of trichomycetes are haploid like those of most other fungi. Duboscq et al. (1948) described the development of resistant spores in *Eccrinidus flexilis* as involving a complicated sequence of steps that would appear to consist of meiotic divisions followed by karyogamy. If this interpretation is correct, the somatic complement

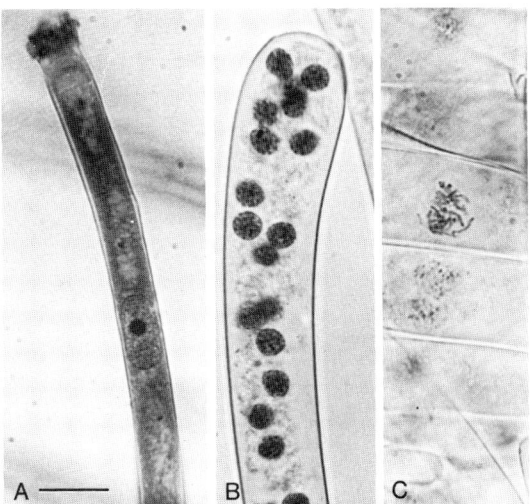

FIG. 7.8. Nuclei of *Enterobryus euryuri*. A. Heidenhain's hematoxylin stain. B, C. Acetocarmine stain. Scale bar = 10 μm for all figures.

would be diploid, but Manier (personal communication) has studied this same species and has not been able to substantiate the kind of nuclear behavior described by Duboscq et al. Moss and Lichtwardt (1977) found four nuclei in the sexual apparatus of *Harpella melusinae* after thallial anastomosis and presumed karyogamy that suggested a meiotic division prior to zygospore formation (Fig. 7.9). If true, this would result in the direct development of haploid thalli from zygospores. Additional studies of sexual processes in these fungi are needed before ploidy can be determined.

Thalli of trichomycetes are conveniently divisible into two forms: branched (Legeriomycetaceae and Asellariales) and unbranched (Harpellaceae, Eccrinales, and Amoebidiales). The branched forms become septate as they develop, though some of the cells may remain coenocytic, whereas the unbranched thalli produce septa only to delimit reproductive cells. One species of Eccrinales, *Ramacrinella raibauti*, produces multiple true branches at the base of the thallus, but the thallus remains coenocytic and develops no septa except during terminal sporangium formation on one or more of the branches. Species of another eccrinid genus, *Palavascia*, produce what appear to be branches from sporangia; however, these are clearly sporangiospores that "germinate" in situ (Manier, 1979a), and not true branches. In a few eccrinids the holdfast may not remain strictly basal due to some peculiar pattern of growth and development, as described in the next section, but these thalli also are not truly branched despite such an appearance in some cases.

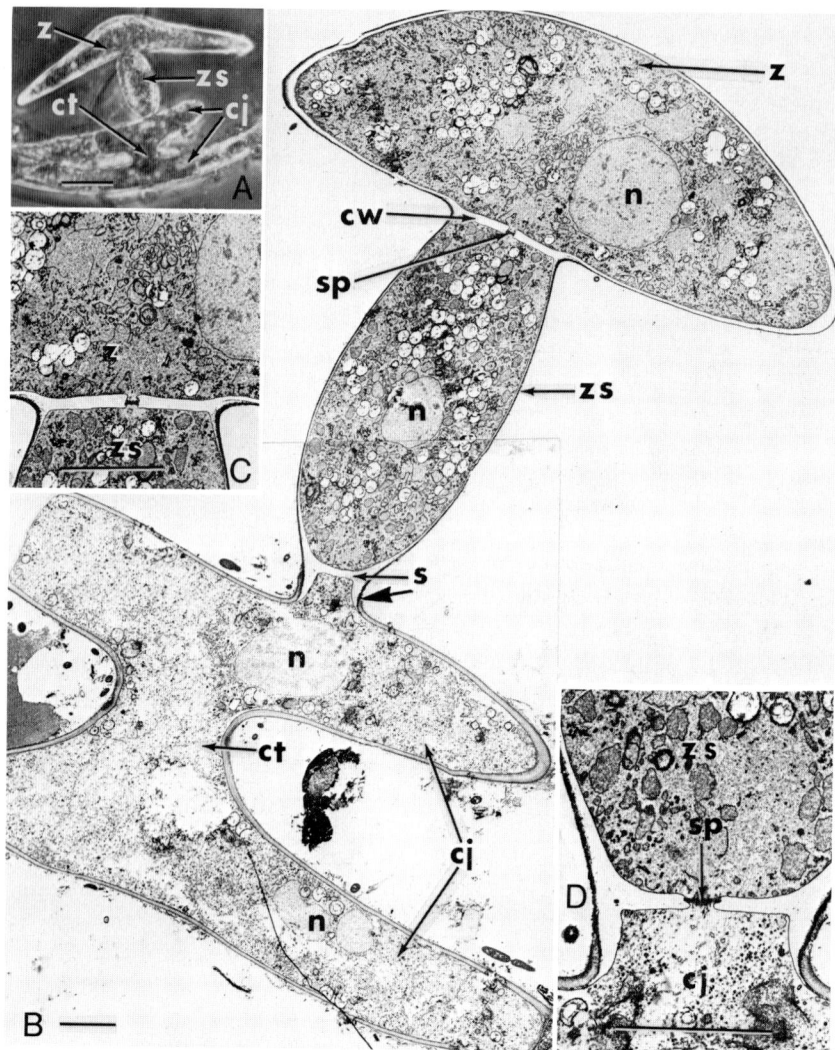

FIG. 7.9. *Harpella melusinae*. A. Mature zygospore in phase–contrast illumination. B. Longitudinal TEM section of an entire sexual apparatus with an immature zygospore in slightly oblique section. Note the four scattered nuclei, compatible with the idea that the zygote may have undergone meiosis and that the nucleus contained in the zygospore is haploid. The zygosporophore wall is seen to have an endogenous origin from the conjugant (arrow). C. Septum with a plugged pore between the zygospore and zygosporophore. D. Septum with a similar structure between the zygosporophore and the conjugant. Scale bars: A, 20 μm; B, C, D, 5 μm. Symbols: *cj*, conjugant; *ct*, conjugation tube; *cw*, crosswall; *n*, nucleus; *s*, septum; *sp*, septal pore; *z*, zygospore; *zs*, zygosporophore. From Moss and Lichtwardt, 1977, by permission of Research Council of Canada.

Holdfasts

A feature common to all trichomycete thalli is their firm adhesion to the substrate by means of a holdfast of some kind (Fig. 7.10). The typical holdfast consists of a rigid structure at the base of the thallus, easily discernible with the light microscope. However, considerable variation in morphology and development exists among the species. As defined here, the holdfast is considered to be either the rigid, acellular structure secreted at the base of an essentially unmodified cell, or a morphologically distinct basal cell (the "holdfast cell") together with whatever substance is secreted to provide adhesion to the cuticle. The holdfast has proved to be sufficiently distinctive in some species or genera as to be of taxonomic value. It is perhaps to be expected that there be some variation in this essential apparatus, considering the morphological heterogeneity found within the fungal class and the diversity of arthropod guts to which the different species must adhere. Nevertheless, as will be discussed later in this section, several recent studies of their fine structure have revealed unexpected differences in holdfasts, even among closely related taxa in some instances.

In the unbranched trichomycetes (Harpellaceae, Eccrinales, and Amoebidiales), the holdfasts are discrete structures that may range from a short cylindrical form, sometimes flared into a broad disk at the base, to longer forms that may be essentially cylindrical, campanulate, or may have other, sometimes less regular shapes. In some Eccrinales the holdfasts are especially large and, although usually much shorter, may measure 50 μm or more in length. Large holdfasts are often longitudinally striated. Intraspecific variation in holdfast size and shape can be found in some Eccrinales, primarily in those species that have a variable thallus morphology.

A curious multiple holdfast system, which characteristically forms in a few species of Eccrinales along with some simple holdfasts, consists of a tuft of individual thalli whose holdfasts are fused into a complex branching structure with a common base of attachment to the host cuticle. In *Enteromyces callianassae* (from the stomach of decapods), this apparently begins as a simple holdfast. Then, as sporulation proceeds, the unbranched thallus curves downward and sporangiospores are released near the basal part of the thallus and become attached to the original holdfast (Fig. 11.35), each new developing thallus contributing more holdfast material to the complex as it grows (Hibbits, 1978). *Enterobryus attenuatus* (in the hindgut of passalid beetles) produces rosettes of individual thalli that arise from a common holdfast system. It has not been observed how the thalli become joined in such a fashion in this species, although it is probably through some kind of deposition of spores as in *E. callianassae*. Manier and Théodoridès (1965) illustrated an unusual multiple holdfast system in an eccrinid from passalid beetles collected in Laos that they tentatively assigned to *E. attenuatus* (but probably is not that species). Their fungus produced

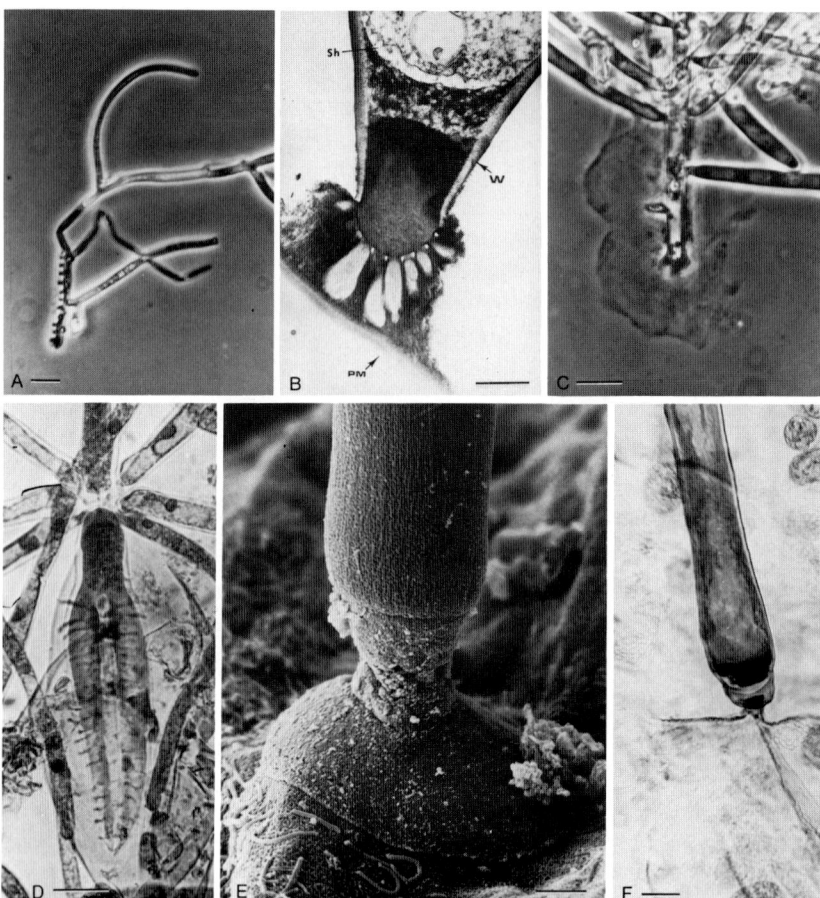

FIG. 7.10. Variations in trichomycete holdfasts. A. *Glotzia ephemeridarum:* row of auxiliary, peglike holdfasts secreted from basal cell above the original holdfast. B. *Harpella melusinae:* cementing substance surrounding finger–like projections from the base of thallus (*PM*, peritrophic membrane; *Sh*, crosswall formed from inner wall layer; *W*, thallus wall). C. *Pennella arctica:* mucilaginous secretion from basal cell. D. *Asellaria aselli:* cleft basal cell with pectinate rows of outgrowths. E. *Enterobyrus* sp.: disklike base of holdfast fitted to contours of cuticle of millipede host (*Californiobolus uncigerus*). F. *Palavascia sphaeromae:* small secreted holdfast firmly attached to a small area of host cuticle. Scale bars: A, C, D, F, 20 μm; B, 1 μm; E, 4 μm. Fig. 7.10B from Reichle and Lichtwardt, 1972, by permission of Springer–Verlag. Fig. 7.10E kindly furnished by R.E. Reichle. Fig. 7.10F from Lichtwardt, 1961b, by permission of J. Elisha Mitchell Sci. Soc.

some very long (up to 70 μm) and narrow holdfasts with a loosely branched appearance when mature. In the several instances cited above, the development of multiple holdfasts is a consistent feature of some of the thalli of those species, and not an abnormal situation such as can be seen on rare occasions in other species in which the holdfasts of two thalli may become fused when they form in very close proximity.

In almost all unbranched trichomycetes the holdfast is located at the base of the thallus in line with its axis. Some exceptions to this have been found in the Eccrinales that were caused by uneven growth of the thallus from the germinating spore or due to other unusual ontogenetic factors. The most extreme example was described by Whisler (1963) in *Enterobryus bifurcatus* from a millipede. In this species the multinucleate secondary infestation sporangiospores have the usual basal rudimentary holdfast upon initial attachment to the cuticle, but the developing thallus bifurcates at an early stage of growth, resulting in a mature nonseptate thallus with a "lateral" holdfast and two divergent arms of approximately equal length. Other examples of holdfasts that are not strictly basal are described in Chapter 11 (e.g., *Enterobryus tuzetae*).

Among the branched trichomycetes (Legeriomycetaceae and Asellariales), one finds several different basic forms of holdfast. The majority of Legeriomycetaceae produce a rigid secreted holdfast not unlike the typical holdfast of the unbranched trichomycetes. Sometimes the holdfast is inconspicuous, or it may be difficult to locate if the mature thallus has no main axis and is prolifically branched near the base (e.g., some species of *Smittium*). Where the thallus contains a prominent main axis (e.g., *Genistellospora homothallica*), the holdfast usually is also prominent.

A different type of holdfast system has evolved in species of *Pennella* and *Stipella*. Here there is a secretion of mucilaginous substance from the basal cell that serves to anchor the thallus, and the basal cell lies parallel to the substrate so as to present a greater adhesive surface. In some species (e.g., *Pennella hovassi*) the basal cell of larger thalli is typically branched and thereby provides an even greater surface area in contact with the host cuticle. *Pennella angustispora* (and possibly also other less studied species) secretes the adhesive substance through the wall onto the outer surfaces of the entire basal cell, not just the lower part of that cell. The cell wall itself is not in direct contact with the cuticle (Mayfield and Lichtwardt, 1980). The mucilage in this type of holdfast system is essentially transparent and is therefore best seen with phase–contrast microscopy. Often bacterial cells and some debris are embedded in or stuck to the mucilage. *Smittium pennelli*, unlike any other described species of this large genus, also produces a *Pennella*-like holdfast with a mucilaginous secretion.

A third type of holdfast in the branched trichomycetes is found in species of Asellariales. It consists of an enlarged or otherwise distinctively modified basal cell, usually without a prominent secreted structure. In some species a small amount of mucilage or a short basal zone of rigid holdfast material can be detected. (*Asellaria armadillidii* is an exception, for a noticeable amount of mucilage is secreted such that the basal part of the thallus somewhat resembles the type found in *Pennella* and *Stipella*.) In most Asellariales the shape of the basal cell is so distinctive that it serves as the principal means by which the various species are distinguished (Manier, 1979). These basal cells are described and compared in Chapter 11.

Modifications can be recognized in the three basic holdfast types in the branched trichomycetes just described. An interesting variation, for instance, is found in *Glotzia ephemeridarum* (Lichtwardt, 1972; Moss, 1979), where the primary holdfast consists of a small, rigid, peglike or disklike secretion. Then, as the thallus become larger, series of auxiliary holdfasts develop along the basal cell to strengthen further the attachment to the cuticle (Fig. 7.10A). Large thalli of *G. ephemeridarum* may have a tapered basal cell with a row of auxiliary holdfasts somewhat resembling the suckers on the tentacle of an octopus.

Light and electron microscopic studies to date, with a few notable exceptions cited in the next paragraph, indicate that holdfasts are entirely superficial on the host cuticle. The holdfasts are in intimate contact with the gut lining and follow the contours, setae, depressions, or other relief features of the lining. Sometimes a slight distortion of the cuticle by the holdfast is evident in whole mounts or sectioned material [e.g., *Genistellospora homothallica* (Mayfield and Lichtwardt, 1980); *Palavascia sphaeromae* (Manier, 1979a)], or, where there is an enlarged and modified basal cell (Asellariales), that cell may grow in such a way that it appears to pinch the cuticle layer. The lack of penetration by the holdfast through the cuticular layers of the gut adds weight to the interpretation that holdfast structures function as anchorage devices and not for absorption of nutrients from the tissues of the host.

However, several exceptions to the strictly superficial attachment of the holdfast are known. Ultrastructural studies of the holdfast of a dozen trichomycete species have revealed that in two Harpellaceae, *Stachylina grandispora* and *Harpella leptosa* (Moss, 1972, 1979; Moss and Lichtwardt, 1980), the secreted holdfast substance may penetrate into the inner layers of the peritrophic membrane immediately adjacent to the holdfast [although another species of *Harpella*, *H. melusinae,* does not appear to do this (Reichle and Lichtwardt, 1972)]. Such penetration should enhance adhesion to the membrane, and it would not be surprising to find this situation in other trichomycetes yet to be studied ultrastructurally. More interesting are three examples of trichomycetes that actually penetrate through the chitinous lining (Fig. 7.11): (1) Several *Stachylina* species have a base that perforates the peritrophic membrane and produces a slightly bulbous swelling underneath the membrane (Lichtwardt, 1984a), or even an outgrowth from this swelling (Moss, 1979), apparently for anchorage

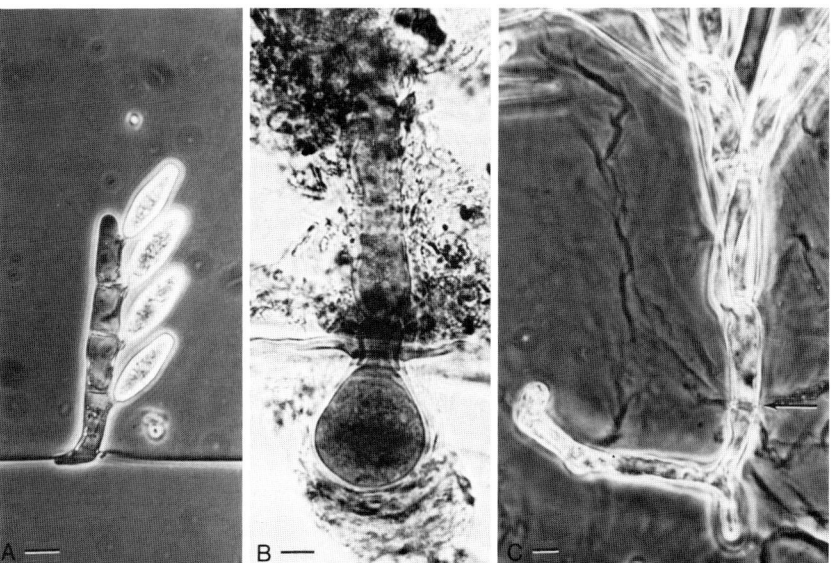

FIG. 7.11. Species of fungi that penetrate through their host's gut lining. A. Footlike base of *Stachylina pedifer* projecting through the peritrophic membrane. B. Cystlike structure produced by *Enterobryus borariae* is located outside of the gut lumen and has germinated through the hindgut cuticle to produce a thallus within the gut. C. Portion of a thallus of *Asellaria ligiae* that has penetrated the hindgut cuticle and produced a rhizoidal growth; the arrow indicates where penetration of the cuticle occurred. Scale bars = 10 μm. Fig. 7.11A from Lichtwardt and Williams, 1983a, and Fig. 7.11B from Lichtwardt, 1958, reprinted by permission of Mycologia, vol. 75: 732 (7.11A) and vol. 50: 556 (7.11B), Copyright 1983 and 1958, R.W. Lichtwardt and the New York Botanical Garden.

and as a substitution for the rigid, secreted type of holdfast that is characteristic of other species of that genus. (2) Peculiar structures in *Enterobryus borariae* were found located beneath the cuticle of its millipede host by Lichtwardt (1958) that consisted of pyriform, cystlike bodies that germinated to produce sporulating thalli of that species within the gut lumen; the origin of these bodies was not determined. (3) A third instance of cuticle penetration has been observed by the author (unpublished) in several specimens of *Ligia* sp. collected in a freshwater stream in Hawaii and containing *Asellaria ligiae:* in heavily infested hosts a number of thalli were seen to have no bulbous basal cell characteristic of this species, but rather had penetrated the hindgut lining and had produced a branching rhizoidal growth underneath the lining. These thalli appeared to be branches of the fungus that had broken off from other thalli in the gut and then had produced this adventitious growth from the proximal end to anchor themselves to the lining.

In none of the cases cited above have the structures in question appeared to penetrate into the underlying cellular tissues of the host. However, Sweeney (1981a) has described penetration of epithelial cells of the midgut of mosquito larvae by *Smittium morbosum*. This apparently unusual occurrence, and other pathological effects by species of *Smittium*, are presented in Chapter 8.

The vast majority of trichomycetes attach to specific and predictable substrates. Some instances of unusual attachment are known (Table 7.2). One of the most interesting is found in *Simuliomyces microsporus* which grows in the hindgut of blackfly larvae. While its thalli usually attach to the gut lining, not uncommonly it is seen in various stages of development, including sporulation, on thalli of *Paramoebidium* sp. (Fig. 7.12A) or *Genistellospora homothallica* that share the same gut, or even on thalli of its own species. In the latter instance, it may require careful observation to see that some of the "branches" arising from thalli are in fact different

TABLE 7.2. Attachment of trichomycete thalli to substrates other than their normal host cuticle.

Trichomycete	Substrate	References
HARPELLALES		
Simuliomyces microsporus	*Paramoebidium* sp.; *Genistellospora homothallica*; *Simuliomyces microsporus*	Lichtwardt, 1972
Smittium culisetae	On external cuticle of mosquito larvae (special culture conditions)	Horn and Lichtwardt, 1981
ECCRINALES		
Enterobyus elegans	Oxyuroid nematodes	Leidy, 1853; Thomas, 1930; Lichtwardt, 1954; Wright, 1979
Enterobryus borariae	Oxyuroid nematodes; *Enterobryus borariae*	Lichtwardt, 1958
Palavascia sphaeromae	*Palavascia sphaeromae*	Lichtwardt, unpublished
AMOEBIDIALES		
Amoebidium parasiticum	Cotton fibers, nylon, silk, hair (special culture conditions); glass	Whisler and Fuller, 1968; Chatton, 1906a
Paramoebidium sp.	*Paramoebidium* sp.	Lichtwardt and Williams, 1983b
	Legeriomyces aenigmaticus	Lichtwardt and Williams, 1983b
	Simuliomyces spica	Peterson and Lichtwardt, 1983
	Glotzia centroptili	Gauthier, 1936
	Spartiella barbata	Moss and Lichtwardt (unpublished)
	Simuliomyces spica	Peterson and Lichtwardt, 1983

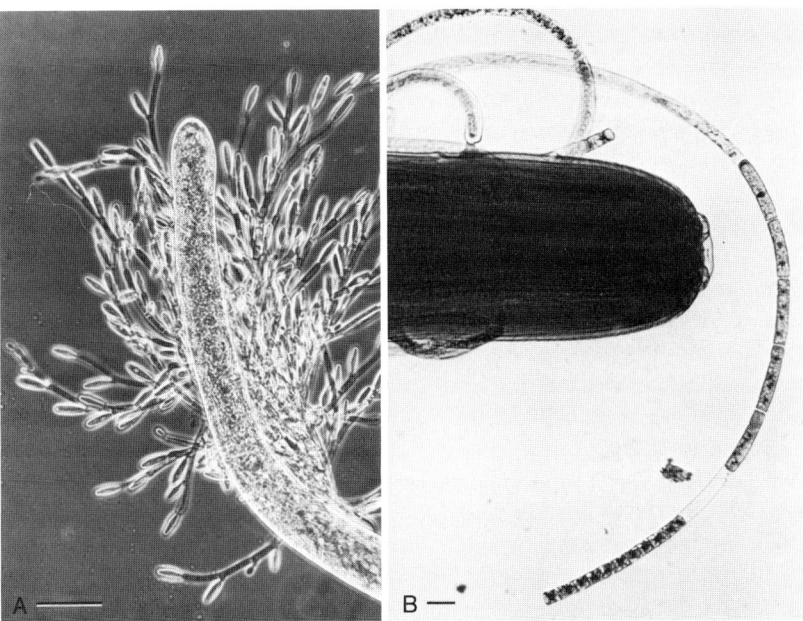

FIG. 7.12. A. Sporulating thalli of *Simuliomyces microsporus* attached to a robust thallus of *Paramoebidium* sp. B. Thalli of *Enterobryus borariae* attached to the cuticle of a gut nematode; one thallus has produced both primary and secondary infestation sporangiospores, one of the latter having emerged from its now empty sporangium. Scale bars = 20 μm. Fig. 7.12B reprinted by permission from Mycologia, vol. 50: 555, Copyright 1958, R.W. Lichtwardt and the New York Botanical Garden.

thalli of the same fungus attached by a rather inconspicuous holdfast. What evidently was *S. microsporus* on *Paramoebidium* spp. was seen and published by several investigators prior to 1972 when *S. microsporus* was described, but it appears to have been misidentified as species of other genera: as *Stipella vigilans* (Manier, 1955a) and as *Smittium* sp. (Ingold, 1967; Moss, 1970).

Enterobryus borariae is another species that usually attaches to the hindgut cuticle of its millipede host but has been found in all stages of development attached to other thalli of its own kind (Lichtwardt, 1958). Frequently it attaches as well to the cuticle of oxyuroid nematodes that are common parasites of millipede hindguts (Fig. 7.12B). The attachment of Eccrinales to nematode surfaces was first mentioned by Leidy in 1849a, and described and illustrated in 1853, in *Enterobryus elegans*. This phenomenon was restudied in the same species by Lichtwardt (1954), who also found a modified spore type that attached to nematodes in abundance just prior to molting of the millipede. Lichtwardt suggested that this attachment to nematodes might be a mechanism to maintain some of the thalli in the gut during ecdysis, because the nematodes, unlike the eccrinids attached to the gut lining, are not expelled with the molt.

The author has seen several other, undetermined species of *Enterobryus* attached to gut nematodes, and other biologists have also observed this phenomenon in *Enterobryus* spp. (Udekem, 1859; Thomas, 1930; Tuzet and Manier, 1952; Dollfus, 1952). Manier (1969b, p. 581) mentioned that *Eccrinidus flexilis* does likewise. Most *Enterobryus* species rarely, if ever, attach to nematodes, however. The frequency of attachment of those that do varies from one millipede specimen to another, but in general it depends on the density of the fungal population: In heavily infested guts, it is not uncommon to find the majority of nematodes carrying some eccrinid thalli. The millipede hosts of *Enterobryus elegans* can have several species of gut nematodes, all of which may carry the fungus, but the percentage of infestation can vary among the species (Wright, 1979). None of the trichomycetes that attach to nematodes, to other trichomycetes, or to their own species penetrates or harms the other organism in any observable way.

The initial attachment of a trichomycete to its substrate is determined by the spore. Except in the Eccrinaceae and Parataeniellaceae, where spore types exist that germinate in the gut immediately on release, spores of trichomycetes pass from the gut and are ingested by a suitable host before they attach and germinate. Alien trichomycete spores, which one can assume are ingested by certain arthropods whose microhabitats overlap those of others (see Chapter 4), apparently pass through the gut without attachment. In fact, recognition of a suitable substrate to which the spore can attach may represent the first level by which host specificity is determined. Once in a suitable host, attachment occurs predictably either on the lining of the foregut, midgut, or hindgut, often in particular regions of those parts of the gut. [Sporangiospores of the ectocommensal *Amoebidium parasiticum* may have a simple adhesive property, which could account for the wide host range and lack of selectivity in that species, for they even attach to various inanimate objects under special conditions (Table 7.2).] Although a few species of trichomycetes can attach to unusual living substrates, there appear to be no instances known where spores attach to any of the many kinds of solid debris that pass through the gut, except by entanglement of trichospore appendages. These observations lead to the conclusion that there is considerable selectivity exercised by the spore. One possibility, not yet demonstrated, is that the spore produces a lectin that binds to specific carbohydrates in the appropriate living substrate. Such a system could operate even if the major component of the substrate were chitin, cellulose, collagen (as in nematode cuticles), etc., provided suitable saccharides were also present.

The process of substrate recognition and initial attachment of the spore could well be a developmental aspect distinct from the subsequent formation of the holdfast proper. The several electron microscopic studies of holdfasts to date have been based mostly on relatively mature holdfasts; as a consequence, developmental features are not well understood. Nevertheless, these studies have revealed some aspects of their probable development as well as a rather surprising difference among the few species studied so far (Table 7.3). There have been no published analyses of the chemical nature of the secreted substance that forms the holdfast. Duboscq et al. (1948) and Manier (1950) stated that, on the basis of its staining properties (particularly with cotton blue), the holdfast in *Eccrinidus flexilis* consists of callose. However, the stains they used are not sufficiently specific, nor is callose a likely candidate in this instance because of its solubility properties. Mayfield and Lichtwardt (1980) reported that the silver–methenamine stain gave no indication of polysaccharides being present in the multiple holdfast of *Enterobryus attenuatus,* but a strong positive test for polysaccharides was demonstrated in the "ring complex" at the base of each thallus through which the holdfast material is extruded.

Holdfast formation commences soon after a spore has become attached, and in some species may continue to develop for some time along with thallial growth and maturation. In others, holdfasts are of limited size, and development appears to cease as soon as firm attachment is insured. Multinucleate secondary infestation sporangiospores of Eccrinales are capable of immediate attachment on release from the sporangium. In some species the end of the cylindrical sporangiospores that will produce the holdfast can be distinguished while still in the sporangium by virtue of a slightly swollen base, and in a few species the spore may even produce a pad of substance resembling a small holdfast before release.

Whisler and Fuller (1968) studied the attachment of cultured sporangiospores of the ectocommensal *Amoebidium parasiticum* to both *Daphnia* and cotton fibers (Table 7.3). (The attachment to cotton fibers occurred when thalli were first transferred to a dilute salt medium but not when kept in their tryptone broth growth medium.) The fusiform spores were found to have numerous conical pits that perforated the wall and were primarily concentrated at each pole, but there were some pits widely dispersed along the lateral wall as well. Within the spores they found many membrane–bound vesicles aggregated mostly near the two poles and containing homogeneous material that was electron microscopically similar in appearance to the amorphous substance of the mature holdfast. Unreleased spores showed no holdfast, but extrusion of the cementing substance apparently occurred rapidly after initial attachment. The concentration of pits at both ends of the spore may be adaptive; either end of the spore can break through the sporangial wall at maturity, with the unexposed end of the spore lying within the sporangium until (under natural conditions) the spore attaches to a passing arthropod on contact and is pulled from the sporangium. Species of the closely related endocommensal genus *Paramoebidium* produce no sporangiospores. However, like *Amoebidium* species, they produce fusiform cystospores that develop from

TABLE 7.3. Principal characteristics of trichomycete holdfasts studied ultrastructurally.[a]

Species	Distinctive features	References
HARPELLALES		
Harpella melusinae	Digital projections radiating from base of thallus and surrounded by a cementing substance that exudes from a basal compartment in the thallus	Reichle and Lichtwardt, 1972
Harpella leptosa	No digital projections; granular holdfast substance exudes from a basal compartment in the thallus and penetrates into the peritrophic membrane	Moss and Lichtwardt, 1980
Stachylina grandispora	Thin pad of granular or fibrous material surrounds base of thallus and penetrates into the peritrophic membrane	Moss, 1972
Genistellospora homothallica	Biconcave, context homogeneous and dense, or fibrous	Mayfield and Lichtwardt, 1980
Pennella angustispora	Amorphous cement, indistinct from outer wall layer, secreted along entire wall of the basal cell which lies parallel to substrate	Mayfield and Lichtwardt, 1980
Smittium chironomi	Thin pad indistinguishable from outer layer of cell wall	Moss, 1972
ECCRINALES		
Eccrinidus flexilis	Granular holdfast secreted through a zone of distinct pores formed in the innermost wall layers	Manier and Grizel, 1972
Enterobryus attenuatus	Compound fibrous holdfast secreted through a reticulate ring complex at the base of each individual thallus	Mayfield and Lichtwardt, 1980
Enterobryus elegans	Large fibrous holdfast containing numerous vertical parallel channels	Wright, 1979; Mayfield and Lichtwardt, 1980
Palavascia sphaeromae	Biconcave, surrounded by a mucous substance	Manier, 1979a
AMOEBIDIALES		
Amoebidium parasiticum	Amorphous cement secreted from one end of the thallus through discrete pores	Whisler and Fuller, 1968; Coste-Mathiez, 1970; Dang, 1979
Paramoebidium curvum	Planoconcave, context homogeneous and dense	Dang and Lichtwardt, 1979

[a] Based on Mayfield and Lichtwardt, 1980. Excluded are some species insufficiently described or documented: *Astreptonema gammari*, *Taeniellopsis susplugasi*, and *Smittium mucronatum* (Grizel, 1971; Manier and Grizel, 1972).

encysted amoebae. Dang and Lichtwardt (1979) reported that at least one end of these cystospores in *P. curvum* contained high concentrations of cylindrical pits and that vesicles were aggregated around the plasmalemma in the same region.

Discrete wall pits have also been found at the base of the thallus in the eccrinid *Eccrinidus flexilis* (Manier and Grizel, 1972), with the pits restricted in this case to a small basal zone on the thallus. An even more specialized area of holdfast secretion has been found in *Enterobryus attenuatus* (Mayfield and Lichtwardt, 1980) that consists of a "ring complex" at the base of each thallus, with the holdfast material apparently being extruded through a network of spaces in the middle of the complex. In *Enterobryus elegans* there are no discrete pores or spaces, but there are thinner areas along the inner wall at the base of the thallus through which the holdfast substance is presumed to pass.

Distinct pits at the base of thalli have not yet been found in the few species of Harpellales studied, but in trichospores of *Legeriomyces ramosus* (Manier, 1973a) and *Genistellospora homothallica* (Moss and Lichtwardt, 1976) there are some peculiar apical channels on the inner spore wall that may well be associated with the process of spore attachment and holdfast formation. These are described in the next section. In the eccrinid *Palavascia sphaeromae* the oval sporangiospores, while still in the sporangium, possess an area of pits in the wall at one end (Manier, 1979a). On release of the spore and attachment to the host cuticle, the holdfast material passes through these pits. The pits can be seen in the more mature thalli in the outer wall layer just above the holdfast; the inner wall layers do not form along this zone of holdfast secretion.

Membrane–bound vesicles present within cells near the region of holdfast formation suggest that these may function as carriers of the holdfast substance or some precursor. The origin of such vesicles is not known for certain, though in *Palavascia sphaeromae* Manier (1979a) implied they were derived from Golgi. Mayfield and Lichtwardt (1980) reported that the membranes of some of the vesicles in *Enterobryus elegans* fused with the plasmalemma and were seen releasing their contents just above the zone where the holdfast material is secreted through the wall, and Manier (1979a) described a similar process in *P. sphaeromae*. The holdfast of *E. elegans* continues to form and enlarge for some time during thallus maturation, in contrast to most of the other species of trichomycetes that have been studied. It is probably necessary to observe earlier stages in most species if it is to be determined whether vesicular transport is a feature common to all types of holdfast formation. It should be mentioned here that vesicles are also involved in cell wall formation, consequently it is conceivable that some (or all) of the vesicles seen in the general zone of holdfast secretion might in fact be active instead in deposition of wall material, at least in the younger stages of thallus development. This is a problem that should be addressed in future studies. However, the observed fusion of some vesicles with the plasmalemma in the immediate area above the holdfast at the base of relatively mature thalli of *Enterobryus elegans* and *E. attenuatus* (Mayfield and Lichtwardt, 1980), in cases where the cell wall seems to be already well developed, gives credence to the concept of involvement of vesicles in holdfast formation.

It is evident from observations on some species of Eccrinales with large holdfasts that the holdfast substance must be at least initially in a fluid or plastic state. For instance, thalli of *Arundinula washingtoniensis* have been seen where the holdfast substance has wrapped around setae on the stomach cuticle of its decapod host (Hibbits, 1978). In other Eccrinales the base of the holdfast is spread into a disk much broader than the zone at the base of the thallus through which the holdfast substance is extruded. Some thalli of *E. elegans* have a large and roughly conical shape with successive tiers of holdfast material heaped one on the other (Dang, 1979), looking as though the lower layers had spread out as the holdfast grew in length.

The holdfast of *E. elegans* is especially perplexing in terms of its development because of the numerous parallel vertical channels (containing a loose fibrous substance) that form within the holdfasts (Fig. 7.13). The upper holdfast region, just below the cell, consists of a relatively dense and homogeneous fibrous material. As one sections the holdfast serially downward, the channels, numbering around 200, become discernible and progressively more numerous. There were about 1800 in the broad base of one large thallus that was studied (Mayfield and Lichtwardt, 1980). Each channel arises from a pitlike depression in the gut cuticle. The parallel channels closer to the periphery of the conical holdfast open as pores on the outer edges, and only the more central ones extend up through the holdfast. Because of their arrangement, it appears reasonable to assume that the channels develop due to some form of condensation (polymerization?) of the holdfast substance well after it has been extruded. Wright (1979) studied this species and made the interesting observation that the holdfast channels are found only in thalli attached to millipede cuticle; those attached to nematodes (*Rhigonema infecta*) had no distinct channels nor any pores opening to the outside of the holdfast. Taking into account that the arthropod hindgut serves to dehydrate intestinal contents by absorption of water, Wright suggested that if the pitlike depressions on the millipede cuticle represent areas of greater permeability, then fluid flow through these points might prevent consolidation of holdfast material above them and result in channel formation in the holdfast; there is no comparable fluid flow through the cuticle of nematodes and consequently there are no channels within holdfasts attached to nematodes. This interesting hypothesis does not account for the channels in the central core of the holdfast that do not open to the outside of the holdfast.

The many variations in holdfast structure that have been detected

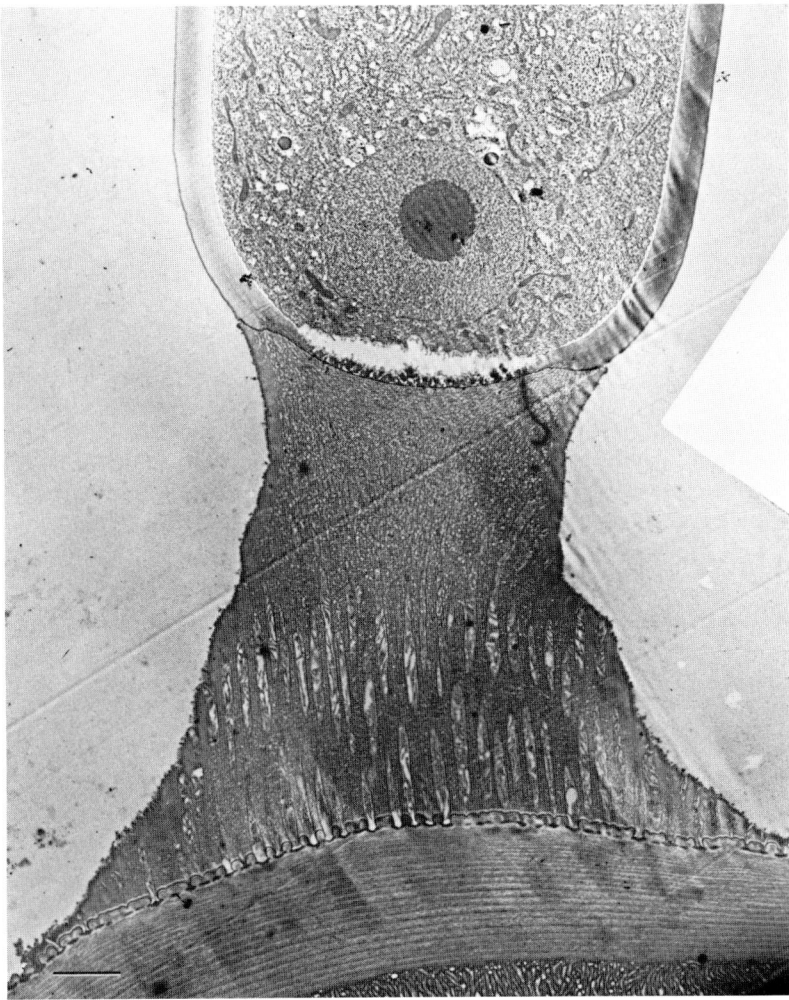

FIG. 7.13. Holdfast of *Enterobryus elegans* attached to the laminated hindgut cuticle of its millipede host. Scale bar = 2 μm. From Mayfield and Lichtwardt, 1980, by permission of National Research Council of Canada.

through electron microscopic studies are not necessarily adaptations induced by substrate differences in the hosts of the respective species they presently occupy. As an example, one has but to compare the substantially different holdfasts of *Pennella angustispora* with *Genistellospora homothallica* growing side by side in blackfly hindguts. Understanding the commonalities as well as the differences that characterize trichomycete holdfasts will have to await further investigations on their development and chemistry.

Asexual Reproduction

Asexually produced spores are the major means of propagation in trichomycetes. Sexual reproduction is apparently lacking in most taxa, or where it does occur it is generally less frequent than asexual reproduction. In *Graminella* spp. a means of vegetative propagation appears to be significant in producing more thalli endogenously in the gut (Léger and Gauthier, 1937; Lichtwardt and Moss, 1981), but trichospores are still formed in relatively large numbers and may function as the major propagule in transmitting the fungus from one individual to another.

Asexual spores develop entirely within the gut lumen in all but a few instances. The exceptions include representatives of three orders. In the Harpellales, *Pteromaktron protrudens* thalli protrude from the anus of mayfly nymphs and produce a brushlike cluster of spores in the external aquatic environment (Whisler, 1963). The larger sporulating thalli of *Zygopolaris ephemeridarum* and *Z. borealis* often project from the anus of their mayfly hosts (Moss et al., 1975; Lichtwardt and Williams, 1984), and Moss (1970) reported that thalli of *Stipella vigilans* occasionally extended beyond the anus in blackfly larvae. In the Eccrinales there have also been reports of spores projecting from the gut. Thalli of *Palavascia sphaeromae* apparently do not sporulate until the tip reaches the anus of its marine isopod hosts. The tip then coils and produces a series of sporangiospores, which can often be seen projecting outside of the gut (Lichtwardt, 1961b). Wolf and Wolf (1947) recorded that tufts of thalli of an eccrinid, possibly *Enterobryus* sp., have been seen protruding from the anal opening of the mud crab *Panopeus herbstii*, but they did not state if any of these projecting thalli were sporulating.

Species of Eccrinales that are capable of producing unusually long thalli probably will be found from time to time protruding from the hindgut, as reported by Hibbits (1978) in *Arundinula haplogaster*. Terrestrial beetles of the Passalidae in the neotropics sometimes contain *Enterobryus compressus,* which Thaxter (1920) described as "growing wholly exposed on the anal plates of a large species of *Passalus.*" The author has made several collections of passalids with this fungus, but has found that the majority of thalli are relatively sheltered within the gut and enclosed within the posterior edge of the elytra most of the time. They are probably exposed only during extension of the anal plates. Nevertheless, their periodic exposure to the air may account for the rather thick walls that develop in most of the thalli and sporangia.

The order Amoebidiales includes the ectocommensal *Amoebidium parasiticum* whose thalli as well as all reproductive stages are, of course, continuously in the external aquatic environment. Other Amoebidiales inhabit the rectum of their hosts, but *Paramoebidium curvum* thalli are sometimes found growing on the retractile anal gills of blackfly larvae, rather than being in their more normal position just inside the anus, and consequently are fully exposed to the outside when the gills are extended. It is possible that a number of trichomycetes in the rectum of aquatic arthropods develop and sporulate with at least a sporadic exposure to the external water, brought about by muscular activity, which may draw water into the rectum.

The remainder of this section is devoted to describing the development, structure, and function of asexual spores in each of the orders.

HARPELLALES - The sole asexual spore of the Harpellales is the trichospore. This structure can be defined as an exogenous, deciduous sporangium containing a single uninucleate sporangiospore and normally having one to several basally attached filamentous appendages (Moss and Lichtwardt, 1976). The term trichospore was first introduced in this sense by Manier and Lichtwardt (1968) [although its first use in the trichomycete literature was by Chatton (1925), who used it once to refer to the elongate spores of *Amoebidium*]. Other terms that have been applied to the trichospore are conidium (Léger and Duboscq, 1929a), azygospore (Poisson, 1936), mastigocyst (Manier, 1955b), trichosporangium (Moss, 1972), merosporangium (Moss and Young, 1978), and sporangiole or sporangiospore (Moore–Landecker, 1982). Thalli of the unbranched Harpellaceae are completely converted to reproductive cells on reaching maturity, whereas in the branched Legeriomycetaceae only part of the thallial biomass normally becomes reproductive. In *Orphella coronata*, the trichospores were described as being produced singly at the tips of branches (Léger and Gauthier, 1931), but in all other species of Legeriomycetaceae they develop in basipetal series from generative cells located at the distal region of branches. An unusual situation can be seen in some strains of *Smittium culisetae* wherein the production of trichospores is not restricted to a short series of generative cells at the tips of branches, as is normal. Rather, many of the thallial cells that normally would remain vegetative become fertile, and in extreme cases virtually the entire thallus becomes reproductive through production of spurious generative cells (Fig. 7.14A). This development may be preceded by septation in some of the longer cells of the thallus. In some species of *Smittium* where cells remain longer than normal they can give rise to trichospores markedly larger than the norm (Fig. 7.14B).

Appendages have not been seen, or are not produced, in 4 of the 17 genera of Harpellales: *Carouxella, Gauthiera, Orphella,* and *Zygopolaris*. In all other genera the number of appendages on trichospores varies, according to the genus, from one to six or seven. Two types of trichospores based on current electron microscopic studies of appendage formation (Moss and Lichtwardt, 1976) are illustrated in Fig. 7.15. Type A consists of one or several appendages that develop in a corresponding number of appendage sacs, which are invaginations of the plasmalemma located in the central region of the distal end of the generative cell. In Type B, the appendages form adjacent to the generative cell wall. The common feature

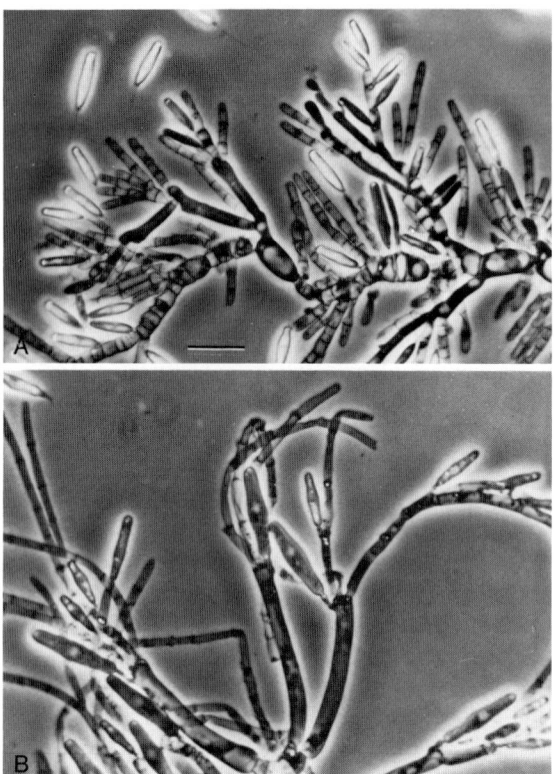

FIG. 7.14. A. Strain of *Smittium culisetae* wherein most cells become reproductive by virtue of spurious generative cell formation. B. A species of *Smittium* (possibly *S. pusillum*) with normal trichospores as well as abnormally large trichospores arising from long cells that became reproductive. Scale bar = 20 μm for both figures.

of all appendages that have been sufficiently studied is that they form outside of the plasmalemma, rather than within the generative cell protoplast. This process involves synthesis and transport of appendage material, or some precursor, from the cell by means of membrane–bound vesicles that fuse with the plasmalemma at specific sites and deposit appendage material to the outer side of the membrane where the appendages are constructed. Trichospores thus appear to be unique fungal propagules in terms of development and fine structure of the appendages. The chemical composition of appendages has not been determined. Descriptions and comparisons of appendage formation in several genera follow.

Stachylina grandispora (Harpellaceae) has been studied by Moss (1972, 1974, 1976). Each generative cell contains a nucleus that divides mitotically. One nucleus migrates into the developing trichospore, and the other remains in the generative cell, after which a perforate septum forms to

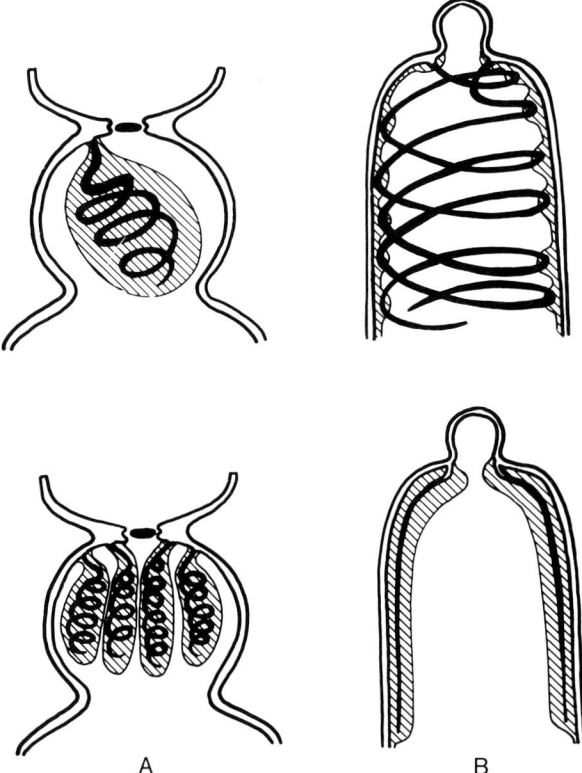

FIG. 7.15. Diagrams of two basic types of appendage formation in Harpellales trichospores as revealed by electron microscopy: Type A, consisting of one to several appendage sacs (invaginations of the plasmalemma), each containing a single appendage; and Type B, consisting of one or more appendages formed adjacent to the generative cell wall and outside of the plasmalemma. From Moss and Lichtwardt, 1976, by permission of National Research Council of Canada.

separate the two cells. Appendage vesicles appear in the cytoplasm of the generative cell, and seem to originate in the perinuclear space and bleb off from the outer nuclear membrane. The vesicles have an electron–opaque core and a less dense peripheral layer. They migrate to the septal region next to the plugged pore and fuse with the plasmalemma, the process eventually causing that membrane to become invaginated. As additional appendage material is deposited near the septum, a larger invagination develops to become the appendage sac containing a single appendage folded several times upon itself. Within the appendage sac is a matrix surrounding the ribbon–like appendage of electron–opaque amorphous material. The initially formed portion of the appendage at the base of the spore, however, is rod shaped. While the appendage is taking shape, the generative cell becomes progressively more vacuolated, and eventually all the cellular organelles appear to degenerate. At maturity the trichospore breaks loose, retaining a small collar of wall material below the septum, and at this moment the appendage unfurls and quickly leaves the generative cell as though forced from the inside. The appendage of the released spore provides no motility. It is about 0.3–0.4 μm wide near the base of the spore and 200–500 μm long, although the length is often difficult to determine because it tapers toward the terminal end and cannot always be resolved optically with the light microscope.

The actual process of appendage formation has not been studied in the genus *Smittium*, but the mature appendage in both *S. mucronatum* (Manier and Coste–Mathiez, 1968) and *S. culicis* (Moss and Lichtwardt, 1976) forms within a large appendage sac that fills the outward extension of the generative cell beneath the trichospore that later becomes the spore collar. Transverse sections of appendages in both of these species appears in electron micrographs as a series of concentric electron–opaque rings. This unusual structure was not found by Preisner (1973) in either *S. culisetae* or *S. simulii*, however, and there may be variation in appendage substructure within the genus.

A single appendage is formed on trichospores of *Smittium* spp., but in *Trichozygospora chironomidarum* there are about 5–7 appendages, each produced in a separate appendage sac located primarily in the collar region but extending into the main body of the generative cell (Moss and Lichtwardt, 1976). *Trichozygospora* appendages consist of an amorphous electron–opaque material surrounded by a lighter matrix that fills the appendage sac, and, as in *Smittium* spp., the appendages are twisted about within the sac.

A study of *Genistellospora homothallica* trichospores by Moss and Lichtwardt (1976) revealed yet another modification of appendage formation (Fig. 7.16). Appendages begin to form as soon as trichospore growth commences. As the developing spore begins to bulge out of the generative cell, membrane–bound vesicles are seen in the apical and subapical regions of the generative cell and appear to be involved both in deposition of wall material in the enlarging spore as well as in formation of appendages at its base. These vesicles originate from areas of convoluted endoplasmic reticulum. Approximately six appendages develop; they initiate adjacent to the generative cell wall in invaginations of the plasmalemma, and continue their development straight downward. The appendage sacs enlarge, and in transverse section a matrix is seen filling the appendage sac and completely surrounding each of the semicircular to lunate appendages.

The most complex appendage substructure occurs in *Harpella melusinae* (Reichle and Lichtwardt, 1972). In longitudinal section the appendage has an alternating light– and dark–banded structure (Fig. 7.17C). The lighter bands are approximately equal in width, but the dark bands are alternately wider and narrower. The periodicity of the banding is about 136 nm. Light

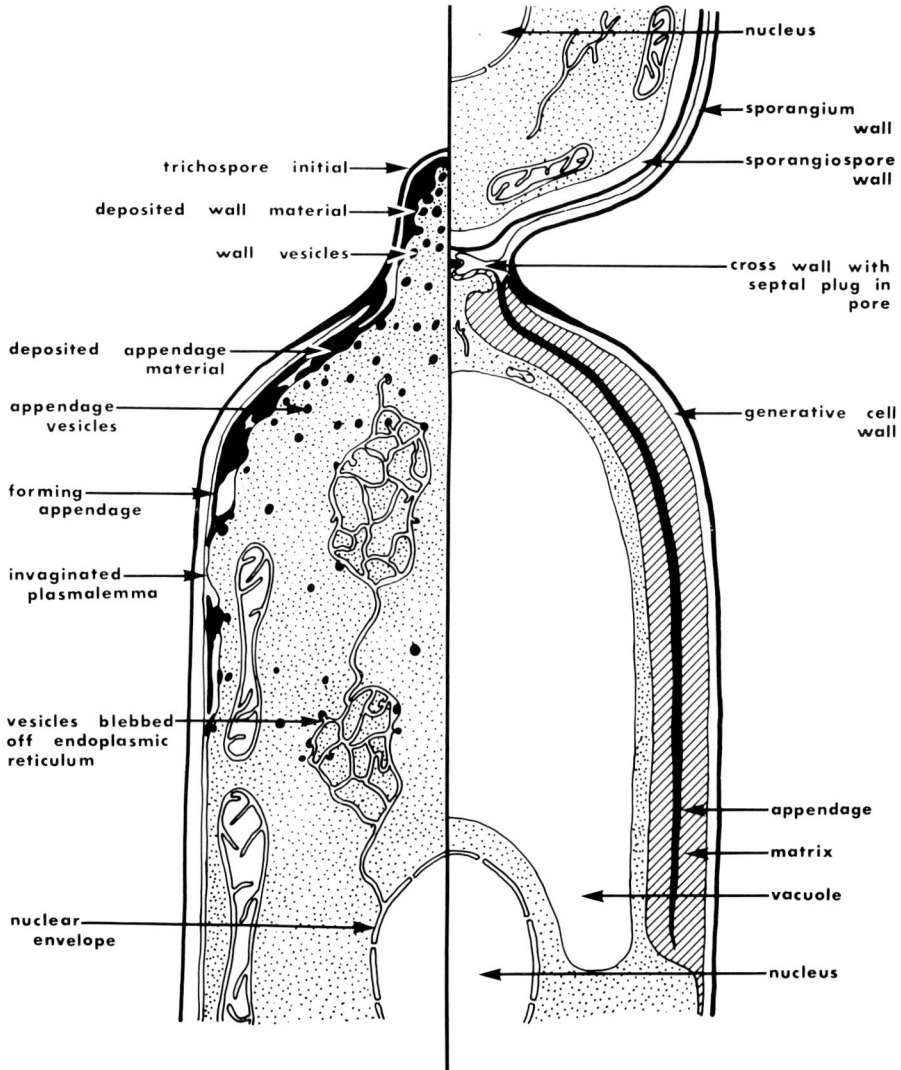

FIG. 7.16. Schematic representation of appendage formation (left) and a formed appendage (right) of *Genistellospora homothallica*. See text for a discussion of the process involved. From Moss and Lichtwardt, 1976, by permission of National Research Council of Canada.

microscopic observations (Lichtwardt, 1967) indicate that there may be a 10–fold longitudinal expansion of the appendages after release from the generative cell. Four appendages normally form at the base of each *H. melusinae* spore, and these are seen more or less tightly spiralled against the inner generative cell wall and outside the plasmalemma (Fig. 7.17A, B). Active fusion of vesicles with the plasmalemma can be seen at certain

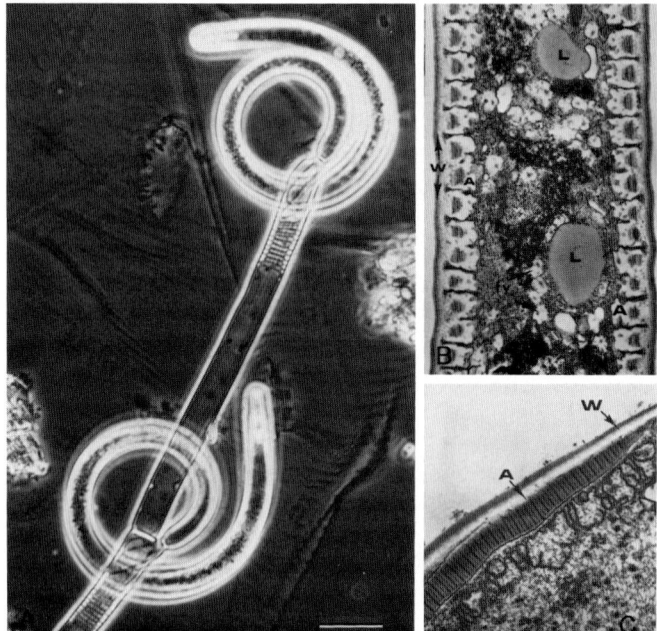

FIG. 7.17. Appendages of *Harpella melusinae*. A. Two sporulating generative cells with appendages lying tightly coiled inside next to the wall, as seen in phase–contrast microscopy. B. Longitudinal electron microscopic section of a generative cell with the forming appendages seen in cross section; note fusion of vesicles with the plasmalemma, which partly surrounds the appendages. C. Longitudinal section of an appendage showing periodicity in the banded substructure. Symbols: A, appendage; *fv*, forming vesicle; L, lipid; W, wall of generative cell. Scale bar = 20 μm. Figs. 7.17B and 7.17C from Reichle and Lichtwardt, 1972, by permission of Springer–Verlag.

stages along the length of the spiral, but the origin of the forming vesicles has not been determined. Nor is it known how these structurally complex appendages are constructed after vesicles fuse with the membrane.

In some genera (e.g., *Legeriomyces, Genistelloides*) appendages begin to form well before the trichospore starts to grow out from the generative cell. Thus, the time at which appendages begin to develop seems to depend on the particular genus or species.

Terminal trichospores in *Genistellospora homothallica* arise by holoblastic outgrowth from the generative cell; that is, the wall of the developing trichospore is an extension of the two wall layers of the generative cell (Moss and Lichtwardt, 1976). Subterminal trichospores develop indirectly from their generative cells: First, an outgrowth develops laterally from the terminal end of the generative cell by production of new wall layers laid down internally, in the same manner in which branches develop in vegetative hyphae of this species. Then, after a short (1–4 μm) extension of this branch, the trichospore develops by holoblastic outgrowth, as in the terminal trichospores. A septum develops at the base of the trichospore, and this is followed by the manufacture within the original trichospore wall of a new wall around the protoplast consisting of two layers of wall material similar to those of vegetative hyphae and generative cells. In young spores the new inner wall is appressed to the outer wall, but the two wall layers are not fused, and sometimes can be seen physically separated. Thus, the trichospore can be interpreted as a monosporous sporangium consisting of a sporangial wall with an inner, separate sporangiospore wall surrounding the spore proper.

Two unusual internal features of trichospores characterize several genera of Harpellales whose mature trichospores have been studied by electron microscopy (Manier, 1973a; Preisner, 1973; Moss and Lichtwardt, 1976). The apex of the inner wall contains an annulus consisting of a thickened structure with several canals that traverse both inner wall layers (Fig. 7.18). Immediately behind this perforated thickening and extending almost to the nucleus are several (1–10) elongate or spherical membrane–bound "apical spore bodies." These undoubtedly correspond to the refringent bodies seen with the light microscope in trichospores of many genera of Harpellales. In *Genistellospora homothallica,* numerous membrane–bound vesicles 30–69 nm in diameter have been found located between and near the apical spore bodies and the annulus. The membranes of these vesicles are sometimes continuous with the membrane of the apical spore bodies or with the plasmalemma lining the annulus. The implication is that vesicles carry material from the apical spore bodies and pass it through the plasmalemma to the canals of the annulus. In *Legeriomyces ramosus,* considerable secreted material appears to accumulate between the annulus

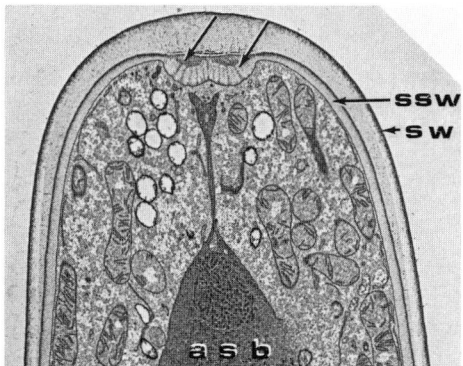

FIG. 7.18. *Genistellospora homothallica* trichospore apex showing section of an annulus with canals (arrows). Symbols: *asb,* apical spore body; *sw,* sporangial wall; *ssw,* sporangiospore wall. From Moss and Lichtwardt, 1976, by permission of National Research Council of Canada.

and the outer trichospore wall (Manier, 1973a). Manier, and also Moss and Lichtwardt (1976), have suggested that this apparatus is involved in release of the sporangiospore from the sporangial wall, or attachment of the spore to the gut lining of the host, or both. Germinated spores of *G. homothallica* and *L. ramosus* have been seen only rarely; therefore electron microscopic studies of these spores during germination have not been feasible. Only future fine-structural studies will indicate whether apical spore bodies and annuli are found in all trichospores.

The appendages at the base of released trichospores appear to have a passive but possibly important function in transmission. They have been seen to entangle with various external objects as well as host fecal material (Fig. 7.19), and it is believed that this helps to prevent spores from floating or washing away, in this way keeping many of them on the substrate of their hosts and increasing their chances of being ingested. Most species of Harpellales grow in insect larvae that inhabit streams with fast-flowing water, and a mechanism to restrain spores from floating downstream may have been critical in the evolution of these fungi. In the case of still waters, where mosquitoes and bloodworms may be found for instance, the entanglement of spores in fecal matter may allow them to settle more rapidly or remain at the bottom where such hosts usually feed.

Species of two monotypic genera apparently have no appendages associated with their trichospores. *Zygopolaris ephemeridarum* (Legeriomycetaceae) trichospores remain attached more tenaciously to their generative cells than do most Harpellales. Occasional loose trichospores have been seen with a small blob of material at the base of the spore that might represent incompletely formed appendage material (Moss et al., 1975), and Moss and Lichtwardt (1976) reported finding in their electron micro-

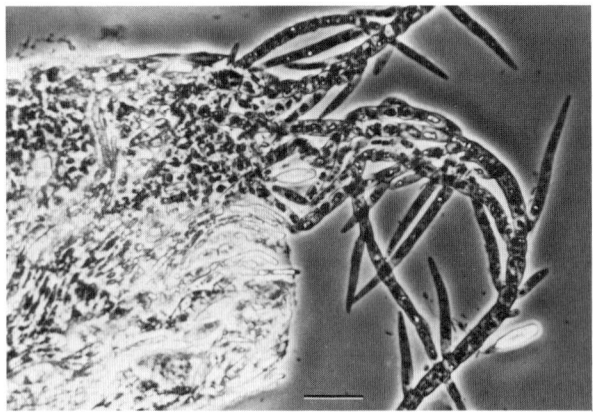

FIG. 7.19. Posterior end of a dissected midge hindgut from which algal filaments are emanating together with trichospores of *Smittium simulii* attached by their appendages to the filaments. Scale bar = 20 μm.

scopic study of this species a rudimentary type of appendage structure at the base of the spore still attached to its generative cell. It thus appears that in *Z. ephemeridarum* the ability to form discrete appendages may have been lost, or at least they are not completely developed in many instances. Tufts of trichospores on the brushlike end of the thallus of this species often project from the mayfly nymph anus, leading Moss et al. to suggest that one method of transmission may be by direct ingestion of trichospores by another nymph, without the spores being first released. *Carouxella scalaris* from ceratopogonid larvae is a curious member of the Harpellaceae in that it has no appendaged trichospore and an unusual type of dispersal. The spores do not detach from their generative cells; rather, the generative cells break apart from each other at maturity and the two structures are disseminated as a unit, referred to as a "diaspore" by Manier et al. (1961). According to these authors, the generative cell on reestablishment in a host produces the holdfast, after which the trichospore, still attached, develops into a new thallus.

Trichospores normally germinate only after being ingested. Occasional spores of several genera have been seen to germinate in water on microscope slides. In some cases this may be due to changes in osmotic conditions or other artificial factors. Germination typically involves an extension of the inner wall resulting in the spore body breaking through a tear in the sporangial wall. This process is described in more detail in Chapter 9 in connection with studies on cultured species of *Smittium*.

ASELLARIALES - The branched thalli of this order reproduce asexually by cells that disarticulate like arthrospores. Immature thalli are septate, the septa being the perforate harpellid type discussed earlier in this chapter (Manier, 1973b; Moss, 1975; Moss and Young, 1978). At this stage the cells typically are relatively long in most species, and contain one to several nuclei. On maturation, and possibly in response to onset of molting in the host (Poisson, 1932a; Manier, 1958, 1964c), nuclei may begin to divide in longer cells, and secondary crosswalls develop to delimit uninucleate segments of more or less uniform length. It is not known whether the secondary septa are perforate or consist of nonperforate walls. This process is essentially basipetal, and in some species (e.g., *Asellaria ligiae, Orchesellaria lattesi*) the complete thallus may disarticulate, leaving only the specialized holdfast cell intact. As a consequence, the molt of the arthropod may contain large numbers of arthrospores, or, if these have already disseminated, only the basal cells remain still firmly attached to the cuticle.

On first consideration this type of asexual reproduction would appear to be quite distinct from that found in other orders of trichomycetes, were it not for the observations (Manier, 1963a, 1969b; Lichtwardt, 1973a) that the arthrospores of *Asellaria ligiae*, when kept in water, "germinate" to produce a lateral cell whose size and shape resembles that of a trichospore (see Fig. 11.26C). Lichtwardt (1973a) called attention to the remarkable similarity of this "germinated" arthrospore to the trichospore–generative cell unit of *Carouxella scalaris* (Harpellaceae) (Fig. 11.3) referred to previously. Unfortunately, material has not been available for an electron microscopic study to ascertain whether in *A. ligiae* (or in *C. scalaris* for that matter) the trichospore–like bodies have the same double wall and internal structures that characterize the typical trichospore of the Harpellales.

Arthrospores of other Asellariales have not yet been seen to develop in this manner. Poisson (1932a) kept arthrospores of *Asellaria cauleryi* in water, but saw no development similar to that found later in *A. ligiae*. Rather, he stated that only on ingestion by its isopod host did the arthrospore of *A. cauleryi* germinate, with one end of the arthrospore giving rise to the specialized holdfast cell while the other grew into the highly branched thallus. Manier (1958, 1964c, 1979b) has described a similar direct germination of ingested arthrospores in several species of *Orchesellaria*.

ECCRINALES - Species of this order reproduce asexually by producing one or more kinds of sporangiospores. These typically develop singly in sporangia that form basipetally. Two basic types can be recognized. One is the *primary infestation sporangiospore* (Fig. 7.20A, B). It is uninucleate

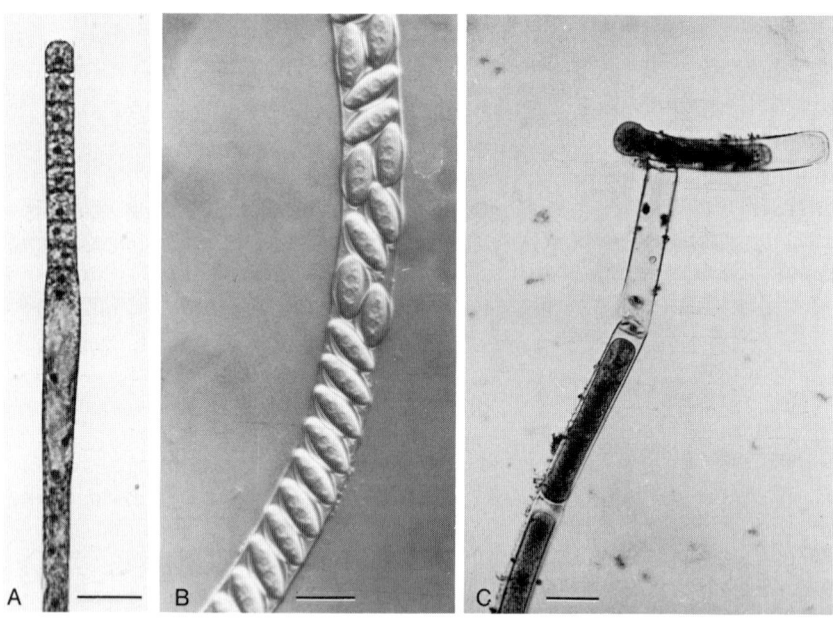

FIG. 7.20. Spore types of Eccrinales. A. Uninucleate thin–walled primary infestation sporangiospores of *Enterobryus moniliformis*. B. Binucleate thick–walled primary infestation sporangiospores of *Parataeniella* sp. C. Multinucleate secondary infestation sporangiospores of *Enterobryus euryuri*. Scale bars = 20 μm. Figs. 7.20A and 7.20C reprinted by permission from Mycologia, vol. 49: 737 (7.20A) and vol. 46: 575 (7.20C), Copyright 1954 and 1957, R.W. Lichtwardt and The New York Botanical Garden.

in most species and bi– or quadrinucleate in others, often thick walled (and presumably "resistant"), and is believed to germinate only after passage from the gut and ingestion by another suitable host. It is a propagule for transmitting infestation from one host to another. The other type is the *secondary infestation sporangiospore* (Fig. 7.20C). This type is multinucleate and capable of immediate germination on release from the sporangium, thereby serving to increase the population of thalli in the already infested gut without requiring the ingestion of additional spores. The two basic spore types can be recognized in most instances by their structure. As used in this treatise, any eccrinid spore that germinates immediately within the gut where it was produced, regardless of its morphology, is considered to be a secondary infestation spore. Reproduction in the Eccrinales is complicated by the fact that several morphological variants of both types exist; furthermore, there are cells produced in some species whose function is not known and which may in fact be nonfunctional. Several of these variants will be described later in this section.

First, let us consider the typical secondary infestation spore which is capable of immediate germination, for in the majority of eccrinid species these are the most commonly encountered reproductive unit during the intermolt period. Prior to their formation in the coenocytic thallus the terminal nuclei become aligned more or less equidistantly, and a septum forms to delimit the most terminal nucleus. This is followed by the production of successive septa in such a way that a series of initially uninucleate cells is formed. While this process goes on, the more terminal nuclei begin to divide mitotically and in synchrony within each cell so that the cells or sporangia usually end up with 4 or 8 nuclei aligned roughly in a row. The number is relatively constant in any given species, but may be as few as 2 (e.g., *Parataeniella* spp.) or as many as 16 (e.g., *Enteromyces callianassae*). Within each sporangium a new wall is laid down around the protoplast to form a single sporangiospore.

The mature spore usually escapes from the sporangium through a hole that develops in the lateral wall of the sporangium at either the distal or proximal end. Presumably the hole is produced by enzymatic dissolution or weakening of the wall material by the spore. The distal or proximal position of the exit hole on the lateral wall is generally constant for a species, but in some thalli it can be reversed. Invariably, however, it is the basal end of the spore—the end that will attach to the cuticle—that emerges first. This can be identified in most species by a slight swelling at that end of the spore or, more rarely, by a rudimentary structure resembling a small holdfast. In some cases spore emergence may entail a tearing of the sporangial wall around the exit hole. After release of the spore the sporangial wall disintegrates, possibly by bacterial decomposition (Fig. 7.21B). (Empty sporangia, in time, almost consistently have bacteria attached to their walls, sometimes in large numbers. It is interesting that bacteria of similar appearance often are seen attached to the walls of immature sporangia and nonreproductive parts of the thallus, but they do

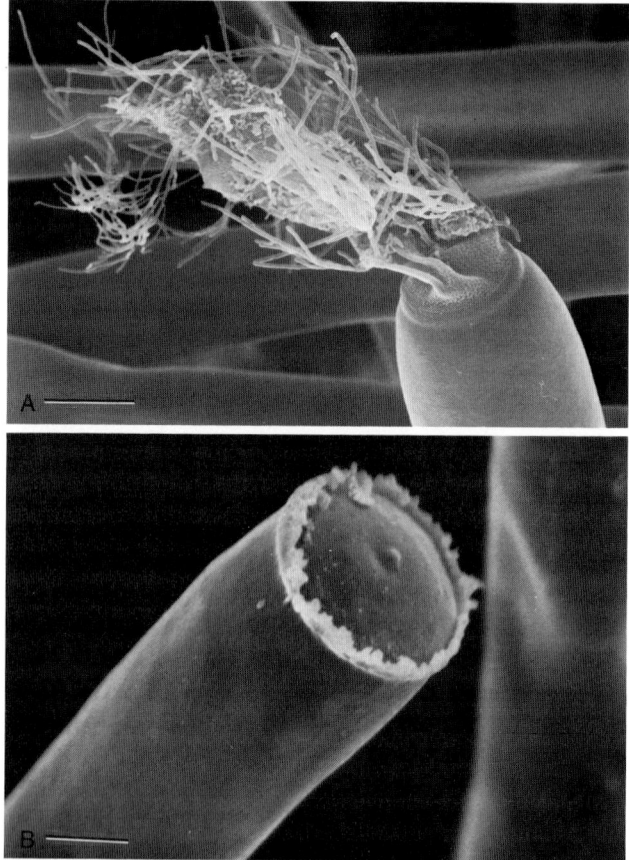

FIG. 7.21. A spore mother–cell at the tip of an *Enterobryus* sp. thallus covered with filamentous bacteria that presumably assist in its eventual complete disintegration. B. Remnant of the wall of an empty sporangium of *Enterobryus* sp. after disintegration by bacteria; note the crosswall with its plugged perforation. Scale bars: A, 10 μm; B, 2 μm. Fig. 7.21A kindly furnished by S.T. Moss, who prepared the material from a millipede (*Thyropygus* sp.) collected in India. Fig. 7.21B kindly furnished by R.E. Reichle, who prepared the SEM from a fiddler crab *(Uca crenulata)* from southern California.

not appear to inhibit development of the eccrinid.) Occasionally it is possible to find all of the sequential stages, from sporangial delimitation to spore release, at one time in individual thalli of some species.

The released multinucleate spore attaches immediately to the host cuticle and begins to germinate. In a few eccrinid species, release of several of the spores can occur within a short space of time with the result that a row of attached spores or developing thalli from one mother thallus can be seen on the host cuticle (e.g., *Arundinula* spp.). Germination commences by extension of the wall at the base of the spore immediately above the holdfast so that the spore maintains a terminal, usually reflexed, position on the elongating thallus, changing little in shape or size as the thallus grows. The fate of the spore mother–cell follows one of several patterns: (1) it remains in its apical position only during the initial stages of thallus elongation; a wall then forms to cut off the spore case, and it disintegrates and disappears (e.g., *Enterobryus* spp., Fig. 7.21A); (2) it is persistent through sporulation of the thallus, eventually deteriorating as the sporangiospores mature (e.g., *Palavascia* spp.); or (3) it is persistent and eventually functions as the terminal sporangium (e.g., *Enteromyces callianassae*). These patterns are more or less uniform in any given genus, and often are useful taxonomic characters. What is presently called the spore mother–cell has been referred to in the literature under various other names: gland, spore case, mother spore, appendage, etc.

The primary infestation spores, which serve as dissemination propagules, are in most instances uninucleate on release and are thick walled in many, but not all, species. Their development within thalli is similar to that just described in the secondary spore type, except that mitotic divisions do not occur in the young sporangium. Frequently the spores assume an oval or ellipsoidal shape on reaching maturity. These spores form only in response to molting in some hosts, as evidenced by the fact that mature ones are found only during ecdysis. The shed molt may contain large numbers of primary infestation spores in such cases.

In some genera *(Astreptonema, Eccrinoides)* the developing primary spores do not, in fact, remain uninucleate, but may be quadrinucleate at the time of release. Or, as in *Eccrinidus flexilis,* there is a further development after release to produce two chambers within the spore, each containing a quadrinucleate protoplast. In *Enterobryus,* as this genus is currently conceived, the primary infestation spores remain uninucleate, but are not thick walled and do not seem to form in response to ecdysis. These examples illustrate that several forms of primary spore exist.

The above outline of the two basic spore types (primary and secondary) is based on the largest of the three eccrinid families, the Eccrinaceae. In the Palavasciaceae, only primary (but multinucleate) infestation spores are formed. In the Parataeniellaceae both types occur; the secondary (binucleate) type is produced individually in sporangia, whereas the primary (uninucleate) type develops in larger numbers within a thallus that functions as one multispored sporangium. In the Parataeniellaceae there appear to be no septa separating the primary spores even in early stages of cleavage, but it should be pointed out that this unusual multispored type of eccrinid sporangium has not been studied at the electron microscope level of magnification. In several Eccrinaceae the primary infestation spores clearly develop individually within sporangia; however, at maturity the crosswalls separating the spores may disappear, leaving the spores loose and scattered in the thallus tip until they are released from the apical end or the remaining thallus wall material disintegrates around them.

In most species of Eccrinales much of the thallus remains nonreproductive. In a few (e.g., *Enteromyces callianassae*) only a short basal portion of the thallus may remain unconverted to sporangia at the end of the reproductive process. Complete conversion of the coenocytic thallus to spores is found in the Parataeniellaceae (primary spore type only), as mentioned above, but apparently no complete conversion is found in the Palavasciaceae and only rarely in the Eccrinaceae. [In *Arundinula washingtoniensis*, Hibbits (1978) reported finding the entire protoplast cleaved into uninucleate spores, and she saw a similar cleavage (but no spores) in *A. haplogaster* on one occasion.] In almost all sporulating thalli of Eccrinaceae, one finds either the primary or the secondary spore type, but occasionally thalli that have begun to form one type may shift to the other (e.g., *Enterobryus borariae*, *Parataeniella* spp.), resulting in a thallus bearing both types simultaneously, as in Fig. 7.12B.

An interesting structural feature has been found in the thick walled, oval, primary infestation spores of several genera of Eccrinaceae, namely that on release some have appendages at one or both ends. These are all associated with molting marine or freshwater Crustacea. One may assume that the appendages serve to entangle the spores in aquatic substrate materials, thus increasing their chances of successful transmission. At least this may be true in those eccrinid spores where the appendages are long and filiform, as in appendages of harpellid trichospores. Eccrinid primary spores with appendages were first reported in *Arundinula capitata* from hermit crabs by Duboscq et al. (1948), who found them in only 3 of 150 molting crabs. They were described and illustrated as being uni– or binucleate with one short appendage emanating from each pole of the oval spore. Duboscq et al. thought the spores developed from a fusion of uninucleate cells functioning as gametes within the thallus. These spores in *A. capitata* have not subsequently been found for restudy. Hibbits (1978) saw similar spores with one long appendage at each pole in *Arundinula washingtoniensis;* however, she did not consider them to have arisen by a sexual process, nor has sexuality been suggested in the formation of appendaged spores in other species.

Primary spores with a single appendage at each pole are now known in *Astreptonema gammari* (Fig. 11.30B) and *A. typica* (Manier, 1964b), and Hibbits (1978) described an unnamed species of *Astreptonema* from *Exosphaeroma amplicauda* that had a single appendage at one pole. (She also found a single appendage on a spore of *Alacrinella sanjuanensis*, but stated that it might have been an artifact of fixation.) In *Taeniella carcini*, Moss (1979) discovered that there are two appendages at each pole of the oval primary spores, and Hibbits (1978) found two attached to only one pole in *T. grandis*. Species of *Taeniella* often have gelatinous masses at the ends of the spore in the sporangium, and it is probable that in species of this genus, and in others as well, additional appendaged primary spores will be found on further study. Appendages unfurl only after release from the sporangium. They are often difficult to see without phase–contrast microscopy or other special techniques, and probably have been overlooked in some cases.

The significant question is whether appendaged spores of Eccrinales and Harpellales have evolved separately through convergent evolution, or whether they are in fact structurally and ontogenetically similar and phylogenetically related (see Chapter 12). At the present time little is known about the development of eccrinid appendages. Moss (1972, 1979) found little similarity between these structures in the two orders based on his investigation of the eccrinid *Astreptonema gammari*. He showed that in this species the appendages are extensions of an outer, mucilaginous sporangiospore wall formed within the sporangium by deposition of Golgi–derived material. Further electron microscopic studies of this kind of spore in the Eccrinales would be most valuable in order to compare their development with the better known trichospore of the Harpellales.

A different form of spore appendage was briefly reported by Lichtwardt (1973a) in *Palavascia sphaeromae*, and is described here in more detail (Fig. 7.22). This widespread species of Eccrinales occurs in the hindgut of marine isopods. A collection of *Sphaeroma serratum* from the Bassin de Thau near Sète, southern France in July of 1968 was well infested with the eccrinid. When thalli bearing mature primary infestation spores were placed on a microscope slide in 50% seawater and gentle pressure was applied to the coverslip to release spores from their sporangia, the spores

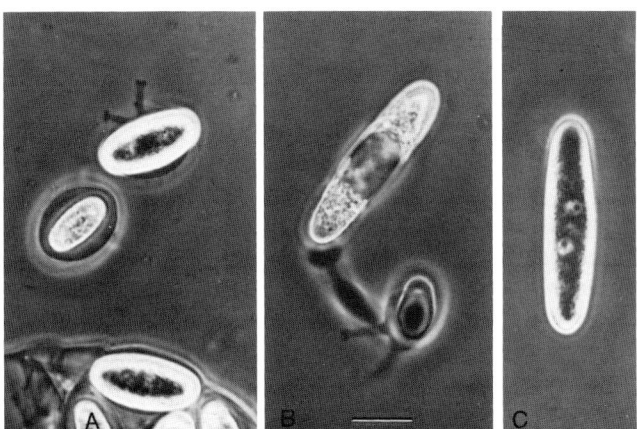

FIG. 7.22. Germination of *Palavascia sphaeromae* primary infestation sporangiospores. A. Released sporangiospore with two hornlike appendages. B. Spore has elongated and burst out of its outer wall, which is now collapsed; a shadowy structure connects the emerged spore with its former outer wall. C. Emerged binucleate spore completely separated from its former outer wall. Scale bar = 20 μm for all figures. Reprinted by permission from Mycologia, vol. 65: 8, Copyright 1973, R.W. Lichtwardt and The New York Botanical Garden.

were seen in phase contrast to bear two adjacent short, hyaline, hornlike gelatinous appendages, each with a small terminal knob, arising from an inconspicuous flat pad on the outer surface of the spore. Most commonly they were attached to the side of the oval spore, but they also occurred in pairs on other parts of the spore. These sometimes adhered to the glass slide. Within an hour or less, many of the released spores elongated and burst forth from a thin, elastic, outer wall that surrounded the spore, and the outer wall immediately collapsed in a pleated fashion. At the base of each of these ejected spores, and attached to the collapsed outer wall, was seen a hyaline body of similar appearance but larger than the outer appendages, possibly consisting of a hygroscopic substance that had swollen to aid in ejecting the spore. The hyaline body and the spore did not remain attached to each other. This process was observed many times on slides prepared from different isopods, and undoubtedly was not an artifact.

Whether spores of all *P. sphaeromae* populations regularly behave in this fashion *in vitro* has not been determined. It is possible that the two hornlike appendages serve to cause the spore to adhere to the substratum until ingested, and that the sudden shedding of the outer wall in natural circumstances occurs only after ingestion and just preceding attachment of the spore to the host cuticle. Manier (1979a) did a study of the fine structure of *P. sphaeromae* collected from the same general site and detected two distinct walls in the sporangiospores within the sporangium, as expected, but she did not see fine-structural evidence in her sections for the appendages on the outer wall of the unreleased spores.

The terminology one finds in the literature in reference to the two basic spore types in the Eccrinales is extensive and confusing. At one time or another they have been called conidia, endoconidia, oidia, etc., as well as sporangiospores, which of course they are. Most often the prefixes "macro" and "micro" have been used to differentiate between the secondary infestation spores and the primary ones, respectively. However, a size difference is not consistent in all species. The thicker walled primary spores have also been called "resistant spores" or "durable spores" by the French workers, and while these designations are descriptive for particular genera of eccrinids, they are not applicable in a general way to all primary infestation spores. The author in past publications has distinguished between the two types by calling them multinucleate and uninucleate sporangiospores, but there are known exceptions to the uninucleate condition in the latter type because of either species differences or the stage of development seen after spore formation. In her 1969 monograph, Manier (1969b) called the secondary infestation sporangiospores "immediate-development spores" and the primary infestation sporangiospores "dissemination spores," thus placing their terminology on a functional basis, which seems preferable in view of the structural variation among the taxa. Despite some evidence to the contrary (Hibbits, 1978), there are no convincing data at present to indicate that the function of each of the two basic spore types is not distinct.

Another complicating factor in studies of reproduction in the Eccrinales, particularly in the Eccrinaceae, is the presence in thalli of some species of cells whose function is not known and which, in fact, may be nonfunctional. It is not uncommon in some Eccrinaceae to find series of cells at the tips of thalli that resemble sporangia, but do not appear to produce spores [Hibbits (1978) called these "partitioned" thalli]. In *Enterobryus* spp., such cells may be larger than normal sporangia and are sometimes slightly swollen, with nuclei that are not arranged in a row as is typical in true sporangia; or they remain uninucleate but are morphologically different from the true uninucleate primary infestation spores. It is surprising that Leidy (1853), who was the first to study eccrinids and a competent microscopist, never found spores in *Enterobryus elegans* and other species he described. He only saw what he called "secondary cells" in six instances out of several thousand thalli from over 150 millipedes collected from early spring until late fall. In *Eccrina longa* (nomen nudum) he found a few such cells detached singly or in groups of two or three, and assumed that reproduction in the eccrinids was by segmentation. Since Leidy never identified nuclei in these fungi, it is not known for certain whether they were a modified type of sporangium or immature ones. The disarticulation he reported has not been confirmed in subsequent studies of some of his same species (Lichtwardt, 1954b, 1957a,b), but one cannot discount this type of occurrence. It has been said to happen in other eccrinids, such as *Eccrinoides (Eccrinopsis) helleriae (nom. nud.)* (Duboscq et al., 1948) and *Trichellopsis schizophylli (nom. dub.)* (Maessen, 1955).

An unusual cell type that has been seen occasionally in a number of species of *Enterobryus* (Lichtwardt, 1954b, 1958, and unpublished) consists of apical cells tightly packed with dozens of nuclei. Thalli that produce them are quite distinct, even before such cells have formed, for they contain more numerous nuclei than are normally found, and these are scattered randomly throughout the thallus. In time, some of the nuclei become densely aggregated at the tip of the thallus, and a septum then forms to isolate them. This process continues until up to about half a dozen such cells are formed. What may be nuclear divisions in some of these terminal cells have been observed on rare occasions, and, judging from observations of the static condition but involving sequentially produced events in cells connected to both ends of those with dividing nuclei, the divisions appear to result in nuclei with about one–half their previous volume. The function of these cells, if any, is not known. Thalli bearing this cell type can be found interspersed with other thalli reproducing normally, and consequently it does not seem that their development is necessarily affected by some unusual environmental factor in the gut.

Another spore type whose function is unknown is frequently found in both species of *Palavascia*. The reproductive unit in the genus, as described previously in this section, is a multinucleate primary infestation sporangiospore. However, some of the many sporangia germinate *in situ*, putting out one or more long, narrow thalli ("microthalli") that resemble a small branch. Manier (1979a) has clearly shown in *P. sphaeromae* that the walls of these narrow thalli are extensions of the spore wall contained within the sporangial wall and are not outgrowths of the outer, sporangial wall itself. After breaking through the outer wall, the narrow thalli elongate considerably and eventually divide into a series of small uninucleate cells. Spores that germinate *in situ* in this manner do not normally emerge from the sporangium, and the uninucleate cells they produce appear to have no function.

Other modified spores of Eccrinales are described in Chapter 11 in the descriptions of individual species.

AMOEBIDIALES - In *Amoebidium parasiticum* the entire protoplast cleaves to produce uninucleate sporangiospores or, on a molting or injured host, uninucleate amoeboid cells. Whisler and Fuller (1968) found no significant difference in development of the two cell types, except that sporangiospores become enveloped in a rigid wall whereas amoebae are merely membrane bound. There is a question remaining as to whether the thallus functions as a single sporangium, or whether spores form in separate, walled chambers within the thallus. The author on many occasions has studied *A. parasiticum* under the phase–contrast microscope and frequently, but not always, has seen what appear to be walls or partitions within emptied thalli. In her 1970 thesis, Coste–Mathiez stated that the empty walls of thalli often show an internal partitioning as though the sporangiospores are isolated individually or in groups within chambers, and two of her electron micrographs of a sporulating thallus (her Plate II, Figs. 11 and 12) clearly show a separate wall within the thallial wall surrounding the spore walls. On the other hand, electron micrographs by Whisler and Fuller (1968) of cultured *A. parasiticum*, and by Dang (1979) of the same species occurring naturally on mosquito larvae, show no internal partitioning of the thallus around the spores. It is possible that the formation of chambers may or may not occur, depending on factors such as the environment during development, thallial size, and possibly strain differences.

It seems clear, however, that amoeba production in both *A. parasiticum* and *Paramoebidium* spp. does not involve internal wall formation, so that the entire thallus functions as a single reproductive structure. This phase of the life cycle is similar but not identical in both genera. The amoebae of the two genera are essentially indistinguishable. (On a few occasions the author has seen thalli of *Paramoebidium curvum* and other undetermined species of *Paramoebidium* release unusually large amoeboid cells that appear to be multinucleate, along with normal amoebae, apparently due to abnormal cleavage of the protoplast.) Manier and Raibaut (1970) observed that the cystospores of *A. parasiticum* emerge from the cyst wall as a compact cluster within a membranous, partitioned sac from which they later are released. Lichtwardt (1973a) found no such membranous sac surrounding cystospores of *P. curvum;* rather, the cystospores are ejected forcibly from individual chambers within the cyst wall. These differences in cyst development, if indeed they exist, may be attributable to the different functions of the cysts: *Amoebidium* cystospores are released in the external environment where they can eventually attach promiscuously to the external cuticle of various arthropods, whereas it is believed the cysts of *Paramoebidium* are ingested and may release their cystospores suddenly and only when they reach the proper location in the arthropod hindgut (Dang and Lichtwardt, 1979).

Sexual Reproduction

Of the four trichomycete orders, only the Harpellales produce structures (zygospores) that appear to arise from a true sexual process. There is no cytological or genetic proof of sexuality currently available. Nevertheless, the sequence of zygospore development, beginning with thallial conjugation, and the thickened walls and storage materials contained within the spore, leave little doubt that they are comparable to zygospores of other fungi, although structurally and ontogenetically different. No zygospores have been found yet in the closely related order Asellariales, but observed conjugations identical to those seen in the Harpellales suggest that in time zygospores will be discovered in that order too (Lichtwardt, 1973a). Poisson (1931b) reported observing fusion of amoeboid cells in *Amoebidium parasiticum* (Amoebidiales), but this has not been verified by subsequent investigators, as mentioned in the first section of this chapter. In the Eccrinales a peculiar form of thallial conjugation and fusion has been seen in several instances in crustaceans that is suggestive of an attempt at sexuality, and this will be described following an account of zygospore development in the Harpellales.

HARPELLALES - Zygospores are known in 14 of the 18 genera of Harpellales, but not in all species of two of these genera. Zygospores may be unknown in some species due to insufficient investigations, or the host specimens may have been collected at the wrong season or stage of maturity. In other more studied species zygospores appear in fact to be rare. For instance, *Stachylina grandispora* has been seen conjugating but never producing zygospores (Lichtwardt, 1972; Moss, 1972). Conjugations are frequent at times in some populations of blackfly larvae containing *Harpella melusinae,* but zygospores have been found in only some seven larvae out of thousands dissected from different climatic regions (Lichtwardt, 1967; Moss and Lichtwardt, 1977; and unpublished). It may be that in *H. melusinae* there is frequently some impotence factor present involving an inhibition or loss of genetic control over the sequence of events necessary for completion of the sexual process. Other possibilities, not mutually exclusive, are that larvae may sometimes develop too rapidly for zygospores to form prior to molting, or that environmental factors within the gut lumen are not favorable for complete sexual development. Fortunately, zygospores are not difficult to find in many other harpellids.

It is clear that in some species of Harpellales zygospores are associated only with the prepupal molting process. Mature aquatic insect larvae about to molt can often be identified externally by the appearance of pupal structures beneath the larval skin; internally, the gut linings become loosened as the epithelial cells lay down new cuticle material prior to ecdysis. Since the last larval stage is normally the longest, it is possible that the development of zygospores just before pupation is due primarily to temporal factors. However, it seems evident in some harpellid species (e.g., *Legeriomyces ramosus*) that there is a rather abrupt morphogenetic shift from asexual to sexual reproduction, with the result that the exuviae may contain for the most part masses of zygospores and no remaining trichospores. In other species (e.g., *Genistelloides hibernus*), when most trichospores are mature but while some are still maturing, clumps of thallial branches begin to conjugate and produce zygospores. In a few species (e.g., *Pennella simulii, Trichozygospora chironomidarum, Genistellospora homothallica*), trichospore and zygospore development can be concurrent. Nevertheless, sexual reproduction in the majority of Harpellales does seem to be influenced by the molting processes of the host or at least is correlated with ecdysis.

Currently there is no understanding of the stimuli that effect the morphogenetic change in these gut fungi. Is it due to the direct influence of molting hormones or other chemicals activated during the molting process, or do these fungi sense less directly the physiological changes occurring in the host? The only practical means of resolving this intriguing biological problem is through the use of axenically cultured fungus material, but unfortunately at present zygospores have not been induced to form in any of the cultured harpellid species.

The onset of sexuality in some species of both families of Harpellales is marked by uneven swellings in some of the cells of the thalli. This is especially noticeable in the unbranched thalli of *Harpella melusinae*, where this initial stage of sexuality can be readily apparent at low magnification in well–infested peritrophic membranes, even before actual conjugations are detected at higher magnification. This suggests that there may be some kind of hormonal interplay that makes certain regions of the thalli receptive to adhesion and fusion. It should be noted that these gut fungi, unlike other sexually reproducing fungi on more stable substrates, are frequently in motion due to peristalsis and the movement of host–ingested particles through the gut. Also, when the gut is dissected and cleansed of debris, the positions of the thalli are considerably disturbed. Consequently, it is difficult to determine whether the swellings on adjacent simple thalli (or branches in Legeriomycetaceae) can develop when they are still separated but in close proximity, or only when in actual contact. Both partners on contact produce a short protuberance contributing to the formation of the conjugation tube, and the walls between them quickly dissolve completely and allow the cytoplasm to intermingle. The nuclei from each of the uninucleate mating cells have been seen juxtapositioned in the conjugation area, and single, presumably diploid, nuclei have been seen (Lichtwardt, 1967), but actual karyogamy has not been observed.

Multiple conjugations between two branched or two unbranched thalli may produce a scalariform pattern, but conjugations also may be promiscuous and involve many branches or thalli. Lichtwardt (1967) reported as many as seven unbranched thalli of *H. melusinae* interconjugated. Analysis of such complexes in *H. melusinae* did not give any indication of a simple mating system, such as the plus–minus type characteristically found in heterothallic Zygomycetes. Some branched thalli of harpellids (e.g., *Pennella* spp.) appear to mate only with other thalli, and therefore may be truly heterothallic, but it is often too difficult to trace the intermingled branches to determine accurately their thallial source (Williams and Lichtwardt, 1971; Lichtwardt, 1972). *Stipella vigilans* is reported to be either "homothallic" or "heterothallic" (Manier, 1963b). On the other hand, *Genistellospora homothallica* differs in that it forms its zygospores without any conjugations.

Unlike the spherical zygospores of other Zygomycotina, those of the Harpellales are biconical at maturity (although spherical very early in their formation) and each develops as an outgrowth from a supporting cell, the zygosporophore. Moss et al. (1975) recognized four basic types (Fig. 7.23):

Type I. The axis of the zygospore lies perpendicular to the zygosporophore, and its attachment is median (midway between the poles). This

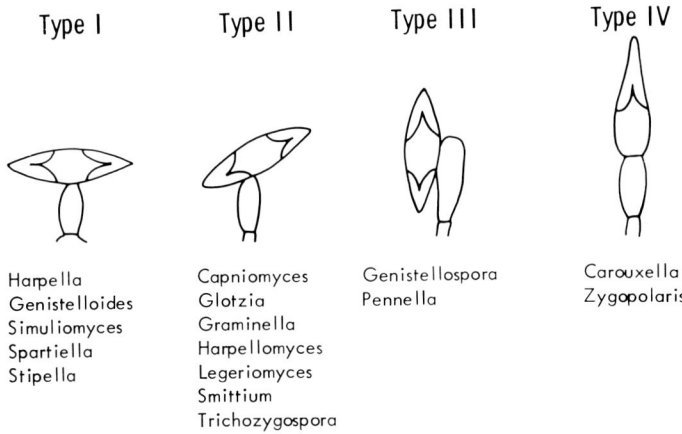

FIG. 7.23. Four basic zygospore types in the Harpellales.

type is found in *Harpella, Genistelloides, Simuliomyces, Spartiella,* and *Stipella.*

Type II. The zygospore position is oblique to the zygosporophore, and its attachment is submedian. Found in *Capniomyces, Glotzia, Graminella, Harpellomyces, Legeriomyces, Smittium,* and *Trichozygospora.* In these genera, released zygospores have a collar and, in some genera, one or many appendages.

Type III. The zygospore lies parallel to the main axis of the zygosporophore, and attachment is median. Found in *Genistellospora* and *Pennella.*

Type IV. Attachment is at one pole, so that the zygospore and zygosporophore are coaxial. Found in *Carouxella* and *Zygopolaris.*

The genera whose zygospores presently are unknown are *Gauthiera, Orphella, Pteromaktron,* and *Stachylina.*

The location of the developed zygospore and its subtending zygosporophore in relation to the zone of conjugation is fairly predictable. In *Capniomyces, Glotzia,* and *Genistelloides* these usually arise directly from the conjugation tube. In other genera these structure arise from one of the conjugated cells, and usually close to the zone of conjugation except in *Pennella, Simuliomyces,* and *Trichozygospora.* Where the conjugations are scalariform, the zygospores normally all arise from the cells of one of the thalli or branches. In a few genera (*Pennella, Stipella, Trichozygospora,* sometimes *Zygopolaris* and others), the fertile cells from which the zygospores arise may become noticeably swollen, whereas the "donor" cells retain their shape. Conjugations may occur between terminal cells and intercalary cells, or (more commonly) two intercalary cells. Williams and Lichtwardt (1971) noted that in *Pennella simulii* donor branches could conjugate with more than one receptor branch, and sometimes a donor branch terminated in the production of an apical trichospore.

Zygospore development in *H. melusinae* (Type I) begins with the budding out of the zygosporophore from one of the conjugated cells (Lichtwardt, 1967). Its two–layered wall originates from new wall layers formed internal to those of the cell from which it grows, in a manner similar to lateral branch formation in the Legeriomycetaceae (Moss and Lichtwardt, 1976, 1977). After the zygosporophore achieves its normal size, it bulges spherically at the top and then the biconical zygospore takes shape. At about this time two crosswalls form, one separating the zygospore from the zygosporophore and the other separating the zygosporophore from the conjugated cell. Both are the typical harpellid perforate septa that become occluded by an electron–opaque plug. An especially interesting feature of the entire sexual apparatus in *H. melusinae* is that at maturity each of the four cells—the two conjugants, the zygosporophore and the zygospore—contains a single nucleus (Moss and Lichtwardt, 1977) (Fig. 7.9). This may indicate that a meiotic division occurs prior to septation and, if true, that the zygospore contains a haploid nucleus. Such a condition would be advantageous, assuming somatic nuclei are haploid, for it would allow the zygospore to be primed for rapid germination, attachment, and thallial development after ingestion by a host, as in trichospores.

In *Trichozygospora chironomidarum* (Type II), the sexual apparatus is somewhat modified. The zygospore is produced some distance from the zone of conjugation and develops at the end of an extended conjugant. The zygosporophore in this species can be identified as a swelling beneath the zygospore, but it is merely a continuation of the conjugant and is not separated from it by a septum as in other harpellids. Moss and Lichtwardt (1977) found in one of their specimens that three of the four nuclei in the sexual apparatus remained in the larger conjugant cell, with the fourth nucleus being located in the zygospore. The numerous appendages (10 or more) in this species develop within the zygosporophore and extend down into the conjugant cell. These appendages do not form in individual appendage sacs, as do the fewer appendages (5–7) of the trichospores of this species. On detachment, a portion of the zygosporophore wall remains with the zygospore to form a collar from within which the numerous appendages trail. In *Genistellospora homothallica* (Type III) and *Zygopolaris ephemeridarum* (Type IV), the zygospores contain a single nucleus when first formed, but Moss and Lichtwardt (1977) did not find any nuclei in the zygosporophores of the limited material available for study.

There appear to be several patterns of zygospore release. This process has not been described for all species, and in some cases only a few zygospores have been seen to detach from thalli when kept in water on microscope slides. Therefore, one cannot always be certain that the observed release is not premature or otherwise abnormal. In *Genistelloides hibernus* and *Simuliomyces microspora* (Type I), detachment usually takes place at the base of the zygosporophore, so that the zygosporophore and the zygospore release as a unit. All Type II zygospores have a small collar when released due to a circumscissile dehiscence of the zygosporophore wall. In *Genistellospora homothallica* (Type III), the zygospore is released with the contents of the zygosporophore attached, leaving behind the zygosporophore wall in place on the thallus (Lichtwardt, 1972; Moss and Lichtwardt, 1977). Zygospores of *Zygopolaris ephemeridarum* (Type IV) have been seen to retain some of the zygosporophore material at their base (Moss and Lichtwardt, 1977), and to produce a skirtlike substance around the zygospore base (Lichtwardt and Williams, 1984).

Before normal release, zygospores begin to develop internal thickenings at their conical ends, and as the zygospores mature the protoplast becomes progressively restricted. In the case of Type IV zygospores the thickenings develop only in the free end of the spores. Completion of this thickening process has been seen to occur also in detached zygospores of many species kept in water mounts. It appears to involve mostly deposition of supplemental wall material, but in addition the protoplasts of some species may develop conical apical caps that aid in the release of the contents on germination. It is believed that normally zygospores are ingested and that germination occurs in the host gut, but germination has been seen only on a few occasions outside the host: in *Legeriomyces* sp. (Whisler, 1963), *Stipella vigilans* (Moss, 1970), and *Glotzia ephemeridarum* (Lichtwardt, 1972). The zygospore contents are apparently under considerable pressure so that on germination the inner structure pushes through one end of the zygospore wall almost explosively. In *G. ephemeridarum* the inner apical caps are quite sharp, and the internal force is sufficient to perforate the lateral wall of the zygospore in cases where the zygospore protoplast is misaligned within the spore, as sometimes happens in this species (Lichtwardt, 1972).

The elongate rather than spherical shape of the trichomycete zygospore appears to be adaptive by providing a means to allow rapid germination or expulsion of the protoplast during its short travel time in the gut, and this may be a major reason for the differences in shape one finds between the biconical zygospores of these gut fungi and the spherical ones of other Zygomycotina.

ECCRINALES - Over the years there have been a number of reports of putative sexual processes in the Eccrinales. Among them are intracellular fusions in *Eccrinidus flexilis* and *Arundinula capitata* (Duboscq et al., 1948); *Enterobryus duboscqui* (Tuzet and Manier, 1948a), and *Palavascia philoscii* and *P. sphaeromae* (Tuzet and Manier, 1948b). These reports have not been substantiated in subsequent studies, and in fact Manier in her 1969 monograph did not acknowledge the presence of sexuality in those species or any other Eccrinales. Maessen (1955) described the fusion of released protoplasts in *Enterobryus flavus (nom. dub.)* and *Trichellopsis schizophylli (nom. dub.)*, but an occurrence such as she illustrated is not likely. More recently Scheer (1977) described *Nodocrinella hylonisci (nom. nud.)* and speculated that there were haploid and diploid phases in that species without offering any good evidence to support his view.

A study of marine eccrinids by Hibbits (1978) merits attention and further investigation. Among the various thallial and spore types in *Enteropogon sexuale* occurring in the anomurid *Upogebia pugettensis*, she described and illustrated parallel thalli from the anterior hindgut with series of scalariform conjugations (Fig. 11.36). Apparently the uninucleate protoplast of one thallus fused with the adjacent uninucleate protoplast of the other thallus, after which the binucleate protoplast moved into the cell of the larger of the two thalli, rounded up, and karyogamy occurred. No conjugation tubes formed in this process; rather, the cell walls simply dissolved between adjacent cells. Loose uninucleate cells (presumed to be diploid) were seen as well, but further development was not observed, nor was it determined if they were functional in the life cycle of the eccrinid. This process was seen only occasionally and mostly in exuviae. If the events as she described them have been pieced together correctly, they would resemble the type of sexual reproduction characteristic of the algal order Zygnematales. Hibbits also saw thalli possibly undergoing scalariform fusion in *Taeniella carcini* from the hermit crab *Pagurus kennerlyi*.

Prior to Hibbits' discovery, Drs. Hiroharu Indoh, Yosio Kobayasi, and the author (unpublished) in May of 1964 collected anomurids [*Thalassina* sp., possible *T. anomale* (Herbst.)] from a mud flat at Usa on the island of Shikoku, Japan. Several of the specimens in their stomachs had an unidentified eccrinid (*Arundinula* sp.?) with a similar type of fusion. In this species, adjacent pairs of cells from two thalli fused so that each pair formed a common chamber containing a binucleate protoplast. Conjugation of two thalli did not necessarily occur at the distal ends of the thalli, and never at their bases. Farther down the series of fused cells, walled protoplasts could be found with a single nucleus. Thus, this is a process, like that described by Hibbits, that is suggestive of karyogamy and sexuality, although possibly vestigial.

Another interpretation of the type of fusion described above is that they merely represent spurious events in a group of eccrinids that often produce aberrant structures. Additional developmental studies are necessary before the presence of sexual reproduction in the Eccrinales is established.

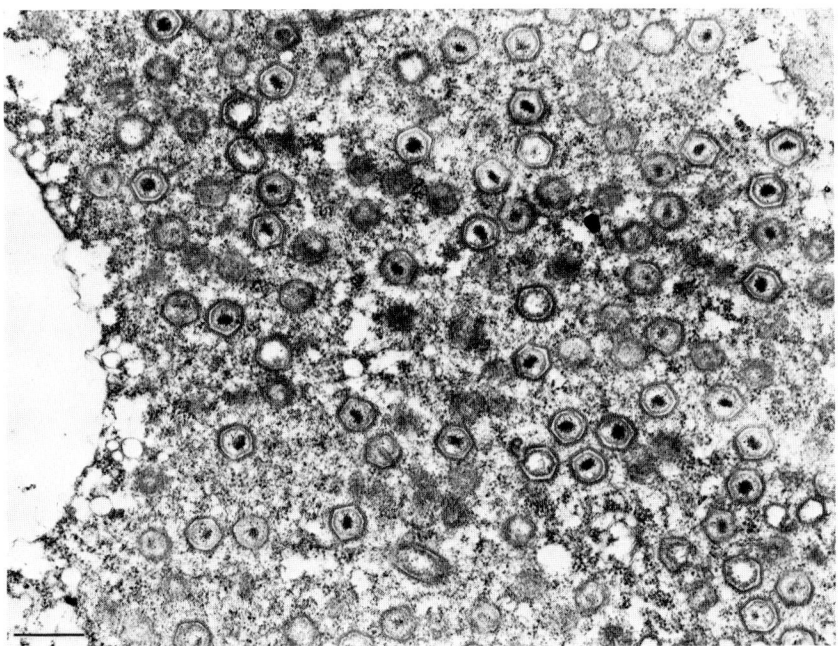

FIG. 7.24. Virus–like particles in the cytoplasm of a *Paramoebidium curvum* thallus. Scale bar = 0.5 μm. From Dang and Lichtwardt, 1979, by permission of National Research Council of Canada.

Viruses

Virus–like particles (VLPs) have been discovered in two species of *Paramoebidium*, *P. arcuatum (nom. nud.)* from the hindgut of a mayfly nymph (Manier et al., 1971) and *P. curvum* from the rectum of a blackfly larva (Dang and Lichtwardt, 1979). They were essentially hexagonal in sectioned material of both species but differed in several other respects. Those in *P. arcuatum* measured 105–110 nm; most particles had an electron–opaque circular core of about 70 nm diameter and were seen partly arranged in paracrystalline bodies in the nuclei of a mature thallus beginning to cleave out amoeboid cells. In *P. curvum*, the VLPs were twice the size (210–220 nm) with an electron–opaque core of about 100 nm in diameter and an outer shell of about 20–30 nm thickness that consisted of two electron–opaque layers separated by an electron–transparent middle layer (Fig. 7.24). These were scatterd throughout a highly vacuolated and membranous cytoplasm, but organelles like nuclei and mitochondria appeared normal.

Unfortunately, no species of *Paramoebidium* has been cultured axenically, and as a consequence it has not been feasible to study the nucleic acid composition of these VLPs or to obtain information on their biological effects on the *Paramoebidium* species. VLPs have been sought in several axenic cultures of *Amoebidium parasiticum* by the author, but were not found, and they have not been reported in any other trichomycete thalli, cultured or not.

Chapter 8
Host–Fungus Relationships

The intimate relationships that have developed between trichomycetes and their arthropod hosts appear to be long–standing ones in terms of evolutionary time. Evidence for such a statement lies in the diversity to be found within the fungal class, the wide range of arthropod types infested, the worldwide distribution of this kind of symbiotic association, and the number of morphological and physiological specializations that these fungi have had to evolve to insure successful adaptation to their respective host types. In the present chapter we will examine these relationships primarily on the basis of studies of uncultured specimens, whereas in the next chapter (Chapter 9) experimental evidence obtained with axenically cultured fungal species will be emphasized.

Much of the older literature, and some of the newer, refers to the trichomycetes as parasites. If one uses this term in a narrow sense to mean that some demonstrable deleterious effect is produced in one of the partners, then it seems clear from currently available information that the relationship is not parasitic in most species. The fungi more properly should be called commensals, with the commensalism obligate on the part of the fungi but not their hosts. Although the commensalistic relationship may apply to most species of trichomycetes, there are several reports of species of *Smittium* that can be lethal to mosquito larvae kept in the laboratory, and it also has been shown that some instars of mosquitoes can be killed by *Smittium* species under certain conditions involving artificial infestation. These will be described later. Alternatively, there are now some data, covered in the subsequent chapter, that *Smittium culisetae* may supply nutrients such as B vitamins and sterols that improve the development of nutritionally stressed mosquito larvae. We thus find—perhaps not surprisingly in view of the variety of partnerships involved—that some trichomycetes can deviate from the normal commensalism with their hosts. Other exceptions are likely to be discovered, but it probably will be necessary in such instances to be able to grow both partners axenically and under controlled conditions before any subtle beneficial or deleterious effects by the fungi can be adequately demonstrated.

Nutritional Relationships

The trichomycetes spend their entire assimilative stage within the gut and therefore obtain all nourishment from substances present in or passing through the gut lumen. As a consequence they could conceivably deprive the arthropod of some nourishment, but there are no data to indicate that this results in any adverse effect. Since most species of trichomycetes inhabit the hindgut where absorption of food is said to be minimal, the deprivation by these fungi may be insignificant. It is often "healthy" populations of arthropods that appear to be most highly infested, but currently there are no studies comparing the condition of naturally infested populations with uninfested ones. Nor would a good correlation of the data necessarily reveal a cause–effect relationship due to the many uncontrolled factors that undoubtedly influence the development of both partners in their natural condition, not the least of which is the presence of other, nontrichomycetous microorganisms within the gut.

There also are no adequate studies concerned with the effects of these gut fungi on reproduction of their hosts. However, observations by the author over the years on different kinds of arthropods do not suggest adverse effects on reproductive capacity. In the field one can observe heavily infested crustacean females carrying apparently normal loads of eggs, and in the laboratory some species of millipedes and isopods that reproduce well in captivity appear to produce offspring just as well if infested as not. Species of Harpellales occur only in larval stages of insects, and therefore if any measurable alteration of the reproductive rate attributable to their gut fungi were to be found it would likely be through nutritional consequences prior to reaching adulthood. Trichomycetes that infest adult arthropods have never been found invading reproductive organs or eggs, and if they were to affect reproduction one would also expect them to do so indirectly by nutritional deprivation or through other effects on host metabolism.

The mandibulate arthropods that have trichomycetes in their guts are either herbivorous or omnivorous. The guts of actively feeding arthropods always contain some undigested or undigestible solid matter throughout the gut, and this often accumulates as a bolus in the hindgut where the majority of trichomycete species grow. It is not known whether trichomycetes are capable of digesting some of this debris, or whether they rely on the digestive processes of the host and perhaps on other host metabolites secreted or excreted into the gut cavity that could be essential to their development. Although several species of trichomycetes have now been cultured axenically, this has not helped to resolve the problem, because the isolatable species show no peculiar nutritional requirements; they may, in fact, represent anomalies when one considers that isolation attempts with numerous other species using the same basic isolation media have failed. The role that nutrition plays in the adaptation of these fungi to such an unusual fungal habitat thus remains to be elucidated.

Pathogenicity

Arthropod hosts of trichomycetes include several groups of economic or medical importance, and it is of interest to know if their fungal symbionts could affect them adversely. Some millipedes, such as *Oxidus gracilis*, and sowbugs (Isopoda) are considered to be pests in greenhouses and gardens when their populations are high because of damage they may do to plants during the course of their feeding activities. Of economic significance are the gribbles (*Limnoria* spp.), small wood–boring isopods that in time can destroy wharf pilings and wooden ships. The importance of mosquitoes as vectors of diseases such as malaria, yellow fever, and dengue fever is well known, and considerable international attention has been given to the blackflies that transmit the filaria of onchocerciasis ("river blindness"), now widespread across central Africa and in some parts of Central and South America. Biting blackflies can also cause considerable loss of livestock (Steelman, 1976). The trichomycetes carried in the adult or immature stage by such arthropod pests and disease vectors cited above in general do not appear to harm them under normal circumstances, and therefore the fungi cannot at present be considered a means of biological control.

At the same time there are several reports that some species of *Smittium* can be lethal to their larval mosquito hosts, suggesting that biological control may occur to some extent in natural populations. During the course of laboratory experiments where mosquito larvae were being infested with a species of *Smittium*, tentatively identified as *S. inopinatum* (= *S. culisetae*), Coluzzi (1966) in Italy found that an average of 27% (12–87%) of the larvae of *Anopheles gambiae* died in 23 infestation studies, whereas larvae of *Culex pipiens* and *Aedes aegypti* were not adversely affected. He attributed the death of the larvae to occlusion of the rectal ampulla, which resulted in unsuccessful metamorphosis. More recently, Dubitskii (1978) in Russia reported up to 80% mortality in laboratory colonies of larval *Culex pipiens molestus* and *Aedes aegypti* by a species he identified as *Smittium culisetae* (but probably not that species, based on his trichospore measurements). He surmised that the fungus was introduced into his laboratory cultures via an infested collection of *Aedes caspius*. The moribund larvae had a black spot visible externally under the abdominal cuticle consisting of dark fungal hyphae, and the larvae were inactive as if paralyzed.

Sweeney (1981a) found a new species, *Smittium morbosum*, infecting *Anopheles hilli* in his Australian laboratory in 1978. Significantly, he was able to obtain axenic isolates of *S. morbosum*, thus making it available for future experimental study as a system for biological control in the field. Sweeney reported 50–90% mortality of larvae in rearing trays, with most of them dying in the 4th–instar. Infection was maintainable *in vivo* by placing infected larvae in trays with uninfected ones. The original inoculum may have been introduced into the laboratory with a previous field collection of *Anopheles annulipes*. Infected larvae of *A. hilli* had a black abdominal spot visible to the unaided eye in the posterior midgut region at the fifth and sixth abdominal segments, and Sweeney attributed this to a possible melanization reaction of the host resulting from restricted penetration by the fungus into the hemocoel. Growth of the fungus established in the anterior hindgut extended forward and penetrated the cells of the midgut epithelium, as well as cells of the Malpighian tubes in some cases. In moribund larvae the fungus seemed to block the hindgut completely, and death also was attributed to the inability of larvae to shed their molts completely. Of special interest is Sweeney's report that hyphal penetration of the midgut epithelium as well as sporulating thalli also occurred in pupae and in adults of both sexes that emerged from trays of infected larvae. This is the first report of Harpellales in pupae or adults of insects, and such a situation could be a means of transmitting the fungus to new breeding sites in this fungal species. *Smittium morbosum* is similar but not identical to *S. culisetae* in trichospore size and shape, and it is possible that the infections Dubitskii and Coluzzi attributed to *S. culisetae* were in fact caused by *S. morbosum* or another as yet undescribed species.

Hindguts of other arthropods are sometimes so packed with trichomycete thalli, most notably millipedes infested with *Enterobryus* spp., that they would appear to occlude passage of materials through the gut. Thalli in some millipedes may be especially numerous near the pyloric sphincter in the most anterior portion of the hindgut. But there are no reports of death resulting from blockage of the gut in these naturally infested arthropods. It is possible to kill mosquito larvae under some circumstances through overinfestation by cultured *Smittium* spp., as described in Chapter 9, but conditions that lead to fatalities of this sort are not likely to arise in the natural environment.

Ecdysis: Effects on Fungi

All trichomycetes, attached as they are to arthropod cuticles, are subject to periodic separation from their living hosts at the time of molting. The linings of the gut are of ectodermal origin and are shed along with the outer cuticle or exoskeleton. Consequently, the adaptive success of trichomycetes includes means by which they can overcome this periodic expulsion. The intermolt period can be as short as 1 day in hosts such as 1st instars of mosquitoes, or up to many months depending on the kind of arthropod, its developmental stage, and environmental conditions. Some Harpellales are subjected to frequent moltings since they occur in aquatic insect larvae with short developmental cycles. Species of Asellariales and Eccrinales are commonly found in both immature and mature forms of their predominantly millipede and crustacean hosts; in these classes of arthropods molting may continue after sexual maturity has been reached, as the adults continue to grow in size. Reinfestation after molting may be accelerated in certain diplopods and crustaceans that habitually consume a portion of their own exuvia.

Due to the absence of cultured material of all but a few trichomycete species and the fact that fungal growth cannot be observed in the guts without dissection, little is known at present about the rate of development of most trichomycetes in their hosts. Williams and Lichtwardt (1972a) demonstrated the *Smittium culisetae* could grow and sporulate in 1st–instar larvae of *Aedes aegypti* within 24 hr, which is the duration of that instar under good rearing conditions. Species of most other trichomycete genera appear to develop spores much more slowly. Juvenile forms of arthropods such as isopods and millipedes can be infested with thalli, but the fungi in their guts may not achieve reproductive maturity before being sloughed off with the molt, consequently, sporulating thalli are seen more commonly in mature specimens. In any event, it is clear that rates of development of all species of trichomycetes must be sufficiently in phase with the molting cycles of their hosts for them to have survived.

The molting process has an effect on the reproductive development of many trichomycetes species. In some Harpellales this is seen as a shift from asexual to sexual reproduction. As a consequence, the shed molt of the last larval stage may contain masses of zygospores and no trichospores, and the same situation can be found if these larval forms are dissected just prior to ecydsis. This abrupt morphogenetic shift can be found in some fungal genera (e.g., *Legeriomyces, Zygopolaris, Simuliomyces, Glotzia*), but not all species of Harpellales respond in the same manner. For instance, thalli of *Pennella simulii* and *Genistellospora homothallica* are found producing both zygospores and trichospores concurrently (Williams and Lichtwardt, 1971; Lichtwardt, 1972), and in these cases it is possible that zygospore production is not so much a response to molting stimuli as it is that the last intermolt period is sufficiently long to accommodate zygospore formation. The production of zygospores in the last larval stage may be of special significance in univoltine species of aquatic insects, assuming that zygospores survive longer than trichospores outside the host and can serve as a source of inoculum for the next generation of larvae.

In the Eccrinales, which have no verified sexual state, the response to molting in some species is the production of thick–walled (resistant) primary infestation sporangiospores. For instance, the resistant spores of some species of *Arundinula, Astreptonema, Eccrinidus, Eccrinoides,* and *Taeniella* seem to be produced only in response to the molting of their respective freshwater, marine, or terrestrial hosts, but the frequency and abundance of resistant spores in the exuviae can vary with the species of fungus, the degree of maturity of the arthropod, and the season when it molts (Manier, 1969b; Hibbits, 1978). Although ecdysis can be recognized in most of the millipede and crustacean hosts just prior to its occurrence, it is not easy to detect externally when the earlier stages of the molting process have begun and when the fungi presumably are stimulated to start producing resistant spores. Consequently, searching the gut cuticle in fresh exuviae is often the most productive way to find resistant spores. In genera that do not have thick–walled primary infestation spores (e.g., *Enterobryus*), the production of thin–walled primary infestation spores seems to be less correlated with molting, or not at all.

The clearest evidence for the effect of host development on fungal reproduction is found in the Amoebidiales. The ectocommensal *Amoebidium parasiticum* produces simple thalli, which convert their entire contents into walled sporangiospores as they mature during the intermolt phase. These sporangiospores serve to establish new thalli on other animals. On molting, injury, or death of the host, developing thalli become converted within minutes or hours into naked amoebae that are released and go through the resistant amoeba–cyst phase of the life cycle. The endocommensal species of *Paramoebidium* have only the amoeba–cyst phase, and amoebae normally develop and are released only when the host molts or on injury such as caused by dissection. The stimulus for amoeba production in pure cultures of *A. parasiticum* has been investigated by Whisler, and his interesting studies are reported in greater detail in the section on amoebagenesis in Chapter 9. Suffice it to say here that Whisler (1966) found he could induce amoebagenesis with an aqueous dyalizate of dried daphnia, and later (1968) determined that amoebagenesis was obtainable in a defined medium so long as the cultures were provided with glucose, amino acids, and relatively high levels of calcium (0.01 M).

One would expect among arthropods to find a common stimulant for amoebagenesis in *A. parasiticum,* considering the lack of specificity of this species (on Cladocera, Copepoda, Isopoda, Amphipoda, and Insecta), if one assumes (but it has not been demonstrated) that all of the various hosts are capable of inducing the amoeba–cyst phase of the cycle. Whisler (1968) found that crab extract and horse serum, if supplemented with calcium, would induce some amoeba release, and Kuno (1973) used a dialyzed homogenate of mosquito larvae. In the more host–specific endocommensal trichomycetes, nothing is known about the chemical nature of zygospore or resting spore induction. Cultures of *Smittium* could provide a convenient system to experiment with, but zygospores at present have not been produced in any of the available isolates, and those most used experimentally (*S. culisetae, S. culicis. S. simulii*) are not known to produce zygospores *in vivo* either, except for one report in *S. culisetae* (Williams, 1983b).

One might expect molting hormones could have a direct effect on the fungus, important as they are in metamorphosis, but the molting process is very complex, involving many changes in metabolism, initiation of cellular growth, production of new enzymes, arrestment of feeding, and absorption of proteins and other substances including calcium (in some Crustacea) from the cuticle. Kuno (1973) attempted to get amoebagenesis in *A. parasiticum* using ecdysterone [as has the author (unpublished)], diosgenin, and juvenile hormone, but without success. Hormones acting directly on the fungi may not be the answer in view of Whisler's (1968) amoebagenesis study on *A. parasiticum* grown axenically on a defined medium, but one cannot necessarily extrapolate from this information and apply it to the endocommensal forms. The natural inducer by which arthropods exert a morphogenetic influence on their gut fungi thus remains an interesting unsolved problem, but not unsolvable provided adequate culture systems become available.

Survival and Distribution Mechanisms

The wide geographic distribution of many trichomycete species and the frequency with which they infest arthropods attest to their successful dissemination. Distribution of trichomycetes is determined by their ability to transmit spores from host to host over short or long distances as well as by the motility and migratory activities of the hosts themselves. Most species of terrestrial trichomycetes (Eccrinales and Asellariales) grow in mature, or both mature and immature, stages of their hosts. Many of these terrestrial arthropods (e.g., millipedes, isopods) overwinter in such stages in temperate regions while carrying the fungi in their guts, obviating the need for resistant spores that can withstand the rigors of the external environment over prolonged periods of time. Distribution of trichomycetes among their one or more species of terrestrial hosts appears to be limited only by sufficient proximity of individuals to enable transfer of inoculum, and the range of the populations of arthropods. At times, terrestrial arthropods along with their gut fungi can be carried over great distances by one means or another, as has presumably happened, for example, with their introduction into Hawaii (see Chapter 5).

Marine arthropods with gut fungi (Eccrinales and Asellariales), most of which are intertidal, are comparable to terrestrial ones in many respects. It is the sexually mature stages that are most highly infested and bear the majority of spore–producing thalli. Although individual populations of given arthropod species along coastlines may be disjunct [e.g., fiddler crabs or ghost shrimps (anomurids) that are restricted to coastal mud flats or estuaries], experience has shown that their respective fungi are present in many populations to one degree or another. Overwintering of the fungi, as in terrestrial arthropods, can occur in the guts of the hosts.

The known distribution of the marine Eccrinales and Asellariales, sketchy as collections have been, suggests that some of these fungi may have a worldwide distribution, but they occur in several different species of their particular host types whose geographic ranges are more limited than their gut fungi. This raises the question concerning how these obligate commensals have become disseminated from one ocean to another. Several examples are given here to illustrate the problem. *Palavascia sphaeromae* has been found in different species of *Sphaeroma* (Isopoda) collected in the Mediterranean, on the east coast of the United States, and in Japan. [In Japan (Hokkaido) *P. sphaeromae* was also found in another isopod, *Tecticeps japonicus,* occurring intermixed with an infested population of *Exosphaeroma oregonensis.*] *Asellaria ligiae* occurs in different species of isopods of the genus *Ligia,* and is known from the Mediterranean, the east and west coasts of the United States, Japan, and in an apparently undescribed freshwater species in Hawaii (see Chapter 4). *Enteromyces callianassae* has a wider host range; it is known to occur primarily in sluggish, mud–burrowing anomurids (*Callianassa* and *Upogebia* species) from Chile, California, and the northwest coast of the United States, but has been found also in true crabs, namely *Uca pugilator* in France and *Hemigrapsus penicillatus* in Japan. In the case of *E. callianassae,* it is tempting to think that dispersal has been via the more active and mobile true crabs, rather than the anomurids.

There is no reason to believe that the distribution of marine trichomycetes has occurred by long–distance dispersal of fungal spores. Rather, it is more reasonable to assume that these gut fungi have dispersed from habitat to habitat over time through migratory activities or accidental displacement of the crustaceans or their molts, and that the fungus has been transmitted from one host species to another while they are living in common habitats. In those trichomycetes (e.g., *Asellaria, Palavascia*) whose range of hosts consists essentially of several species of one genus, it is conceivable that the fungi originally became established in the more primitive forms, and that the fungi later became distributed as host speciation evolved and the derived species of hosts dispersed.

The most puzzling questions relating to survival and distribution of gut fungi are found in Harpellales and *Paramoebidium* spp., which live in freshwater larval insects but not in their respective adults. Some species can be found in permanent lakes or ephemeral pools, but the majority live in rapidly flowing streams or rivers. Infested lotic hosts have been found to be common in the headwaters of scattered river systems in the holarctic. In the United States, they are prevalent on both sides of the continental divide in the small mountain streams that initiate major river systems. The larval insects are incapable of migrating any significant distance upstream and, intentionally or not, often drift downstream during the course of their feeding and development (Davies, 1976; Hynes, 1976). Yet, infestation by trichomycetes has been found over a period of years in some of the same upstream sites. Headwater habitats at high altitudes and latitudes may contain no larval hosts during the winter months because of unsatisfactory environments for larval survival (freezing, drying) or, more commonly, the cyclic nature of their development. Maintenance of trichomycetes in the same site year after year can be accounted for either by resistant spores left in the site for renewed infestation, or by recruitment of new inoculum from other localities. In the event of a resident inoculum, one still has to invoke some mechanism of transport over distance to bring about initial establishment of the fungi, and the same situation applies to the many other river habitats where the insects have multivoltine cycles or extended larval development such that overlapping larval stages occur throughout the year.

A few examples will illustrate the degree to which some Harpellales and *Paramoebidium* spp. can be distributed. *Harpella melusinae* is one of the most widespread species of trichomycetes, and attaches only to the peritrophic membrane of Simuliidae. In the author's collecting experience, it can be found in the majority of larval blackfly populations in Europe, the United States, and Japan. Likewise, few streams with stonefly, mayfly, and blackfly larvae do not contain *Paramoebidium* spp. The successful dissemination of these fungi, both among and within populations of larvae, has been well established. *Smittium culicis* can be found in many mosquito breeding sites in southern France (Tuzet et al., 1961). On the North Island of New Zealand and especially in eastern Australia, *Smittium culisetae, S. simulii,* and *Stachylina grandispora* are common in their respective endemic dipteran larval hosts (Lichtwardt, unpublished). The occurrence of species of *Smittium* in mosquito populations in other parts of the world may be more sporadic (Williams and Nagel, 1980; Lichtwardt, unpublished). Nevertheless, when they are found they may occur in rather unexpected places. In Hawaii (Oahu) the author isolated *S. culisetae* from three of three larval sites examined: in discarded beer bottles and cans (containing *Aedes albopictus*), in a temporary roadside ditch, and in a small rock hole within a forest (both containing *A. vexans*). In Japan the author has isolated *S. culisetae* from mixed species of mosquito larvae in bamboo and pottery flower vases in a Buddhist cemetery, and *S. simulii* (normally found in larval blackflies and chironomids) in mixed species of mosquito larvae in a garden watering can. Almost all mosquito larva populations in rock holes along stream banks in New South Wales and Queensland, Australia examined by the author proved to be infested with *S. culisetae*.

These examples of widely dispersed or unusual habitats for trichomycetes raise the question of how the inoculum was originally introduced. Trichospores would not be expected to become airborne except under unusual and unpredictable circumstances. Conceivably birds could transport spores or infested larvae from one feeding site to another. A more logical explanation is that female insects are capable of carrying spores within or on their bodies and can deposit them in a site during egg laying. However, at present there are no records that adult insects are capable of doing this (Williams, 1983a), with the exception of the pathogenic species *Smittium morbosum* (Sweeney, 1981a). It is not inconceivable that a small but significant percentage of females are able to carry some kind of inoculum, and perhaps the appendages on trichospores assist in the process (Lichtwardt and Williams, 1983a). Whatever the propagule may be, it would have to exhibit some degree of resistance to desiccation if it is located outside the host. Zygospores have not been seen in any of the three species of *Smittium* mentioned in the preceeding paragraph, either *in vivo* or *in vitro* [with one exception (Williams, 1983b)], and those of *Harpella melusinae* are very rare. Cysts of *Paramoebidium* spp. are a normal part of the life cycle, however.

The mechanisms of survival and distribution among habitats of these aquatic fungi thus remain an unsolved problem. Once they are established in one or more larval hosts, transmission from one individual to another over the growing season is more understandable, particularly if they are gregarious. It is interesting that trichospores and zygospores generally fall within the size range and shape of naviculate diatoms that many of these aquatic insects ingest.

CHAPTER 9

Experimental Studies on Cultured Species

Several species of trichomycetes are available in axenic culture, and this has permitted the design and analysis of various experiments, some strictly *in vitro*, others *in vivo*. Mosquito larvae have been the host of choice for studies with *Smittium* spp. because of the relative ease with which the larvae can be manipulated and raised axenically in the laboratory, their short developmental cycles, and the background of information on their growth and physiology available to investigators. The fungi used in experimental studies have been *Amoebidium parasiticum, Smittium culisetae, S. culicis, S. simulii, S. mucronatum,* and several isolates of unidentified *Smittium* species. Currently available cultures are listed in Appendix C, and methods for their isolation and maintenance in culture are given in Chapter 3.

The first of these fungi to be isolated axenically, by Whisler (1960), was the ectocommensal *Amoebidium parasiticum*. This feat was followed by the isolation of two species of *Smittium* by Clark et al. (1963) from the guts of mosquito larvae, and Lichtwardt's (1964) isolation of species of the same genus from larval mosquitoes and blackflies. Since that time a large number of isolates of *Smittium* spp., and some of *A. parasiticum* and new endocommensal genera, have been obtained from a variety of hosts and a range of geographic localities.

It has been stated previously, but bears stressing here, that the nutrients and culture conditions that satisfy the currently available isolates have proved to be unsatisfactory for the *in vitro* isolation or stable growth of a large number of other trichomycetes. All species of *Smittium* are not amenable to the isolation techniques that have been employed with the culturable species. Further, the author has been unable to make isolations from some natural populations of those same species currently in culture, leading to the conclusion that in some cases there may be cultivation idiosyncrasies among strains.

Nutrition

Whisler (1962) found that *A. parasiticum* grew well in a shaken liquid medium consisting of 1% Bacto–tryptone, 0.3% glucose, and inorganic salts. The addition of thiamine (200 μg/liter) to the semidefined medium increased growth more than sixfold, and it was later found to be the only vitamin required. A defined medium was devised by substituting methionine (0.01%) for tryptone and making some adjustments in the levels of ammonium and phosphate. This defined medium provided thallial dry weights of only 21–55 mg/50 ml medium, as compared with maximum dry weight levels of about 150 mg/50 ml medium with the tryptone–containing formula, but it permitted a more precise determination of the organism's nutritional requirements. Replacement of methionine with cystine or sulfate was not successful, nor did nitrate serve as a nitrogen source in place of ammonium. Mannose supported good growth as did also glucose, and the fungus was able to utilize fructose fairly well as a carbon source. From Whisler's study it is clear that *A. parasiticum* has nutritional requirements that do not differ from those of many aquatic phycomycetes and other organisms, and therefore its natural growth on living aquatic arthropods does not appear to be determined by nutritional factors alone.

The type isolate of *Smittium culisetae* (COL–18–3) has been used almost exclusively as a model for studies on nutrition of endocommensal trichomycetes, as well as for other biological parameters to be covered in subsequent sections of this chapter. This fungus grows well on a tryptone–glucose medium (TGv) containing 2% tryptone, 0.5% glucose, inorganic salts, and vitamins (see Chapter 3 for the formula), a modification of Whisler's (1962) medium. Dry weight yields of around 170 mg/50 ml medium were obtainable in shaken cultures (Williams and Lichtwardt, 1972b). A satisfactory defined medium capable of yielding good growth has not been devised yet, for there are unidentified components in tryptone that greatly stimulate growth of *S. culisetae* (Williams, 1971). Significantly, Bacto–tryptone (Difco) in the TGv medium provided almost two and one–half times the amount of growth as did Fisher tryptone, and more than eight times the growth obtained with BBL trypticase. According to a quantitative analysis provided by Difco, Bacto–tryptone (a pancreatic digest of casein) contains 18 amino acids and five vitamins, but growth was not satisfactory when tryptone was reconstituted from the given ingredients. The addition of thiamine in concentrations as low as 10 μg/liter to the basal TG medium with 2% tryptone stimulated growth considerably. The growth stimulant in tryptone can be eluted with warm ethanol and apparently is not a vitamin; ethanol–washed tryptone yielded inferior growth even when supplemented with thiamine, and growth levels were restored by the addition of air–dried ethanol extract of tryptone to the washed medium.

Williams (1971) obtained no positive effects on growth of *S. culisetae* when he added an auxin (naphthylacetic acid), a cytokinin (*N*–6–benzyladenine) and gibberellic acid (GA_3) in various concentrations at which these plant hormones are used with higher plants. At relatively high concentrations, the hormones suppressed both dry weight yields and spore production.

Glucose produced better growth in *S. culisetae* than 18 other carbohydrate sources (Williams and Lichtwardt, 1972b). In these tests tryptone was reduced to 1% in the medium, and dry weights were compared against the baseline growth level without carbohydrate supplementation. The fungus did not ferment glucose. Glycerine, mannose, and fructose were assimilated but were less satisfactory carbon sources. Trehalose, an important blood sugar in insects, supported almost no growth, as was true of soluble starch even though hydrolysis of starch by *S. culisetae* was demonstrated on starch plates. Synergistic effects on growth were not found when the other carbohydrate sources were combined individually with glucose.

Smittium culisetae was found to utilize ammonium compounds and urea when grown on TGv medium containing 0.1% tryptone (to provide at least minimal growth). Cultures with nitrate and methionine as the major nitrogen supplements produced dry weights only slightly greater than the controls. Asparagine was not utilized, and nitrite was inhibitory to growth. Ammonium and urea are commonly excreted into insect guts (Chapman, 1966), and therefore may be major sources of nitrogen for the gut fungi *in vivo* as well.

The studies cited show that the nutrition of *A. parasiticum* and *S. culisetae* is very similar with respect to carbon, nitrogen, and vitamin requirements. The growth of both species is greatly stimulated by tryptone. Unknown components of tryptone are necessary for growth of *S. culisetae* in an otherwise defined medium, whereas methionine can replace tryptone with *A. parasiticum* to provide satisfactory, but less, growth. These species will grow well on a number of other undefined media commonly used with fungi (Lichtwardt, unpublished). Dilute brain–heart infusion (BHI/10) is one of the best, but has been used mostly as a primary isolation and culture maintenance medium.

Growth Dynamics and Culture Conditions

Growth rates of cultured trichomycetes compare favorably with many other fungi. Under good conditions, growth rates as high as 220 mg dry weight/day (150 ml medium in 500–ml flasks) have been measured for *S. culisetae* (Williams and Lichtwardt, 1972b), and 120 mg/day (50 ml medium in 125–ml flasks) for *A. parasiticum* (Whisler, 1962). Maximum dry weights were attained in about 4 days for both species. As is to be expected, growth rates and dry weight yields can vary among species of *Smittium*

(Fig. 9.1), just as there are measurable species and strain differences in other physiological properties in this genus. Whisler reported satisfactory growth of *A. parasiticum* over a temperature range of 15 to 30°C. Farr and Lichtwardt (1967) grew *S. culisetae* (COL–18–3) in stationary cultures and obtained best growth rates between 22 and 28°C; however, maximum dry weight yields were produced at 10°C (Fig. 9.2), with satisfactory growth occurring at the more extreme temperatures of 7 and 32°C. Chapman (1966), also using unshaken cultures, found a generally similar pattern of growth as a function of temperature with another strain of *S. culisetae* (HAW–5–7) and with *S. simulii* (JAP–51–5), but *S. culicis* (WYO–51–11) produced considerably slower although steady growth at 12, 18, and 26°C with the growth curves still rising after 27 days and with the best growth at the lowest temperature. (This unorthodox isolate of *S. culicis* came from a larva of *Aedes sticticus* living in water at 8°C in a small pool at 2750 m altitude.) Studies of several strains of *Smittium* in shaken culture corroborate the general pattern of slower growth rates but with higher dry weight yields at lower temperatures (Williams and Lichtwardt, 1972b; El–Buni and Lichtwardt, 1976a).

The shaking of cultures in flasks increases both the rate and the amount

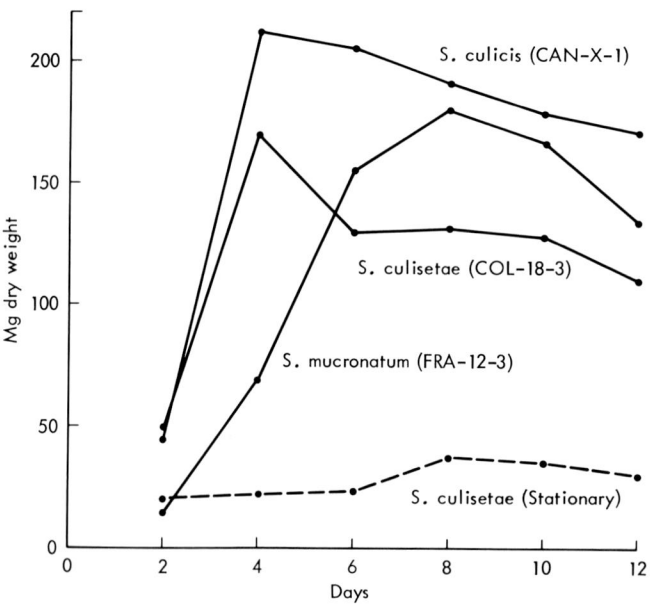

FIG. 9.1. Growth of three species of *Smittium* at 24°C in shaken 250–ml flasks containing 50 ml of tryptone–glucose (TGv) medium and *S. culisetae* (COL–18–3) in stationary culture. Data reprinted by permission from Mycologia, vol. 64: 810 and 68: 565, Copyright 1972 and 1976, R.W. Lichtwardt and The New York Botanical Garden.

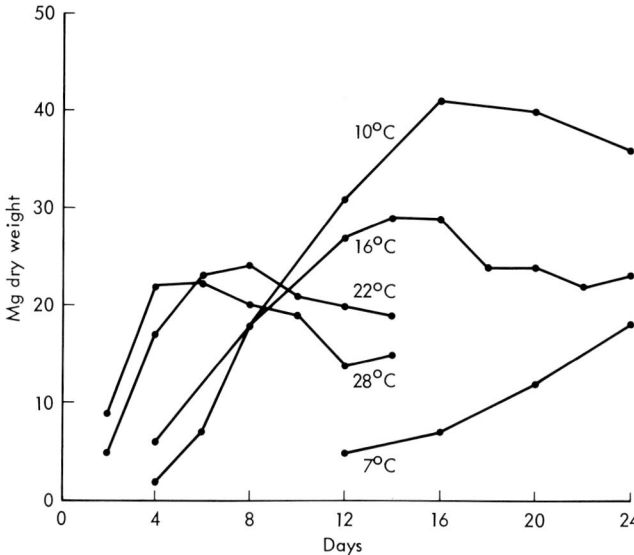

FIG. 9.2. Growth of *Smittium culisetae* (COL–18–3) at different temperatures in stationary 125-ml flasks containing 50 ml of TGv medium. Data reprinted by permission from Mycologia, vol. 59: 174, Copyright 1967, R.W. Lichtwardt and the New York Botanical Garden.

of growth substantially. Whisler (1962) reported increased growth of *A. parasiticum* about 162% over stationary cultures when he used a rotary shaker. With *S. culisetae* the results have been even more dramatic (Williams and Lichtwardt, 1972b): in 250-ml flasks with 50 ml of TGv medium, growth was almost 360% greater in shaken flasks (Fig. 9.1), and in 500-ml flasks with 150 ml of the same medium, shaking resulted in more than 450% increase in dry weight. Maximum yields were obtained in 4 days in both flask sizes that were shaken, and in 8 days in stationary flasks. It would therefore appear that *Smittium* species growing in the guts of larvae, where oxygen potentials can be expected to be low, do not fulfill their growth potentials. However, prolific vegetative growth of thalli in the guts would not be necessarily advantageous to the fungi and could even be lethal to the larvae.

The hindgut contents of mosquito larvae tend to be slightly acidic (pH 6.4) to alkaline (pH 8.0) (Clements, 1963), depending on the species and probably many other factors. *In vitro* stationary cultures of *S. culisetae* (COL–18–3) grown by Farr and Lichtwardt (1967) using 10 different initial pH values ranging from 4.0 to 9.6 that were unadjusted during growth in TGv medium, produced better growth in those media with initial values on the alkaline side, with the best growth occurring at pH 8.3. No growth was produced at either extreme of that pH range. A subsequent study of the same isolate by Williams and Lichtwardt (1972b) indicated that sta-

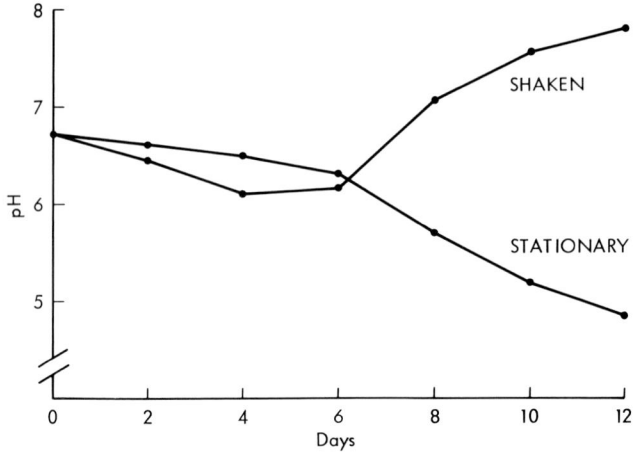

FIG. 9.3. Changes in pH values of TGv medium over time during growth of *Smittium culisetae* (COL–18–3) at 24°C in shaken and unshaken 250–ml flasks with 50 ml of medium. Data reprinted by permission from Mycologia, vol. 64: 810, Copyright 1972, R.W. Lichtwardt and The New York Botanical Garden.

tionary culture conditions lead to a considerable drop in pH of the TGv medium over time (Fig. 9.3), which might account for better growth at the initially higher pH values in stationary cultures. Shaken cultures, on the other hand, became more alkaline after a slight drop in pH. They found that significantly more growth was obtained in shaken TGv medium (initial pH 6.7) when pH values were allowed to change autonomously over time than when the values were maintained constant at pH 6.5, 7.5, and 8.5; least growth was measured at the highest pH value. With *A. parasiticum* grown in shaken flasks at constant pH values in a similar medium, optimum growth was obtained at pH 7.0, and the growth range was approximately pH 6.0–9.0 (Whisler, 1962).

It would appear that pH is not a limiting factor to vegetative growth of either *A. parasiticum* or *S. culisetae* in nature, because of their tolerances. How these fungi respond to their immediate natural environment with respect to pH influences is not clear, however, since the data obtained experimentally are quite artificial. But the data do suggest that those fungi within the confines of the gut might exert some influence on the acidity of the lumen if they are growing luxuriantly.

Sterol and Lipid Production

Starr et al. (1979) did a study of 14 isolates of *Smittium* spp., two strains of *A. parasiticum*, and two Kickxellales *(Dipsacomyces acuminosporus* and *Linderina pennispora)* to determine and compare their ability to syn-

TABLE 9.1. Analysis of free sterols produced by *Smittium culisetae* (COL–18–3) grown on a sterol–free medium.[a]

Days of growth	Total sterol (mg/g mycelium dry wt.)	Desmosterol	Ergosterol	Cholesterol
3	0.767	+	+	−
7	3.38	+	−	+
14	0.129	−	−	+

[a]Excerpt from Starr et al., 1979.

thesize sterols. They were grown for 7 days in shaken TGv medium, which was determined to be sterol free, and mycelial extracts were analyzed by gas chromatography and mass spectroscopy. The major free sterol component of the *Smittium* species was found to be desmosterol, a rare sterol in fungi. A few of the isolates also produced ergosterol, and some contained one of two unidentified sterols that had trimethysilyl ethers with parent ions at either m/e 470 or 498. Strain COL–18–3 of *S. culisetae* was analyzed after harvesting at days 3, 7, and 14, and it was shown that the sterol content varied both quantitatively and qualitatively during growth (Table 9.1). The increase in total sterols in the thalli from day 3 to day 7 coincides with increases in spore production after maximum dry weight has been achieved (Figs. 9.1 and 9.4), and the decrease in sterols on day 14 occurs at the time culture dry weights are decreasing and lytic activities are presumably in progress. Cholesterol was found only in this one strain of *Smittium*, but it was present, along with ergosterol, in the two strains of

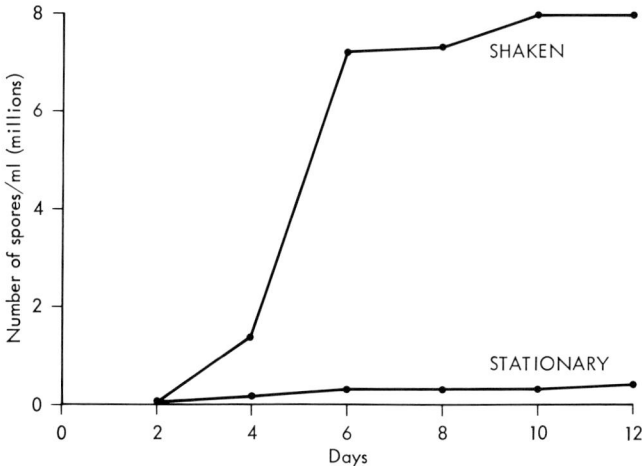

FIG. 9.4. Trichospore production of *Smittium culisetae* (COL–18–3) in shaken and stationary cultures grown at 24°C in 500–ml flasks containing 150 ml of TGv medium. Data reprinted by permission from Mycologia, vol. 64: 810, Copyright 1972, R.W. Lichtwardt and The New York Botanical Garden.

TABLE 9.2. Lipid components of *Smittium culisetae* (HAW–13–2) grown for 10 days on dilute brain–heart infusion medium.[a]

Lipid class	Per cent distribution of total combined lipids
Triglycerides	26.6
Free fatty acids	11.4
1,3–diglyceride	9.9
1,2–diglyceride	7.6
Steroids	12.9
Steroid ester + hydrocarbons	7.2
Phosphatidyl ethanolamine	3.4
Phosphatidyl serine	2.6
Phosphatidyl choline	17.8
Unidentified phospholipid (ninhydrin positive)	Trace

[a]From Patrick et al., 1973.

A. parasiticum. Cholesterol was the sole sterol produced by the two Kickxellales. Only *Smittium* spp. produced desmosterol. The Kickxellales (Zygomycetes) were included in the study because they have been shown to have a slight serological relationship with *Smittium* spp. (see later section on serology). The sterol content of *Smittium* spp., based on the dry weight of fungal tissue, was as high as 0.34%. Sterols are an essential nutrient of insects, and the possible role of these gut fungi in furnishing sterols and vitamins to nutritionally deprived mosquito larvae is presented in the next section of this chapter.

The lipid components of *S. culisetae* were studied by Patrick et al. (1973), who found that 12.9% of the extractable lipid (about 1.28% of the total mycelial dry weight) consisted of unspecified steroids. They grew their fungus on dilute brain–heart infusion (Difco) which, according to a later analysis by Starr et al. (1979), contains 0.715 mg of cholesterol per gram of dry medium. As a consequence, it cannot be determined whether the steroids in the thalli were totally synthesized or partly incorporated from the medium. The lipids extracted by Patrick et al. from 10–day–old cultures of the fungus accounted for 9.9% of the total weight, and consisted of 76.3% neutral lipids and 23.7% polar lipids, with triglycerides comprising 26.6% of the lipid classes (Table 9.2). Of the fatty acids, palmitoleic was present in unusually high concentrations (38.7% of total fatty acids), followed by palmitic (34.3%) and oleic acids (16.3%).

Effects on Host Development

The common occurrence of trichomycetes in such a wide range of arthropods tempts one to postulate that these commensals are not always completely neutral with respect to the welfare of their hosts, even though their effects may not be readily observed. Death of infested mosquito larvae has been recorded in several instances, as described in Chapter 8. In this section the possible beneficial aspects will be considered. The only experimental evidence along these lines was obtained by Horn (1980) and published by Horn and Lichtwardt (1981) using larval *Aedes aegypti* to study the nutritional relationship with cultured *S. culisetae* (COL–18–3). The nutritional requirements of these mosquito larvae have been studied (e.g., Trager, 1935; Lea et al., 1956; Singh and Brown, 1957; Akov, 1962; Lang et al., 1972). The larvae can be raised axenically simply by surface–sterilizing the eggs.

Experiments were conducted using a semidefined phosphate–buffered medium consisting of purified egg albumin, vitamins, cholesterol, RNA, and minerals. Additional controls were run concurrently with each experiment on an undefined (yeast–pork liver) medium, which provided assurance that the larvae were capable of development at an optimal rate. Vitamins and cholesterol were deleted individually from the semidefined medium to compare the rate of development and survivability of larvae infested with *S. culisetae* versus uninfested ones raised on the deficient media. Infestation was accomplished by adding a large measured number of washed fungal spores to test tubes containing 1st–instar larvae. Larvae were kept singly in tubes in order to record the time of molting and mortality, if it occurred.

In the absence of riboflavin in the medium only a few uninfested larvae survived to the 4th instar, and none pupated, whereas almost 50% of the infested larvae were alive at the 4th instar and almost half of these pupated (Fig. 9.5). In pyridoxine–deficient medium, no uninfested larvae survived beyond the 2nd instar, while a few of the infested ones reached the 4th instar but did not pupate. Without nicotinamide the development was similar, except that only a small percentage of infested larvae reached the 3rd instar. In these vitamin–deficient media, the infested larvae thus were able to attain one or two developmental stages beyond what the uninfested ones attained, presumably due to the presence of the fungus. No such improvement in infested larval development was found when thiamine or calcium pantothenate was omitted from the medium.

When cholesterol was omitted, none of the uninfested larvae pupated, whereas 1 out of 45 larvae with *Smittium* did. It was further found that when lipid extracts of 3–day–old *S. culisetae* cultures equivalent to 3.1 and 6.2 mg dry weight of mycelium were provided for each larva, the rate of development of those larvae was as good as that on the complete medium with cholesterol. Extracts equivalent to 37.2 mg mycelium per larva inhibited pupation, however, but this is not unexpected since it is known that excessive amounts of lipids or cholesterol can be toxic to larvae (Golberg and De Meillon, 1948; Akov, 1962). Purified desmosterol, the principal sterol synthesized by *Smittium* spp., when substituted for cholesterol in the medium provided rates of development comparable to cholesterol through the four larval stages; however, only 63% of the larvae pupated

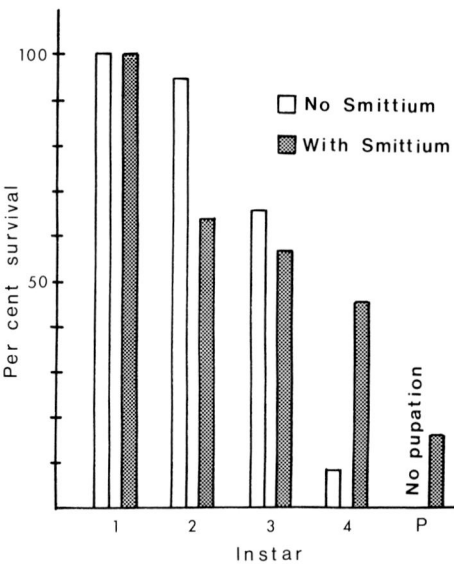

FIG. 9.5. Survival of *Aedes aegypti* larvae through pupation on a riboflavin–deficient medium with and without infestation by *Smittium culisetae*. Reprinted by permission from Mycologia, vol. 73: 731, Copyright 1981, R.W. Lichtwardt and The New York Botanical Garden.

with desmosterol as compared to 97% with cholesterol. When heat–killed thalli of *S. culisetae* were added to sterol–free medium (4.4 mg/mosquito), the rate of development was greater than the controls and more than 50% of the larvae achieved pupation; those not fed dead thallial material did not develop beyond the 3rd instar. Some of those larvae whose development had stalled in the 3rd instar eventually pupated after dead *Smittium* thalli were added to the medium.

Comparison of infested and uninfested larvae raised on complete semi-defined medium showed that about 26% of the infested ones died before reaching the 2nd instar. The same general level of mortality also occurred after the 1st– or 2nd–instar stages more or less consistently in all of the experiments with vitamin– or sterol–deficient media (see Fig. 9.5). This was attributable in most cases to excess growth of the fungus in the larval gut (Fig. 9.6A), presumably caused by the ingestion of too many spores and resulting in unsuccessful ecdysis. On the undefined (yeast–liver) medium, this mortality did not occur in infested larvae fed similar amounts of inoculum. Later instars rarely became overinfested, and in fact infestations were not always maintained through the 4th instar. Williams and Lichtwardt (1972a) also found mortality (5–10%) in artificially infested 1st–instar larvae raised on Akov's (1962) defined medium.

Trichospores of the *S. culisetae* isolate used (COL–18–3), unlike most isolates of this genus, are capable of germinating after some time *in vitro*.

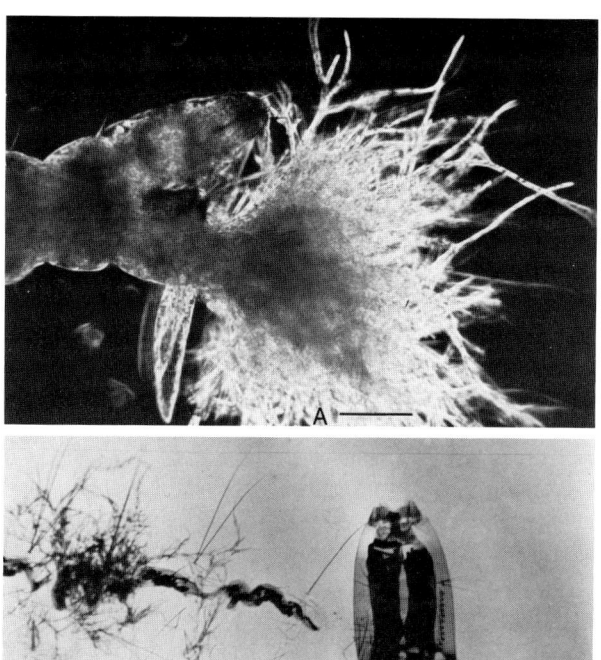

FIG. 9.6. A. First-instar larva of *Aedes aegypti* killed by excessive growth of *Smittium culisetae* in the gut, with the fungus protruding from the rectum; photograph courtesy of M.C.Williams. B. Thalli of *S. culisetae* attached externally to the 4th–instar molt of an *A. aegypti* larva that had survived for several weeks in a semidefined medium lacking riboflavin; from Horn, 1980, with permission. Scale bars = 100 μm.

Sparse thallial growth was observed by Horn and Lichtwardt in tubes with larvae whose development became stalled and which were held for up to 21 days. Larvae apparently ate some of this fungal material, and it is not known to what extent this may have influenced their nutrition in the later stages of those experiments. An unusual observation was made: Thalli of *S. culisetae* could attach to the external cuticle of some larvae that survived several weeks when reared on the semidefined medium lacking riboflavin, pyridoxine, nicotinamide, or sterols (Fig. 9.6B). This is the first instance of external attachment reported in any Harpellales.

These experiments with *S. culisetae*, although not unequivocal, offer some support to the hypothesis that mosquito larvae growing in vitamin-

and sterol–deficient media may develop more favorably if infested with *Smittium*. In natural environments, it is more likely that the deficiencies would be only partial, and it is probable that the amounts of inocula would not be so high as to produce the kind of mortality observed in the experiments. If the results are corroborated by other similar investigations, and if one extrapolates this to other gut fungi, one could envisage some trichomycetes living commensalistically within their hosts under conditions of good nutrition, but assuming a more mutualistic relationship when the arthropods become nutritionally stressed. From an evolutionary point of view, this would suggest that those populations of arthropods that are infested with trichomycetes would, over time, have some selective advantage.

Sporulation

Adequate rates of spore production in gut fungi that are frequently expelled from their hosts due to ecdysis are probably more critical to their survival than is the amount of vegetative growth they produce. The rate of spore production in the holocarpic thalli of *A. parasiticum* in culture is equal to the generation time, but in *Smittium* spp. there normally is appreciable growth before sporulation commences, and spores will be formed at the tips of some branches more or less continuously as the thalli grow. The author has seen in some isolates of *Smittium* the most abbreviated structural form of sporulation possible: A spore will germinate some time after release in culture and the extruded inner cell directly becomes a generative cell that produces a trichospore (Fig. 9.7). Apparently this kind of de-

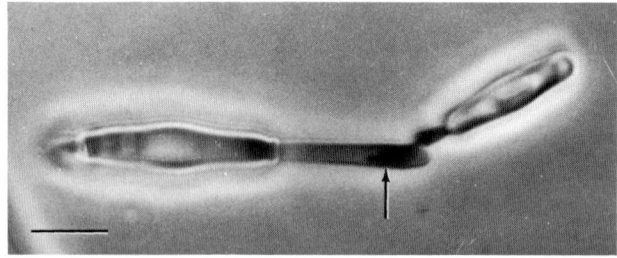

FIG. 9.7. A germinated trichospore of *Smittium simulii* (JAP–33–2) in culture whose extruded inner spore has become generative and has produced a trichospore immediately. Note the appendage within the new trichospore's generative cell (arrow). This unusual abbreviated type of development appears to occur only *in vitro* and under conditions not conducive to vegetative growth. Scale bar = 10 μm.

velopment does not occur regularly but only under cultural conditions that are inimical to vegetative growth.

A study by El–Buni and Lichtwardt (1976a) of ten isolates of *Smittium* grown in shaken TGv medium at 22°C showed that three isolates of *S. culisetae* used were more prolific (5.0–7.5 × 10^6 spores/ml) than any of the isolates of *S. culicis*, *S. simulii*, or *S. mucronatum* (0.5–2.5 × 10^6 spores/ml). Several cultural parameters affecting sporulation were studied, including medium composition. Glucose was inhibitory to sporulation at all levels above 0.2% in TGv medium, even at concentrations that produced maximum dry weight (Fig. 9.8). Glucose proved to be the best carbohydrate source of 12 tested (at 0.5% w/v) for spore production in *S. culisetae*, *S. culicis*, and *S. mucronatum*. Considerably different effects were detected depending on the temperature used. For instance, with *S. mucronatum* fructose provided almost as many spores at 14°C as glucose did, but relatively fewer spores developed with fructose at 22°C, at which temperature other carbohydrates induced more spores. When tryptone concentration was varied in TGv medium, a substantial increase in spore

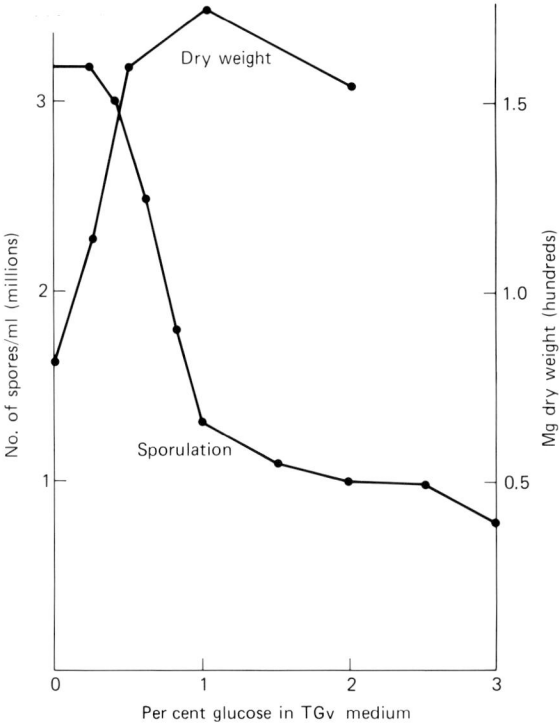

FIG. 9.8. Effect of glucose concentration on sporulation and dry weight of *Smittium culisetae* (COL–18–3). Reprinted by permission from Mycologia, vol. 68: 569, Copyright 1976, R.W. Lichtwardt and The New York Botanical Garden.

production in *S. culisetae* occurred in 2% tryptone, the increase coinciding with a sharp rise in final pH of the medium (Fig. 9.9). It was found that brain–heart infusion prepared as recommended by the manufacturer (3.7%) almost totally inhibited spore production in three species of *Smittium;* all three sporulated best when the medium was diluted to 0.74% or one–fifth of the recommended concentration. The addition of 20 mg/liter of sitosterol acetate and β-sitosterol to an undefined medium increased sporulation in *S. mucronatum* twofold or more, but ergosterol and cholesterol were both inhibitory to spore production.

El–Buni and Lichtwardt also showed that optimum temperatures for sporulation can be quite different among species. In TGv medium, *Smittium culisetae* had maximum sporulation at 22°C, and *S. culicis* and *S. mucronatum* produced most spores at 14°C (Fig. 9.10). Thus, while vegetative growth is good at room temperature in all three species (Fig. 9.1), cultures must be grown at considerably lower temperatures in the laboratory if maximum spore production is desired in the latter two species.

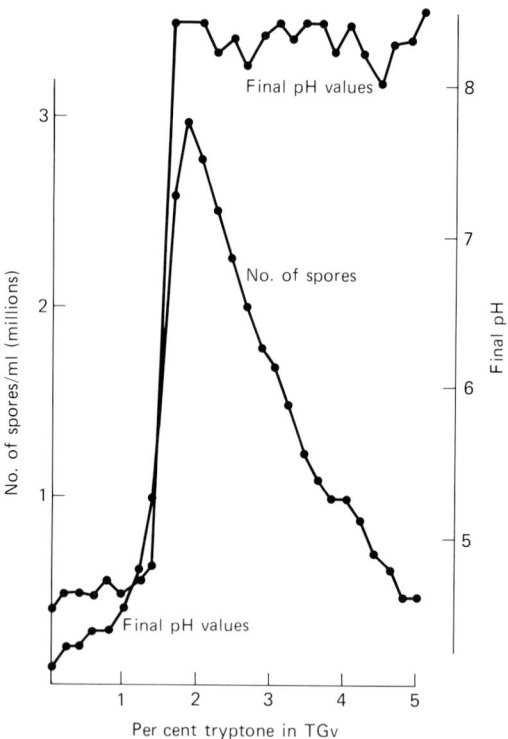

FIG. 9.9. Effect of tryptone concentration on sporulation of *Smittium culisetae* (COL–18–3), and final pH values of the medium at time of harvesting (initial pH 6.7–6.9). Reprinted by permission from Mycologia, vol. 68: 569, Copyright 1976, R.W. Lichtwardt and The New York Botanical Garden.

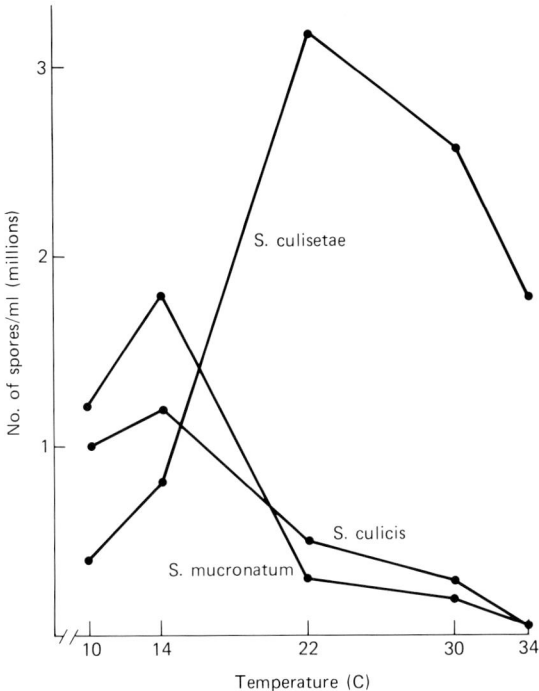

FIG. 9.10. Spore production as a function of temperature in three species of *Smittium*. Reprinted by permission from Mycologia, vol. 68: 565, Copyright 1976, R.W. Lichtwardt and The New York Botanical Garden.

These same three isolates were grown at 22°C at different initial pH values spanning the range that has been reported in mosquito hindguts (Clements, 1963), and all three had maximum sporulation at pH 6.9 or slightly lower (Fig. 9.11), indicating that unadjusted TGv medium (initial pH 6.7) is satisfactory for spore production. Just as aeration in shaken cultures increases vegetative growth *in vitro* (Fig. 9.1), so does it increase spore production (Fig. 9.4). When *S. culisetae* was shaken in cotton–plugged flasks (the usual procedure) and compared against similar shaken replicates with rubber stoppers, the spore numbers were almost fivefold greater in the cotton–plugged flasks. It was not determined what role carbon dioxide accumulation may play in reducing spore production.

The greater number of spores produced by *S. culisetae* isolates does not necessarily mean that this species puts more of its resources in terms of biomass into reproduction, because the spores of *S. culisetae* are smaller than those of any of the other *Smittium* species used in these sporulation experiments. The results obtained in the experiments are useful to the investigator in selecting appropriate media and culture conditions to obtain adequate numbers of spores. How the data apply to these same species

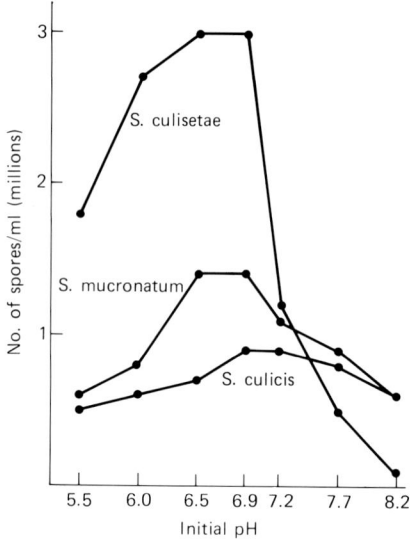

FIG. 9.11. Spore production at 22°C in three species of *Smittium* over a range of pH values reported to occur in mosquito larva hindguts. Reprinted by permission from Mycologia, vol. 68: 567, Copyright 1976, R.W. Lichtwardt and The New York Botanical Garden.

growing in the gut, where the rate and quantity of spore production are undoubtedly of importance to their transmission and survival, is less clear. Since the hosts are poikilothermic, the temperatures at which the fungi grow are close to those of the environment in the hosts' aquatic habitats. Temperature tolerances for spore production may well be a major factor in the degree to which infestation is sustained in any given population of suitable arthropod, and could account for shifts in the percentage of hosts infested as temperatures rise or fall in aquatic habitats during different seasons of the year.

Spore Germination and Longevity

Germination of trichospores is an especially sensitive adaptive process in the life cycle of Harpellales, for it must occur within the short time span that the spore moves through the gut. Trichospores in natural situations do not normally germinate in the outer aquatic environment, for such an occurrence would be counterproductive to survival. When ingested, they apparently are stimulated by still unknown factors in the digestive tract to initiate germination. Williams and Lichtwardt (1972a) determined that trichospores of *S. culisetae* can germinate and attach to the gut cuticle in *Aedes aegypti* larvae in as little as 30 min.

The germination process differs substantially from that in spores of other fungi by virtue of the trichospore being a deciduous sporangium. As germination commences, the inner, walled protoplast—the spore proper—begins to elongate and pushes through the outer, sporangial wall at the apical end (or, rarely, at the base next to the collar). The inner spore often vacates the sporangial wall completely, or its base may remain partially or entirely encased within that wall. The emergence phase may last but a few minutes, sometimes seconds. Further development of the newly emerged thallus *in vitro* seems to depend on the isolate of the fungus and the culture conditions. Williams (1983b) published a set of excellent photographs of spore germination in *Smittium culisetae*, and provided evidence that the chain of apical bodies in the tip of the spore are involved in initial attachment of the spore to the substrate (Fig. 9.12). The juvenile thallus may extend considerably before branching commences, or branching may occur soon after emergence.

Holdfasts do not form under ordinary culture conditions, but in some situations the first end of the spore to emerge may produce a swelling surrounded at times by a secreted substance barely visible with phase microscopy, or a narrow rhizoidal extension may be seen, which branches or coils. In some cultured strains capable of germinating there is no further *in vitro* development into thalli, whereas a few produce highly branched

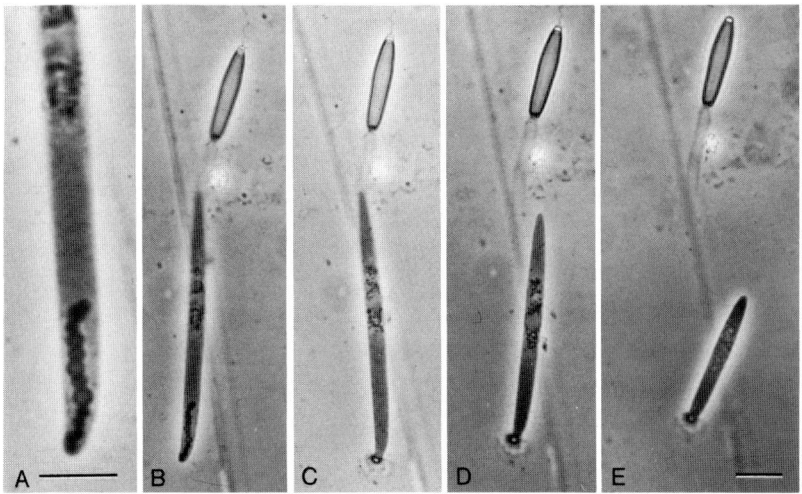

FIG. 9.12. Germinated trichospore of *Smittium culisetae*. A. Apical spore bodies at tip of emerged spore. B. Emerged spore and empty sporangium (outer trichospore wall). C. Extrusion of holdfast material at tip (base), apparently derived from apical spore bodies. D, E. Shrinkage of spore. Scale bars = 10 μm. Reprinted by permission from Mycologia, vol. 75: 253, Copyright 1983, M.C. Williams and The New York Botanical Garden.

fertile thalli with the old trichospore wall sometimes remaining at the point of origin. Shifts in developmental polarity apparently can occur immediately after germination, the best evidence being found in the juvenile thalli that produce trichospores almost immediately: In most instances the trichospore forms at the apex that emerges first (Fig. 9.7), but sometimes a trichospore from the same culture will develop at the opposite end instead.

The vast majority of isolated strains of *Smittium* do not germinate *in vitro* for all practical purposes. There is unquantified evidence that on initial isolation none of the strains at present in culture exhibited a high percentage of germinability; those that now do apparently have been selected for germinability through repeated transference over time in the laboratory. Thus, while selective pressures may not favor germination in the natural environment outside the gut, the opposite can be expected in those new laboratory cultures that have, initially at least, a small percentage of spores that germinate and produce fertile thalli.

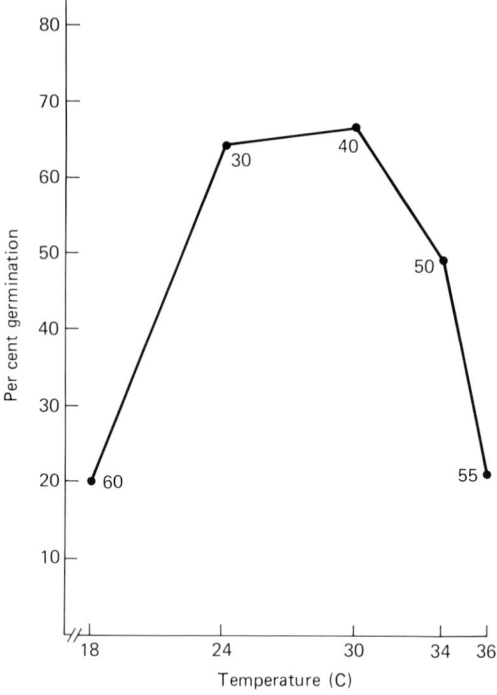

FIG. 9.13. Germination of *Smittium culisetae* (HAW-5-7) trichospores in stationary TGv medium at different temperatures. The numbers on the curve represent time in hours at which maximum germination was reached. Reprinted by permission from Mycologia, vol. 68: 580, Copyright 1976, R.W. Lichtwardt and The New York Botanical Garden.

El–Buni and Lichtwardt (1976b) found five strains of three species in the culture collection of *Smittium* that had better than 20% spore germination. The best germination occurred in three strains of *S. culisetae*, with COL–18–3 topping the list by producing up to 82% germination. TGv medium was superior to BHI as a germination medium, and incubating the spores in still cultures resulted in better germination than shaken cultures in almost all tests. They found that 24–72 hr were required for spores to germinate, depending on various culture parameters; thus, germination *in vitro* takes considerably longer than *in vivo*. Temperatures between 24 and 30°C produced maximum germination in *S. culisetae* (HAW–5–7), as well as the shortest times to achieve those maxima (30–40 hr) under the experimental conditions used (Fig. 9.13). At least 10% germination in strain COL–18–3 occurred over a range of pH values from 5.5 to 8.5, with an optimum (77%) near pH 7.2 (Fig. 9.14). This is slightly more alkaline than the optima for growth and sporulation cited previously. El–Buni and Lichtwardt also found that germination was dependent on spore concentration to some extent, the germination percentages decreasing more or

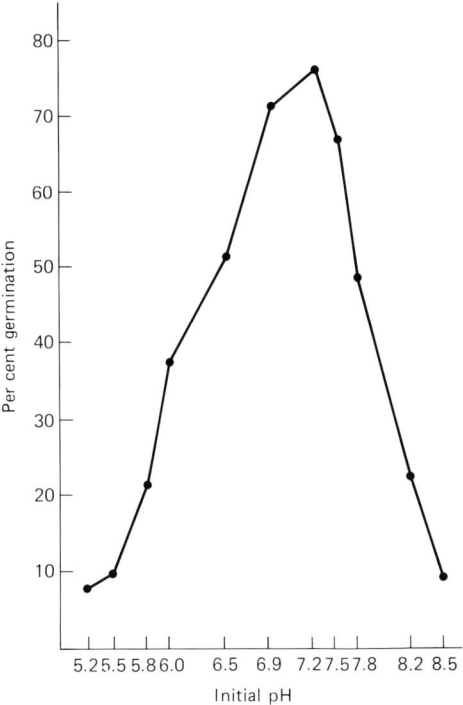

FIG. 9.14. Germination of *Smittium culisetae* (COL–18–3) trichospores at 22°C in TGv medium having different initial pH values. Reprinted by permission from Mycologia, vol. 68: 580, Copyright 1976, R.W. Lichtwardt and The New York Botanical Garden.

less linearly in TGv medium from a maximum of about 79% germination with 0.2×10^6 spores/ml to 31% with 4×10^6 spores/ml.

The duration of spore viability is of interest both because of the need for maintaining cultures in the laboratory and the implication such data may have for understanding survival of spores in nature as a source of inoculum. Freeze–drying has not succeeded as a method for long–term laboratory storage, but storage in liquid nitrogen appears to be satisfactory with all isolates of *Smittium* and *Amoebidium*, provided the cultures are initially cooled down slowly (1°C/min) to below freezing before plunging the sealed ampules into liquid nitrogen (El–Buni, 1975).

It has been noted from the time axenic isolates were first obtained that the longevity of refrigerated cultures varies considerably among the isolates. Vegetative cells of thalli seem to die much sooner than trichospores, as expected, and it is conceivable that the apparent lack of viability in stored cultures sometimes may actually be the result of spores not germinating, although viable, when cultures are transferred to new medium. This possibility has not been thoroughly tested, although El–Buni (1975) fed mosquito larvae with 2–year–old cultures of four species of *Smittium*, and obtained no infestation. In further studies on storability, El–Buni and Lichtwardt (1976b) harvested new trichospores from six isolates of four *Smittium* species and fed them to different instars of *Aedes aegypti* before and after the trichospores were refrigerated in distilled water at 5°C for 8 and 12 months. Spores from *S. culisetae* strains COL–18–3 and HAW–5–7 infested mosquito larvae well to moderately well after 12 months of storage, but spores of strains CAN–X–1 and WYO–51–11 *(S. culicis)* and JAP–33–2 *(S. simulii)* did not survive either storage period at 5°C. FRA–12–3 *(S. mucronatum)* infested no larvae, even before refrigeration.

In the case of *S. simulii* and *S. mucronatum*, the lack of germination in the guts may have been due to the fact that *A. aegypti* is not a suitable host for those particular strains (see Table 9.3). The indications are, however, that strains of *S. culicis* have much shorter viability time spans than *S. culisetae*. El–Buni and Lichtwardt conducted further tests on *S. culisetae* (COL–18–3) using *in vitro* germination, and found that when spores were stored in water at 5°C, germination decreased linearly from 80% to 14% over an 18–month period. At –5°C, germination dropped abruptly to 20% in 2 weeks, and at 4 and 8 weeks it decreased to 10% and 8%, respectively. Williams (1983a) demonstrated that trichospores of *S. culisetae* placed in moist soil and stored in a freezer (~–10°C) for 5 months were capable of infesting 71% of *A. aegypti* larvae, whereas lower infestation resulted from subsets stored in moist soil that had been refrigerated (~4°C) or kept at room temperature. These data suggest there might be considerable loss of viability in overwintering spores, although natural substrates obviously were not duplicated in these experiments.

Experiments such as those described above have shown that not only

TABLE 9.3. Ability to infest larval *Aedes aegypti* by various axenic isolates of *Smittium* spp.[a]

Species and isolate no.	Original larval host	1st Instar	4th Instar
S. culicis			
WYO-51-11	Mosquito	Good[b]	Good
CAN-X-1	Blackfly	Good (slow)[c]	Good
S. mucronatum			
FRA-12-3	Chironomid	Slight[d]	None[e]
S. simulii			
CLW-2-44	Blackfly	Good	Good
JAP-32-4	Chironomid	Slight	None
JAP-33-2	Chironomid	Slight	None
JAP-51-11	Blackfly	Slight	None
S. culisetae			
COL-18-3	Mosquito	Good	Good
HAW-5-7	Mosquito	Slight	Good
HAW-13-2	Mosquito	Good	Good
HAW-14-5	Mosquito	Good	Good
JAP-77-9	Mosquito	Good	Good

[a] The names of the original hosts, where known, and their geographical locations are given in Appendix C. Data are from Williams and Lichtwardt, 1972a.
[b] Good: Young thalli observed at 2–4 hr with abundant growth and sporulation evident in or previous to the molt.
[c] Slow: Few to no thalli at 2–4 hr but abundant growth and sporulation evident in or previous to the molt.
[d] Slight: Only 1 to 3 immature thalli with sparse growth observed but with no sporulation.
[e] None: No thalli observed when at least 10 larvae were dissected in each of two separate trials.

do trichospores germinate in considerably less time in the host than *in vitro*, but many can be induced to germinate in the host whereas they will not do so under culture conditions. Spores of Harpellales and other trichomycetes apparently are stimulated in some way to germinate almost immediately after ingestion. Chapman (1966) tried treating trichospores of several *Smittium* species in various ways to see if germination could be triggered *in vitro*. Included in her studies were treatment of spores with enzymes (protease, trypsin, amylase, maltase, invertase), buffer solutions at different pH values (5.0, 6.0, 7.0, 8.0), embedment in agar, osmotic shock, horse dung extract, detergent (1:1000 Tween 80), citric acid (0.1 M, pH 6.1), different bile concentrations, anaerobic conditions, high carbon dioxide tensions, ethylenediamine tetraacetate (EDTA), amino acids (alanine, serine, valine), and an aqueous extract of mosquito larvae. None of these stimulated germination, yet spores of all species were shown to germinate and establish thalli in the guts of larval *Aedes triseriatus*. One cannot conclude that these kinds of *in vitro* spore treatments are necessarily ineffective, however, because adequate concentrations, time of exposure, or even sequential combinations of exposures to the inducers could be critical.

Host Inoculations

The availability of many axenic *Smittium* isolates obtained from different dipteran hosts and geographic localities has provided the opportunity to study selected representatives for their ability to infest a "foreign" host. This was first done by Chapman (1966), who fed sporulating thalli (and spores alone of one isolate) to larvae of *Aedes triseriatus*. She found that all of the tests resulted in infestation of that larval species. Her studies were expanded and refined by Williams and Lichtwardt (1972a), who used spores of 12 isolates and four species of *Smittium* and determined their infestibility in two instar stages of *A. aegypti* (Table 9.3). The results showed that the isolates that originated from mosquito larvae infested *A. aegypti* well, but so did two isolates from blackflies, CAN–X–1 and CLW–2–44. *Smittium culisetae* and *S. culicis* almost always have been found infesting species of mosquito larvae, but exceptions are known (see Appendix C). *Smittium mucronatum* has been found only in chironomid larvae, and while the cultured isolate did not infest *A. aegypti* well in these experiments, Coste–Mathiez (1970) reported some growth of *S. mucronatum* in *Culex pipiens* after infested chironomid larvae were placed in the same containers with the mosquitoes (see Chapter 6). In nature *S. simulii* has been found mostly in chironomid and blackfly larvae, but one of the isolates, CLW–2–44, readily infested *A. aegypti*. These cross–infestation experiments, together with data on field isolation of species from different families of hosts, show quite clearly that species of *Smittium* are not highly host specific, and therefore their taxonomic determinations cannot be based on the host in which they are found.

Amoebagenesis

Amoebagenesis is the term used by Whisler for the process of inducing thalli of *Amoebidium parasiticum* to release amoeboid cells instead of sporangiospores. Several earlier investigators had noted that amoebae were normally produced only when the host molted or was injured. Whisler's earlier study (1966) was concerned with exploring the nature of the chemical inducer and environmental conditions that influence amoebagenesis *in vitro* using axenic isolates. He developed a quantitative assay by growing cultures of *A. parasiticum* in shaken tryptone–glucose broth, harvesting the thalli during the logarithmic phase of growth by centrifugation, washing them in a dilute salts solution, then keeping the thalli in shaken dilute salts for 16 hr. This provided a homogeneous culture of starved young thalli against which various substances could be tested to record the percentage of thalli that released amoebae. Out of a long list of defined and undefined substances tried, all but one showed no more than a trace of amoebagenic activity. Only an extract of dried *Daphnia* (commercially available as a fish food) produced appreciable (17%) release. He was able to obtain very high release of amoebae (80–100%) within 16 hr when he used a dialysate of the daphnid extract, and showed that the release was proportional to the concentration used. The inducing substance was heat stable, had a relatively small molecular size, and was not soluble in ether, acetone, or chloroform. He further showed that amoebagenesis in his system was optimal near pH 8.0 and 30°C.

In 1968 Whisler was able to identify more precisely the amoebagenic factors by fractionating the daphnid concentrate through a Sephadex G–10 column. This, together with other tests, implicated calcium, glucose, and various amino acids as essential for amoebagenesis. Histidine and methionine promoted amoeba formation more than other individual amino acids, but highest release was obtained with a combination of 10 amino acids. Calcium was optimal around 0.01 M, and he speculated that this relatively high concentration might somehow interfere with wall synthesis around the cleaved protoplasts within the thallus such that amoebae rather than sporangiospores result.

Wall Composition

Studies on chemical components of trichomycete cell walls have been few, and consist mostly of tests performed on uncultured species (Table 7.1). In this section two analyses of wall material from cultured species will be described. Sangar and Dugan (1973) used a strain of *Smittium culisetae* (HAW–13–2), and isolated wall fragments mechanically. They found glucosamine to be the predominant component (35%) in hydrolyzed walls, with glucose (13%), mannose (5.5%), and galactose (4%) comprising the other monosaccharides. The overall analysis indicated 65% carbohydrate, 15% protein, 13% lipid, and 3% ash. The presence of chitin was substantiated by digesting the acid– and alkali–resistant portion of wall material with chitinase, which released N–acetylglucosamine, and by infrared spectroscopy. The presence of cellulose was not detected using the standard $IKI–H_2SO_4$ test.

A study by Trotter and Whisler (1965) on *Amoebidium parasiticum* revealed neither chitin nor cellulose in the walls. Their analysis gave 30% galactosamine, 10% galactose, 3% xylose, 30% protein, and 4% ash. Acid hydrolysates of the wall material gave no evidence of glucose or glucosamine. The presence of hemicellulose was indicated. Thus, it appears that *A. parasiticum* not only differs from *S. culisetae* in its wall composition, but the apparent lack of chitin and cellulose raises the possibility of a lack of relationship of *A. parasiticum* with other groups of fungi as well.

Serology

In 1972 Sangar et al. published the results of an immunological study on trichomycetes and Zygomycetes designed to test their serological relatedness. Antigens were extracted from 21 isolates of *Smittium* representing what were later determined to be four species, from one isolate of *Amoebidium parasiticum*, and from six Zygomycetes: *Linderina pennispora* Raper & Fennell, *Dipsacomyces acuminosporus* Benjamin, *Syzygites megalocarpus* Ehrenb. *ex* Fr., *Entomophthora apiculata* (Thaxter) Gustafsson, *E. virulenta* Hall & Dunn, and *Basidiobolus ranarum* Eidam. Antisera were raised in eight rabbits against seven of the *Smittium* isolates (one duplicate), and the eight antisera were tested against all 28 antigens by immunoelectrophoresis and immunodiffusion. The data were processed with a computerized, numerical taxonomic program that included cluster and principal component analyses.

Perhaps the most surprising result of the immunodiffusion test was that the antigens of *L. pennispora* reacted strongly with the antibodies of one isolate of *S. simulii* (93%, based on 100% reference reaction), and *D. acuminosporus* reacted fairly strongly with five of the eight antisera (22–52%). These same two species of Kickxellales produced lesser reactions against *Smittium* antibodies in the immunoelectrophoretic tests (15–18% against three antisera and 15–22% against five antisera, respectively, for *L. pennispora* and *D. acuminosporus*). The other Zygomycetes gave no, or very low, reactions with the *Smittium* antisera in both kinds of tests. *Amoebidium parasiticum*, significantly, reacted likewise, producing no precipitates.

In general, immunoelectrophoresis resulted in more distinct groupings in the analyses than did immunodiffusion. Figure 9.15 shows a projection of the first three principal components of variation derived from the correlation matrix for all 28 cultures using immunoelectrophoresis. *Smittium culisetae* clustered farthest from the other *Smittium* species, and is also morphologically most distinctive. *Smittium simulii* clustered into what Sangar et al. interpreted as two closely related groups, but they can be considered one species inasmuch as the morphological differences among those isolates also are very slight. The immunoelectrophoretic tests were not able to distinguish between the six *S. culicis* isolates and the single isolate of *S. mucronatum* (for which no antiserum was developed); the latter species is close to *S. culicis* on a morphological basis as well. In Fig. 9.15 the nontrichomycete species are clustered together, with the two Kickxellales paired and slightly apart from the others; this clustering was considered to be due to an artifact resulting from the failure of those antigens to react consistently and strongly with any of the antisera, and perhaps due to the limited number of antisera available for comparison with a large number of antigens. Ideally, antisera would have been obtained

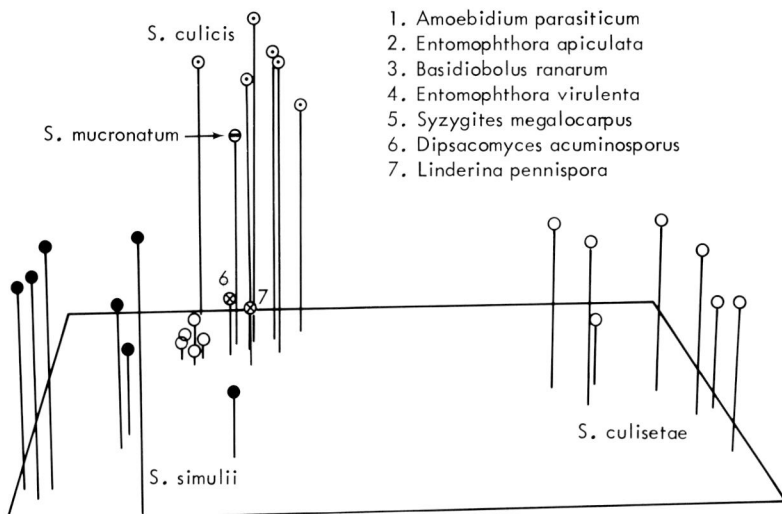

FIG. 9.15. Three–dimensional projection of the first three principal components of variation derived from the correlation matrix computed from immunoelectrophoretic tests on 21 isolates of *Smittium, Amoebidium parasiticum,* and six cultures of Zygomycetes. The five clear circles clustered near circles 6 and 7 represent the first five species in the list. See text for explanation. Reprinted by permission from Mycologia, vol. 64: 349, Copyright 1972, R.W. Lichtwardt and The New York Botanical Garden.

against each of the 28 antigens in order to do reciprocal serological tests throughout, but this was not practicable at the time.

The study by Sangar et al. showed a greater serological affinity among isolates within each *Smittium* species than between species (except for *S. mucronatum*), as expected. But unexpected was the serological relatedness, though at a low average level, of the two Kickxellales with some *Smittium* isolates, and the virtual lack of common antigens between *Amoebidium* and *Smittium*. The possible phylogenetic significance of these and other data are discussed in Chapter 12.

RNA Molecular Weights

Porter and Smiley (1979) undertook a study of ribosomal RNA molecular weights using a relatively large number of isolates in an attempt to clarify phylogenetic relatedness of cultured trichomycetes and a wide representation of Zygomycetes. They used nine isolates of *Smittium* (4 species), one isolate of *A. parasiticum,* 15 species of Mucorales representing six families, three Kickxellales, and one Entomophthorales. The molecular weights for the heavy (25S) and light (18S) rRNA constituents were determined by their relative electrophoretic mobility in polyacrylamide gels. The average rRNA molecular weights for all *Smittium* isolates (calculated from their tables) were 1.47 (1.41–1.51) × 10^6 and 0.75 (0.72–0.78) × 10^6 for the 25S and 18S molecules, respectively; for *A. parasiticum,* they were 1.36 and 0.69 × 10^6; and for all the Zygomycetes, they averaged 1.29 (1.26–1.33) × 10^6 and 0.70 (0.67–0.73) × 10^6. Thus, each group of species appeared to be quite distinct on this basis. As the rRNA molecules are considered to be conservative in terms of evolution, Porter and Smiley concluded that the genus *Smittium* is not phylogenetically related to either *A. parasiticum* or the Zygomycetes. An overall evaluation of phylogenetic relationships is given in Chapter 12.

Abnormalities in Cultures

Like most fungi maintained in culture, trichomycetes may produce aberrant structures not normally seen when they are in their natural environment. In *Smittium* isolates this may include abortive trichospores, chlamydosporelike vegetative cells, and other thallial deformities. Such abnormalities are more frequent when the fungi are grown on agar media with liquid overlayers, and predominate in older cultures.

One interesting variant was found by Chapman (1966). She noted in a petri dish of BHI/10 agar with a water overlayer and seeded with *S. culisetae* isolate HAW–14–5 that numerous small colonies of the *Smittium* were developing around a fungus contaminant growing under the agar layer. Examination of the *Smittium* revealed that the numerous colonies came from germinating trichospores (spores of this isolate had not been found to germinate *in vitro*), and that the trichospores were virtually all asymmetrical, often bent to one side. Colonies that developed from the abnormal, germinated trichospores produced spores with similar bizarre shapes. The contaminant was isolated and identified as *Aspergillus amstelodami* (Mangin) Thom & Church. Normal subcultures of the *S. culisetae* isolate remained normal when grown adjacent to the contaminant species in test plates of BHI/10 medium and when grown on medium that had previously supported growth of the contaminant. A subculture of the abnormal strain of *S. culisetae* was fed to larval *Aedes triseriatus,* and the resulting thalli in the gut produced spores predominantly, but not entirely, of the normal type. The abnormal strain was kept in culture with repeated transfers, and after many years it eventually reverted to producing mostly normal spores. The possibility of some type of viral infection transmitted from *A. amstelodami* was not investigated.

The author has found that some strains of *Amoebidium parasiticum* also may produce aberrant structures. These are more pronounced in still cultures than shaken ones, but the precise conditions that lead to such growth have not been investigated. One form of abnormality involves the formation of giant, thick–walled cells resembling cysts. These apparently do not arise from the amoeba–cyst phase; rather, they seem to be vegetative cells (thalli) that enlarge abnormally and become thick walled. Although these are usually spherical, in some instances they may have irregular shapes. Another form of giant thallus seen in isolates contains two or more generations of cells within a thin outer wall. In this instance it appears that sporangiospores are not released from a mother thallus, and these in turn enlarge and produce unreleased sporangiospores, and so on. Meanwhile, the older thalli keep enlarging to accommodate the internal generations.

The author has had several strains of *A. parasiticum* that originally grew well *in vitro* but began to decline over time and eventually died out. During this process some abnormally shaped cells were observed as well as cells that appeared to be lysed. That such cultures might have had viral infections is a possibility, in view of the presence of virus–like particles in *Paramoebidium* spp. (see Chapter 7). Current cultures of *A. parasiticum* have been examined by the author, but no virus–like particles were detected with the electron microscope.

Part III Systematics

CHAPTER 10

Taxonomic Problems

The taxonomy of trichomycetes is replete with the same kinds of classificatory and nomenclatural problems found in many other large, variable, and understudied groups of organisms. The problems with trichomycetes, however, are compounded by the habitation of the fungi within arthropod guts. This presents some peculiar situations not found in other taxa of fungi. Part of what follows is paraphrased or quoted (with permission) from Lichtwardt, 1978a.

Many species of trichomycetes have well–defined characters with narrow ranges of variation and therefore are readily described or identified. At the other extreme are some species and even genera that have such a variable morphology that their description or identification is difficult unless sufficient material is available in the right stages. Unlike many fungi whose developmental stages can be induced in culture or where visual selection of particular structures can be made directly from their natural substrates, the trichomycetes, with relatively few exceptions, are not currently culturable and are not visible until the host is dissected. The stage of fungal development one finds may depend on the season of the year and the phase of the host at the time of degutting (see Chapter 7). Dissection kills the arthropod and inhibits further development of the fungus in most cases.

It is recognized that species of some genera of Eccrinales *(Alacrinella, Astreptonema, Enteromyces, Ramacrinella)* are dimorphic. That is, the thalli fall into two size categories. However, in other genera one finds some species whose many individual thalli exhibit extensive morphological variation within one gut. In fact, using taxonomic criteria applied to many other species of trichomycetes, we arrive at the initially disturbing conclusion that, in comparing some Eccrinales, intraspecific morphological variation can be greater than interspecific differences. This is not a taxonomic absurdity if one recalls that we are considering the expression of only a small part of the total genotype. Somewhat analogous morphological situations occur in other groups of fungi. For example, single species of *Fusarium* may produce conidia ranging from small, oval, and unicellular to large, lunate, and multiseptate. At the same time, considerably more subtle spore differences are used to separate species in other related genera, and even to make generic distinctions. We simply accept the fact that variability of spore morphology in *Fusarium* species is a characteristic of that genus but not others.

In general, the larger and more complex the gut, the more variable is the trichomycete species within it. A large gut such as that of some millipedes, crustaceans, and beetles may contain a range of microhabitats. Certainly the physiological conditions near the anus would be different from those near the Malpighian tubes. One can measure differences in pH values and dryness, and one can observe that other organisms present in the gut frequently are more or less compartmentalized. Species of Eccrinales that are restricted to particular regions of the gut often show less variability than those which grow throughout the length of the gut. If the variability is very great or disjunct, it may be difficult to determine if the morphologically different thalli represent one species.

A case in point is *Enterobryus borariae,* one of the variable species of Eccrinales studied by the author. The hindgut of its millipede host, *Boraria carolina,* is relatively long and has several distinct structural regions often containing bacteria, nematodes, and protozoans with more or less restricted localization within the hindgut. The fungal thalli can be attached to the gut cuticle from the anterior to the posterior end, or even to the surface of nematodes in the gut. The mature anterior thalli may be as narrow as 3.5 μm, whereas farther back they may be up to 50 μm wide. Various kinds and sizes of spores have been observed (see Chapter 11), but because thallial forms more or less intergrade, it seems preferable to regard these as an assembly of one species that is subjected to various environmental conditions differentially influencing thallial development and morphology. Not all morphological variants are to be found in any single gut. Obviously, identification of this species, as well as other species of Eccrinales with variable morphology, may be difficult if based on examination of only one or two guts. It is difficult to write diagnoses for such variable species according to conventional standards, not to mention designing practical identification keys. See also the description of *Enterobryus tuzetae* and the note under *E. isoporostrepti* in Chapter 11 that relate to the problem of identifying species of *Enterobryus.*

Other examples of extreme variation within species of Eccrinales have been reported. *Enteropogon sexuale* has four thallus types with different spore morphologies, according to Hibbits (1978). The prevalence of particular morphological forms within the long hindgut of any given specimen of its marine crustacean host *(Upogebia pugettensis)* can depend on the season when it was collected. To complicate matters further, the same host may be infested with *Enteromyces callianassae* in the cardiac and pyloric stomach (foregut). *Enteromyces callianassae,* in turn, has a much wider host range including both anomurans and decapods. As another example, passalid beetles are hosts to *Enterobryus attenuatus, E. compressus,* and an unnamed (but good) species of *Enterobryus* described by Heymous and Heymous (1934). All three have been found by the author (rarely) in single passalid specimens, and are distinguished by their location in the long and complex beetle hindgut and by their distinct morphologies.

Most hosts of Eccrinales, however, seem to contain but one trichomycete species. This generalization is not true for some larval aquatic insects, which commonly contain two or more species of Harpellales and Amoebidiales (see Chapter 6). While the trichomycetes in larval insects are usually quite distinct and do not show the range of variation exhibited by some Eccrinales, the proliferation of growth and mixture of species may make it difficult at times to associate zygospores with their related trichospores.

Experimental methods using axenic cultures and host inoculations can aid in the resolution of taxonomic problems, but often they are not practicable or possible. Furthermore, culture conditions do not replicate conditions in the gut, and even if cultures of some of the more variable species were available, they would not likely produce the full range of variation found in natural development. In the absence of such experimental methods, several criteria may be used to help in taxonomic determinations of trichomycetes:

(1) Host taxa alone should not be used to distinguish trichomycete species; notwithstanding, host identification can be a useful primary consideration in restricting the options for identifying a given trichomycete on hand.

(2) Studies should include reasonably large samples of hosts so as to comprehend the fungal variations possible within species.

(3) Collections should preferably be made at different seasons of the year and at different stages of host development (immature, adult, or molting individuals).

(4) If any given trichomycete species appears to have a wide latitude of morphological features within one host species, one can suspect that, if it does occur in other hosts, it may be variable there also, and this should be taken into consideration when making taxonomic assessments.

CHAPTER 11

Taxonomic Treatment

Keys to Orders and Families of Trichomycetes

1. Spores (trichospores) produced exogenously, usually bearing one or more basal appendages; zygospores produced in many genera .... **Harpellales** 2
1'. Spores (sporangiospores) or amoeboid cells produced endogenously, or thallus breaking up into arthrospores; zygospores not present 3
    2(1). Thallus unbranched, usually attached to the peritrophic membrane of the midgut (less commonly to the hindgut cuticle) .... **Harpellaceae** (p. 139)
    2'(1). Thallus branched, attached to the hindgut cuticle .... **Legeriomycetaceae** (p. 152)
3(1'). Thallus branched and septate, producing arthrospores .... **Asellariales;** single family **Asellariaceae** (p. 200)
3'(1'). Thallus unbranched, or branched only at base; nonseptate; producing sporangiospores or amoeboid cells 4
    4(3'). No amoeboid cells; sporangiospores usually produced singly in basipetal series of terminal sporangia .... **Eccrinales** 5
    4'(3'). Amoeboid cells produced at some stage; entire thallus functioning as a sporangium, releasing spores or amoeboid cells more or less simultaneously ....
        .... **Amoebidiales;** single family **Amoebidiaceae** (p. 262)
5(4). Thalli producing directly only one type of spore; sporangia sometimes germinating *in situ* .... **Palavasciaceae** (p. 255)
5'(4). Thalli usually producing at least two types of spores; sporangia do not germinate *in situ* 6
    6(5'). Primary infestation spores produced in thalli that become converted entirely or partly into multispored sporangia .... **Parataeniellaceae** (p. 258)

6′(5′). Primary infestation spores produced singly in series of
terminal sporangia .... **Eccrinaceae** (p. 210)

Trichomycetes

Obligate symbionts of arthropods, attached by a holdfast structure to the lining of the hindgut, midgut, or foregut, or to the exoskeleton. Thalli simple or branched, vegetative parts coenocytic or uninucleate, septate or nonseptate. Reproducing asexually by sporangiospores, deciduous appendaged sporangia (trichospores), arthrospores, or amoeboid cells and cystospores; or sexually by biconical zygospores.

Consisting of the orders Harpellales, Asellariales, Eccrinales, and Amoebidiales.

Harpellales

Lichtwardt & Manier, 1978

Thalli unbranched or branched, producing basipetal series of trichospores. Zygospores biconical. Attached to the gut lining of aquatic larvae of Insecta or (rarely) Isopoda.

Consisting of the families Harpellaceae and Legeriomycetaceae.

The ordinal name was first used, but not validly published, by Duboscq, Léger, and Tuzet in 1948, and included two families of unbranched Trichomycetes, the Harpellaceae and Palavasciaceae. The latter family is now considered to belong to the Eccrinales. The first use of the name Harpellales in the present sense was by Manier (1962b), but she did not formally establish the order at that time.

Harpellaceae

Léger & Duboscq, 1929

Thalli unbranched, attached to the peritrophic membrane (or rarely to the hindgut lining) of the host. Four genera.

Type genus: *Harpella* Léger & Duboscq.

—Key to Genera of Harpellaceae—

1. Trichospores cylindrical and curved to coiled (rarely straight), with about 4 appendages .... *Harpella*
1′. Trichospores more or less oval, with 3, 1 or no appendages 2
    2(1′). Trichospores with 3 appendages; sporulating thalli attached to hindgut cuticle .... *Harpellomyces*
    2′(1′). Trichospores with 1 or no appendage; thalli attached to peritrophic membrane 3

3(2'). Trichospores with 1 appendage; spores release from generative cells .... **Stachylina**

3'(2'). Trichospores with no appendage; spores remain attached to generative cells which disarticulate .... **Carouxella**

Carouxella

Manier, Rioux & Whisler ex Manier & Lichtwardt, 1969 (1968)

[= *Carouxella* Manier, Rioux & Whisler, 1961, *nom. nud.*]

Modified trichospores remain attached to generative cells, which disarticulate like arthrospores. Zygospores attached at one pole. On peritrophic membrane or hindgut cuticle of larval Ceratopogonidae (Diptera). Monotypic.

Type species: *Carouxella scalaris* Manier, Rioux & Whisler ex Manier & Lichtwardt.

The original publication by Manier et al. (1961) had neither a Latin diagnosis nor a type for the species. These same authors in 1965 provided a genus–species description in Latin. The genus and species were validated by Manier and Lichtwardt in 1969 by citing a nomenclatural type (Art. 37).

□ *Carouxella scalaris* Manier, Rioux & Whisler ex Manier & Lichtwardt, 1969 (1968)

[= *Carouxella scalaris* Manier, Rioux & Whisler, 1961, *nom. nud.*]

Thalli unbranched, attached to the peritrophic membrane by a holdfast located inside the incurved basal cell. Producing 5–30 generative cells from which develop cylindrical trichospores slightly swollen in the middle, 25–(32)–37 μm long by 4.5–(5.3)–7 μm diam. Trichospores remain attached to the disarticulated generative cells. Appendages unknown. Scalariform conjugation of two thalli gives rise to elongate–pyriform zygospores attached at the broader end and produced laterally from one of the conjugated pairs. Type species.

Illustrations: Figs. 11.1, 11.2, 11.3.

Host: On peritrophic membrane or hindgut cuticle of aquatic larvae of *Dasyhelea lithotelmatica* Strenzk and *Dasyhelea* sp. (Diptera, Ceratopogonidae) living in troughs, potholes, and rocky fissures containing algae and decomposing vegetation.

Distribution: Departments of Hérault and Pyrénées–Orientales, France; New South Wales, Australia.

The disarticulating generative cell–trichospore unit of this unusual species has a remarkable resemblance to the arthrospores of *Asellaria ligiae* with their subsequently developing trichospore–like outgrowths (Manier, 1969b; Lichtwardt, 1973a). This reproductive unit in *C. scalaris* is apparently the primary means of dissemination from host to host. In their original paper of 1961, in which the species was invalidly published, Manier

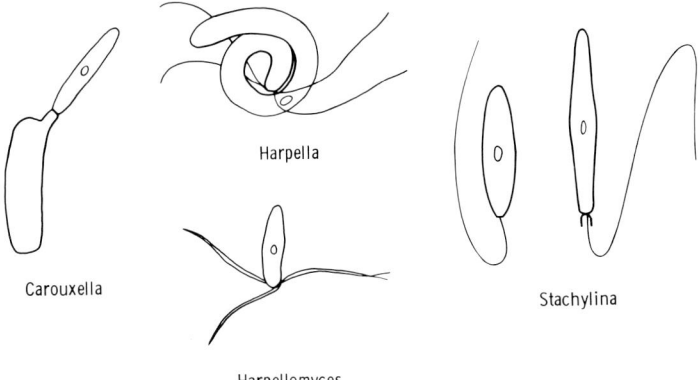

FIG. 11.1. Harpellaceae trichospores: distinguishing generic characters.

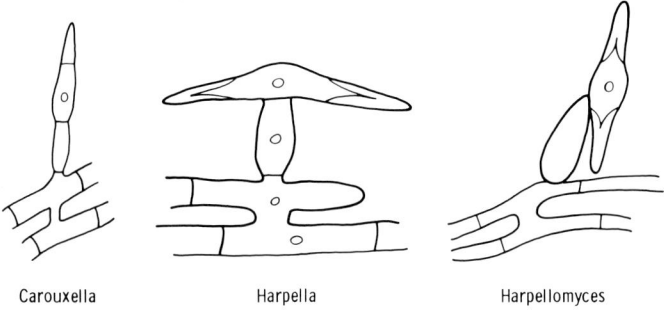

FIG. 11.2. Harpellaceae zygospores: distinguishing generic characters. Zygospore type unknown in *Stachylina*.

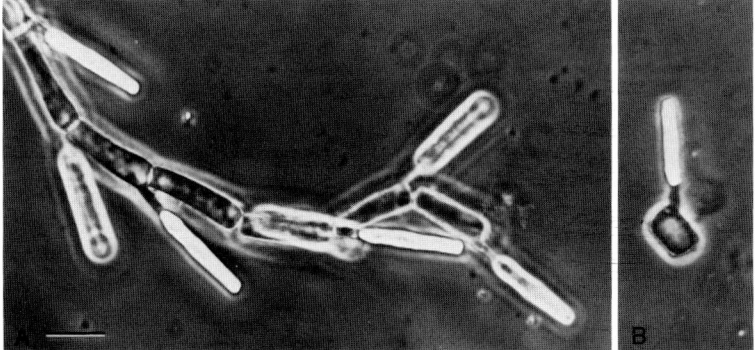

FIG. 11.3. *Carouxella scalaris*. A. Sporulating unbranched thallus with disarticulating generative cells bearing trichospores. B. Disarticulated generative cell with its attached trichospore. Scale bar = 20 μm for both figures.

et al. suggested that the released generative cell–trichospore units (diaspores) may also develop *in situ* without passing out of the gut. Thalli producing trichospores and zygospores simultaneously have been observed. The zygospores are similar to those of *Zygopolaris* (Type IV).

According to Manier et al. (1961), the peritrophic membrane of *Dasyhelea lithotelmatica* extends well into the hindgut, and the thalli of *C. scalaris* attach to the membrane in the region of the Malphighian tubules and develop within the anterior hindgut. The author in 1983 (unpublished) collected larvae of an unnamed and probably endemic species of *Dasyhelea* (not *D. lithotelmatica*) in potholes along the Georges River south of Sydney, Australia. Some of these larvae were infested with a fungus identical to *C. scalaris,* but reproducing only asexually. The Australian *C. scalaris* was attached to the hindgut cuticle, and not to the peritrophic membrane. Some of these larvae, like *Dasyhelea* spp. in France, also had *Smittium culisetae* growing in the hindgut. Another point to note is that Manier (1969b, p. 642) stated that *D. lithotelmatica* is a carnivore, a questionable statement which I have not been able to confirm in the literature. The Australian *Dasyhelea* sp. larvae were clearly feeding on algae, and gave no indication of predaceousness or of having had a carnivorous diet.

References: Manier et al., 1961; Manier et al., 1965; Manier and Lichtwardt, 1968 (1969); Lichtwardt, 1973a; Manier, 1969b (1970b).

Harpella

Léger & Duboscq, 1929

Cylindrical trichospores curved to coiled, sometimes straight, without a collar; appendages (3–)4. Biconical zygospores attached perpendicularly to zygosporophores, which arise from one of the two conjugated cells. On peritrophic membrane of larval Simuliidae (Diptera). Two species.

Type species: *Harpella melusinae* Léger & Duboscq.

—Key to *Harpella* species—

1. Thalli 4–6 μm diam; trichospores about 4.5 μm diam ...... *H. leptosa*
1′. Thalli 6–10 μm diam; trichospores 6–10 μm diam .... *H. melusinae*

□ *Harpella leptosa* Lichtwardt & Moss, 1980 (*in* Moss and Lichtwardt, 1980)

Immature thalli often sinuous. Mature thalli 250–450 μm long and 4–6 μm wide, producing 2–5 coiled or curved trichospores 110–150 μm long and ~4.5 μm diam with 3 (or 4?) appendages. Basal generative cell tapered only immediately above the short holdfast. Zygospores unknown.

Illustrations: See Moss and Lichtwardt, 1980, Figs. 1–3, 5–7.

Hosts: On peritrophic membrane of aquatic (lotic) larvae of Simuliidae (Diptera), including *Simulium canonicolum* (Dyar & Shannon) and *S. venustum* complex.

Distribution: Northwest Montana, U.S.A.

The smaller dimensions of the thalli and spores, as well as the more abrupt tapering of the thallus base just above the short holdfast, distinguish *Harpella leptosa* from the much more common and widespread type species, *H. melusinae*. Electron micrographs have shown that there are considerable differences in the fine structure of the trichospore appendages and the holdfasts of the two species, but the light microscope is sufficient for identification. A few individual larvae have been found with both species attached simultaneously to the hosts' peritrophic membrane. It is probable that *H. leptosa* is more widely distributed geographically than the present records indicate. No zygospores have been found in *H. leptosa*, although conjugations between thalli have been seen. Trichospores with 3 appendages have been observed, but it is possible the normal number is 4.

Reference: Moss and Lichtwardt, 1980.

□ *Harpella melusinae* Léger & Duboscq, 1929b

Thalli slightly arched or undulate, up to 750 μm long and 6–10 μm diam, narrowing at the base to form a tapered holdfast that flares at the area of contact with the peritrophic membrane. Trichospores curved to tightly coiled, occasionally straight, 2–10 per thallus, approximately 150 μm long and as wide as the thallus, with 4 very finely transversely banded appendages. Biconical zygospores attached medially and perpendicularly to zygosporophores, which arise from one of the conjugants opposite the zone of conjugation. Zygospores 38–97 μm long by 10–15 μm diam. Zygosporophore 16–25 × 9–15 μm. Type species.

Illustrations: Figs. 7.9, 7.10B, 7.17, 11.1, 11.2, 11.4.

Hosts: On peritrophic membrane of many species of aquatic (lotic) larvae of Simuliidae (Diptera). Host genera include *Simulium*, *Prosimulium*, *Austrosimulium*, and *Cnephia*.

Distribution: Probably worldwide wherever Simuliidae occur. Reports (some unpublished) include lotic stations throughout the U.S.A.; eastern Canada (Newfoundland, Labrador); Europe from Abisko, Sweden to the Mediterranean; England; New Zealand; Australia; and Japan.

Harpella melusinae is one of the most widespread species of Harpellales. The author has found it in the majority of populations of blackflies examined. Although the trichospores are normally curved to coiled, in larvae of some populations they are occasionally perfectly straight. Sometimes mature thalli with straight spores are found side by side with thalli bearing coiled spores. The fine structure of the holdfast is unusual, consisting of a chamber at the base of the thallus that excretes a cementing substance around some finger–like projections or digits (Fig. 7.10B). These digits can be discerned with the light microscope in good preparations. The appendages also are unique: They have a very fine, regularly recurrent banding pattern barely visible with the light microscope, but clearly seen

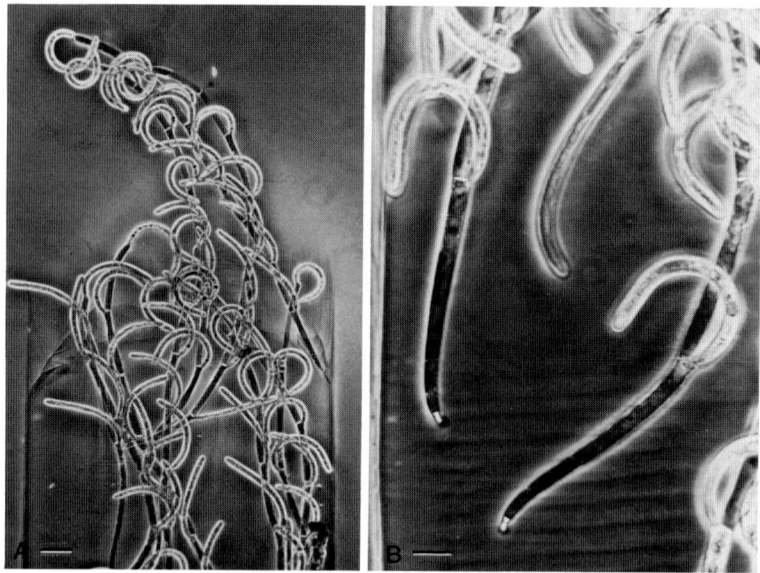

FIG. 11.4. *Harpella melusinae*. A. Posterior portion of a peritrophic membrane within which are attached sporulating thalli. B. Tapered holdfasts of two thalli. Scale bars: A, 40 μm; B, 20 μm.

in electron micrographs (Fig. 7.17C). This periodic banding is not found in *H. leptosa*.

Rarely have zygospores been seen in *H. melusinae*, perhaps only in about 6 or 7 larvae. Yet, conjugations of thalli are very common in some populations of larvae (see Chapter 7). The length of the zygospores appears to vary considerably, but measurements are based on few specimens. The shorter lengths are measurements of zygospores that developed in water on slide cultures (Lichtwardt, 1967), and may be somewhat abnormal; the more usual length may be 60–95 μm. The side of the biconical zygospore where attachment to the zygosporophore occurs is often flattened. Sometimes the arms of the zygospore are bent slightly toward the zygosporophore.

Tuzet and Manier (1955a) once used the name *H. melusinae* var. *eyziesi* (*nom. nud.*) for some specimens they studied, but this was later not recognized as a distinct variety (Manier, 1969b).

References: Léger and Duboscq, 1929b; Léger and Gauthier, 1935b; Manier, 1950 (1951), 1969b (1970b); Tuzet and Manier, 1955a; Lichtwardt, 1967; Moss, 1970; Frost and Manier, 1971; Brassard et al., 1971; Reichle and Lichtwardt, 1972; Crosby, 1974; Moss and Lichtwardt, 1977.

Harpellomyces

Lichtwardt & Moss, 1984b

Thalli unbranched, producing distally a series of collarless trichospores with 3 appendages, or conjugating and producing zygospores attached obliquely and submedially to the zygosporophore. Attached to hindgut cuticle, and possibly the peritrophic membrane, of Thaumaleidae (Diptera) larvae. Monotypic.

Type species: *Harpellomyces eccentricus* Lichtwardt & Moss.

□ *Harpellomyces eccentricus* Lichtwardt & Moss, 1984b

Unbranched thalli up to 1 mm or more long, generally 6–8 μm diam, but with segments sometimes as narrow as 1.5 μm diam. Holdfast a short basal disk. Distal fertile tips producing series of a few or up to more than 30 generative cells of variable lengths. Trichospores oval to slightly ovoid, 20–25 × 6–8 μm, attached eccentrically to the generative cell, on release having 3 broad appendages usually 40–60 μm long, sometimes split at their ends. Zygospores biconical, 48–58 × 10–11 μm, attached obliquely and submedially to a zygosporophore 20–30 × 11–12 μm, both remaining together upon release. Type species.

Illustrations: Figs. 11.1, 11.2, 11.5.

Hosts: In hindgut of larval T*haumalea* sp. and another Thaumaleidae species (Diptera), and possibly growing but not sporulating within the peritrophic membrane of the midgut.

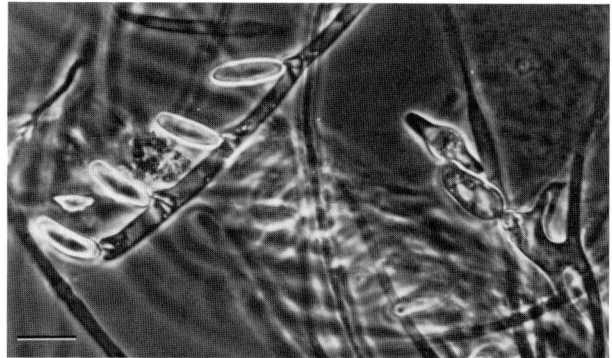

FIG. 11.5. *Harpellomyces eccentricus*. Trichospores and an immature zygospore. Scale bar = 20 μm. From Lichtwardt and Moss, 1984b, by permission of Mycotaxon.

Distribution: In a small stream in northern Sweden and from seeping waters on a steep coastal cliff in Wales.

This species of Harpellaceae, although unbranched, grows and sporulates in the hindgut of its larval hosts and produces spores only on the distal portion of the thallus, and thus has some characters found in the Legeriomycetaceae.

Reference: Lichtwardt and Moss, 1984b.

Stachylina

Léger and Gauthier, 1932

Unbranched thalli cylindrical to fusiform. Trichospores oval to almost biconical, (with or) without a collar, bearing a single appendage. Zygospores unknown. On peritrophic membrane (base of thallus sometimes penetrating the membrane) of larval Chironomidae and Psychodidae (Diptera). Eleven species.

Type species: *Stachylina macrospora* Léger & Gauthier.

—Key to *Stachylina* species—

1.	Base of thallus penetrating the peritrophic membrane; no obvious secreted holdfast structure 2
1'.	Base of thallus attached to inner surface of peritrophic membrane by a small secreted holdfast disk 4
2(1).	Trichospores produced on a narrow cylindrical outgrowth from the generative cell; terminal cell sterile .... *S. minuta*
2'(1).	Trichospores produced directly from the generative cell; terminal cell fertile 3
3(2').	Base of thallus which penetrates the peritrophic membrane shaped like a foot; trichospores 20–30 μm long .... *S. pedifer*
3'(2').	Base of thallus which penetrates the peritrophic membrane bulbous; trichospores 30–50 μm long .... *S. penetralis*
4(1').	Base of trichospores with a collar 5
4'(1').	Trichospores without a collar 7
5(4).	Trichospores less than 22 μm long, collar small and sleevelike around the base of the appendage .... *S. manicata*
5'(4).	Trichospores longer than 22 μm; collar more obvious 6
6(5').	Trichospores 40–72 μm long .... *S. grandispora*
6'(5').	Trichospores 25–35 μm long .... *S. euthena*
7(4').	Trichospores less than 6.5 μm diam 8
7'(4').	Trichospores more than 6.5 μm diam 9
8(7).	Trichospores 35–49 × 4–5 μm .... *S. chironomidarum*
8'(7).	Trichospores 25 × 5–6 μm .... *S. longa*
9(7').	Trichospores 40 μm or more long .... *S. macrospora*
9'(7').	Trichospores usually less than 40 μm long 10
10(9').	Thallus cymbiform to fusiform, producing 2–4 trichospores; appendage ribbon–like near the trichospore base; in larval Chironomidae .... *S. nana*

10′(9′). Thallus more cylindrical, producing usually 4–8 trichospores; appendage not ribbon–like; in larval Psychodidae .... *S. lotica*

☐ *Stachylina chironomidarum* Lichtwardt, 1972

Mature thalli usually 180–250 μm long by 4–5 μm diam, attached to peritrophic membrane by an inconspicuous disklike holdfast, producing 2–6 (or more) generative cells that are usually longer than the trichospores. Trichospores long–ellipsoidal, 35–49 × 4–5 μm, with a collar 1–2 μm long and a single appendage. Zygospores unknown.

Illustrations: See Lichtwardt, 1972, Figs. 63–65.

Host: On peritrophic membrane of unidentified lentic bloodworms (Diptera, Chironomidae).

Distribution: In a pond near Gothic, Gunnison Co., Colorado, U.S.A.

The very slender, collared trichospores of *S. chironomidarum* distinguish this *Stachylina* species from all others.

Reference: Lichtwardt, 1972.

☐ *Stachylina euthena* Manier & Coste, 1971

Thalli 130–150(–200) μm long by 7–8 μm diam, attached to peritrophic membrane by a small disklike holdfast. Trichospores 8 per thallus, fusiform with a median swelling, 25–35 × 7–8 μm, with an ephemeral collar 2.5–3 μm long and a single prominent appendage. Zygospores unknown.

Illustrations: See Manier and Coste, 1971, Figs. 18–21.

Hosts: In *Chironomus plumosus* complex and *Psectrotanypus varius* Fabr. larvae (Diptera, Chironomidae) collected in roadside ponds and watering places.

Distribution: Department of Hérault, France.

Stachylina euthena was described and illustrated in Coste–Mathiez's 1970 thesis under the name *S. chironomi (nom. provis.)*, prior to its valid publication. The species most closely resembles *S. grandispora*, but has shorter trichospores.

References: Manier and Coste, 1971; Coste–Mathiez, 1970.

☐ *Stachylina grandispora* Lichtwardt, 1972

Mature thalli less than 100 μm to more than 250 μm long by 6–10 μm diam, attached to peritrophic membrane by an inconspicuous holdfast. Trichospores 2–16 per thallus, long–ellipsoidal, 40–72 × 6–10 (or more) μm, upon release having a collar 1–3 μm long and a single, prominent and very long appendage. Zygospores unknown.

Illustrations: Fig. 11.6A; also see Lichtwardt, 1972, Figs. 59–62.

Hosts: Various lotic or lentic Chironomidae (Diptera) larvae, including species of *Tanytarsus, Tendipes, Paratendipes, Polypedilum, Chironomus,* and *Cricotopus.*

148　11. Taxonomic Treatment

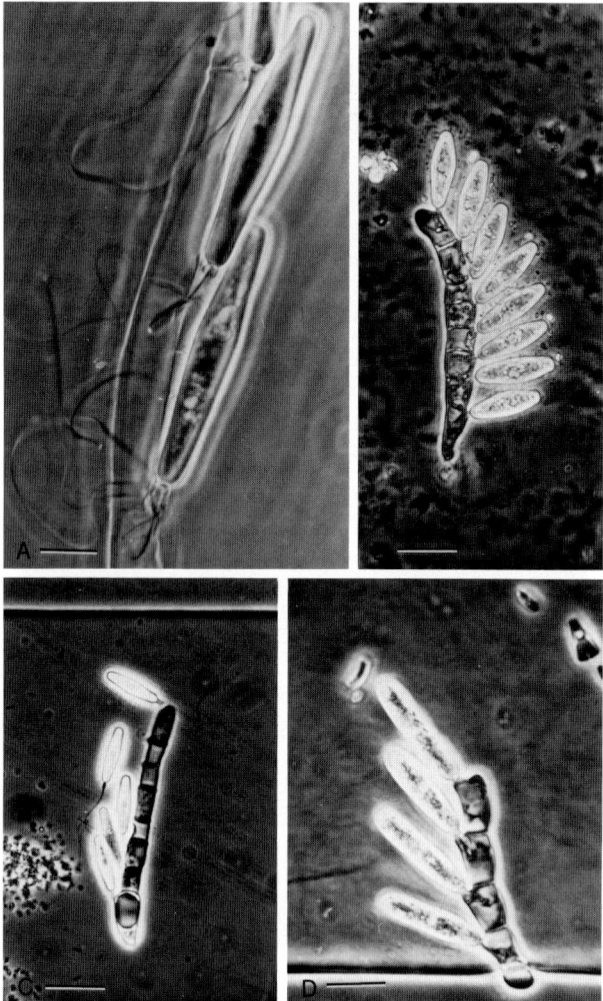

FIG. 11.6. Representative species of *Stachylina*. A. *S. grandispora*. B. *S. lotica*. C. *S. manicata*. D. *S. penetralis*. Scale bars = 20 μm. Reprinted by permission from Mycologia, vol. 76: 206 (11.6B), 205 (11.6C) and vol. 65: 16 (11.6D), Copyright 1984 and 1973, R.W. Lichtwardt and The New York Botanical Garden.

Distribution: Several states of the U.S.A., including Hawaii; England; Sweden; Australia; New Zealand.

Although the thallial and trichospore lengths are variable, *S. grandispora* is easily identified by its large collared trichospores and the long appendages measuring up to 200 μm or more in length. The fine structure of the appendage and septum has been studied in this species (Moss, 1972,

1976; Moss and Young, 1977, 1978). Moss (1972) was able to infest species of seven additional chironomid genera by exposing them to naturally infested larvae (see Chapter 6). Conjugations between thalli have been observed (Lichtwardt, 1972; Moss, 1972), but not zygospores.

The author in 1983 (unpublished) found *S. grandispora* growing in the endemic New Zealand bloodworm, *Chironomus zealandicus* Hudson, as well as in an undetermined species of *Chironomus* in Queensland, Australia. A bloodworm host species from Hawaii, *Chironomus hawaiiensis* Grimshaw, may be endemic to that island group (Lichtwardt, 1963, unpublished). These data indicate that *S. grandispora* has a remarkable record of dispersal.

References: Lichtwardt, 1972, 1976; Moss, 1972, 1974, 1976, 1979; Moss and Young, 1977, 1978.

□ *Stachylina longa* Léger & Gauthier, 1932

Thalli 100 μm or more long by 9–10 μm diam, bearing 6–8 trichospores 25 x 5–6 μm, without a collar. Zygospores unknown.

Illustration: None available.

Host: On peritrophic membrane of larval *Tanytarsus* sp. (Diptera, Chironomidae).

Distribution: Alpine streams, France.

Stachylina longa was originally poorly described, along with the type species, *S. macrospora,* when the genus *Stachylina* was established. In 1961 Gauthier provided a slightly more complete—but not very satisfactory—description. Neither publication presented illustrations, thus leaving in question the form of the thallial base, trichospore shape, range of trichospore lengths, and other such features. To date this has presented no problem, since all subsequently described species have different trichospore dimensions.

References: Léger and Gauthier, 1932; Gauthier, 1961.

□ *Stachylina lotica* Williams & Lichtwardt, 1984

Mature thalli up to 100 μm long by 8–10 μm diam, producing up to 8 trichospores. Holdfast generally with a smaller diameter than the thallus. Trichospores ellipsoidal to subbiconical, 24–32 x 8–10 μm, with appendage up to 10 times spore length, without a collar. Generative cells commonly 8–10 μm long. Zygospores unknown.

Illustration: Fig. 11.6B.

Host: On peritrophic membrane of larval *Maruina* sp. (Diptera, Psychodidae).

Distribution: In stream just south of Glacier National Park, Montana, U.S.A.

This is the only species of *Stachylina* known in Psychodidae larvae. All other species are in Chironomidae.

Reference: Williams and Lichtwardt, 1984.

☐ *Stachylina macrospora* Léger & Gauthier, 1932
[= *Stachylina intermedia* Poisson, 1936, *nom. nud.*]

Mature thalli usually 100–200 μm long by 8–10 μm diam, attached to peritrophic membrane by a broad holdfast disk; bearing (2–)4–8 long–ovoid trichospores 40–50 × 7–8 μm, upon release having no collar and a single appendage several times the length of the trichospore. Zygospores unknown. Type species.

Illustrations: See Poisson, 1936, Figs. 1, 2.

Host: On peritrophic membrane of larval Diamesinae (Diptera, Chironomidae), including *Diamesa* sp. and *Syndiamesa macronyx* Kieff.

Distribution: In alpine streams and fountain waters in France; possibly also in streams near Peters Lake and a pond near Barrow, Alaska.

Stachylina macrospora and *S. longa* were cursorily described by Léger and Gauthier in 1932 when they established the new genus *Stachylina*. Manier and Lichtwardt (1968) selected *S. macrospora* as the type species, because it was the only species of the two to be illustrated (later) by Léger and Gauthier (1935b). Gauthier in 1961 published a somewhat better description of *S. macrospora*. The species description above is a slight modification of Gauthier's so as to accommodate Poisson's (1936) description of *S. intermedia* Poisson *(nom. nud.)*, which appears to be a synonym of *S. macrospora* (Manier, 1969b; Kobayasi et al., 1969; Lichtwardt, 1972). The original collections were from larvae of *Diamesa* sp. in alpine streams, presumably near Grenoble, France, whereas Poisson's collections came from larval *Syndiamesa macronyx* collected in fountains at Besse–en–Chandesse, Puy–de–Dôme, France.

Kobayasi (in Kobayasi et al., 1967, 1969) reported finding *S. macrospora* in unidentified chironomid larvae in Alaska. But the basal cell of his specimens, as drawn, are tapered and almost pointed, a feature not in accord with either Léger and Gauthier's (1935b) or Poisson's (1936) drawings, which leaves a question as to the correct identification of the Alaskan specimens. Coste–Mathiez (1970) and Manier and Coste (1971) tentatively identified as *S. macrospora* a *Stachylina* sp. from larvae of *Syncrotopus rufiventris* Meig., but it appears from their photomicrographs and descriptions to be some other, peritrophic membrane–penetrating species, and not *S. macrospora*.

References: Léger and Gauthier, 1932, 1935b; Poisson, 1936; Gauthier, 1961; Kobayasi et al., 1967, 1969; Manier, 1969b (1970b); Manier and Lichtwardt, 1968; Manier and Coste, 1971; Coste–Mathiez, 1970.

☐ *Stachylina manicata* Williams & Lichtwardt, 1984

Thalli up to 100 μm long by 6.5–8 μm diam, producing up to 8 trichospores. Holdfast disklike, with a diameter smaller than the thallus. Trichospores ellipsoidal to subbiconical, 17.5–20 × 4.5 μm, with a small sleevelike collar 1–1.5 × 0.5 μm surrounding the single appendage, which narrows toward the distal end and is generally at least 5 times the spore length. Generative cell usually shorter than the trichospore. Zygospores unknown.

Illustration: Fig. 11.6C.

Host: On peritrophic membrane of larval *Polypedilum* sp. and/or *Pseudochironomus* sp. (Diptera, Chironomidae).

Distribution: In a cattail swamp bordering Flathead Lake, Montana, U.S.A.

Reference: Williams and Lichtwardt, 1984.

□ *Stachylina minuta* Gauthier ex Lichtwardt, 1984a
 [= *Stachylina minuta*, Gauthier, 1961, *nom. nud.*]

Fusiform thalli 30–55 μm long by 5–6 μm diam, attached by a bulbous base that penetrates the peritrophic membrane, bearing 1–4 trichospores; terminal cell sterile. Trichospores ellipsoidal, 15–16 × 5–6 μm, produced on a narrow cylindrical outgrowth from the generative cell, upon detachment having a single short appendage and no collar. Zygospores unknown.

Illustrations: See Lichtwardt, 1984a, Fig. 32.

Host: On peritrophic membrane of Tanytarsini larvae (Diptera, Chironomidae).

Distribution: In small streams around Grenoble, France.

Among the distinctive features of *Stachylina minuta* are the sterile terminal cell or compartment and the long, narrow outgrowth from the generative cell on which the trichospore is produced.

References: Lichtwardt, 1984a; Gauthier, 1961.

□ *Stachylina nana* Lichtwardt, 1984a

Thalli cymbiform to fusiform, 60–80 μm long, attached by a small holdfast disk. Producing 2–4 oval to biconical trichospores, (25–)30(–40) × (7–)8.5(–10) μm, with a single ribbon–like (near the base) appendage and no collar. Zygospores unkown.

Illustrations: See Lichtwardt, 1984a, Figs. 36–38.

Hosts: On peritrophic membrane of unidentified Chironomidae (Diptera).

Distribution: In the Dranse River east of Thonon–les–Bains near Lake Léman, France.

Reference: Lichtwardt, 1984a.

□ *Stachylina pedifer* Williams & Lichtwardt, 1983 (*in* Lichtwardt and Williams, 1983a).

Mature thalli up to 110 μm long by 7–12 μm diam, with a footlike extension of the basal cell that penetrates through the peritrophic membrane. Trichospores ovoid with a slight median swelling, usually 2–8 per thallus, 20–30 × 7–10 μm, with one long, narrowing appendage but no collar.

Trichospore on terminal generative cell usually produced ⅓ to ½ the distance down from the cell apex. Zygospores unknown.

Illustration: Fig. 7.11A.

Host: On peritrophic membrane of larval *Boreoheptagyia lurida* (Garrett) (Diptera, Chironomidae) living in fast–flowing streams.

Distribution: In and near Glacier National Park, Montana, U.S.A.

Stachylina pedifer occurs in the same host species as *Smittium dimorphum* (located in the hindgut). The most noticeable features of *S. pedifer* are the footlike base of the thallus, which penetrates through the peritrophic membrane, and the commonly subapical placement of the terminal trichospore.

Reference: Lichtwardt and Williams, 1983a.

☐ *Stachylina penetralis* Lichtwardt, 1984a

Thalli 70–180 × 8–10 μm, attached by a bulbous base that penetrates the peritrophic membrane. Producing (2–)4–12 long–ellipsoidal trichospores, 30–50 × 8–12 μm, with a single appendage and no collar. Zygospores unknown.

Illustration: Fig. 11.6D.

Host: On peritrophic membrane of *Diamesa* spp. larvae (Diptera, Chironomidae).

Distribution: In small streams, waterfalls, and a water tank; Island of Honshu, Japan, and in French and Swiss Alps.

Like *Stachylina pedifer* and *S. minuta*, the base of the thallus of *S. penetralis* penetrates the peritrophic membrane of the host. Each of the 3 species, however, is morphologically distinct.

References: Lichtwardt, 1973a, 1984a.

Legeriomycetaceae

Pouzar, 1972

= Genistellaceae Léger & Gauthier, 1932

Thalli branched and septate, eucarpic, attached to the hindgut lining of the host. Sixteen genera.

Type genus: *Legeriomyces* Pouzar (= *Genistella* Léger & Gauthier, 1932).

Pouzar erected the new family name, Legeriomycetaceae, because the former type genus, *Genistella*, was a later homonym of *Genistella* Ortega 1773, a genus that Ortega had assigned to the legume family Papilionaceae Giseke.

—Key to Genera of Legeriomycetaceae—

1.	Trichospores curved or horseshoe–shaped	2
1'.	Trichospores straight	3
	2(1). Trichospores curved, fertile branches in cymelike arrangement. In Plecoptera larvae	**Orphella**

Harpellales

        2′(1). Trichospores horseshoe-shaped, fertile branches not in a cymelike arrangement. In Ephemeroptera larvae .... ***Gauthieromyces***

3(1′). Trichospores with no distinct appendage; zygospores attached at one pole with axis in line with that of the zygosporophore (in Ephemeroptera larvae) .... ***Zygopolaris***

3′(1′). Trichospores normally having 1 to many appendages; zygospores not attached at one pole 4

        4(3′). Trichospores with a single appendage 5

        4′(3′). Trichospores with more than 1 appendage 8

5(4). Trichospores cylindrical, each developing from one of a unilateral series of generative (subsidiary) cells that grow laterally from a fertile branch (in Ephemeroptera larvae) .... ***Pteromaktron***

5′(4). Trichospores ovoid, obpyriform, or ellipsoidal (rarely almost cylindrical but with a median swelling), developing directly from generative cells of a fertile branch 6

        6(5′). Trichospores obpyriform; zygospores attached perpendicularly and medially to the zygosporophore (in Ephemeroptera larvae) .... ***Spartiella***

        6′(5′). Trichospores not obpyriform; zygospores attached obliquely to the zygosporophore 7

7(6′). Trichospores without a collar, ovoid; zygospores with an oblique median collar and no appendage. In Ephemeroptera larvae .... ***Graminella***

7′(6′). Trichospores with a collar, ellipsoidal (or subellipsoidal) to almost cylindrical; zygospores with an oblique submedian collar and 1 appendage. In Diptera larvae .... ***Smittium***

        8(4′). Trichospores with 2 appendages 9

        8′(4′). Trichospores with more than 2 appendages 10

9(8). Zygospores attached submedially and obliquely to the zygosporophore; upon zygospore release having a collar and 1 appendage. In Ephemeroptera larvae .... ***Legeriomyces***

9′(8). Zygospores attached medially and perpendicularly to zygosporophore; no collar or appendage. In Plecoptera larvae .... ***Genistelloides***

        10(8′). Trichospores with a collar; zygospores with a collar and numerous appendages (in Chironomidae larvae) .... ***Trichozygospora***

        10′(8′). Trichospores without a collar; zygospores without appendages 11

11(10′). Zygospores oriented parallel to zygosporophore 12

11′(10′). Zygospores not oriented parallel to zygosporophore 13

        12(11). Zygospores produced without conjugation; trichospores ovoid with about 6 appendages (in Simuliidae larvae) .... ***Genistellospora***

        12′(11). Zygospores produced from conjugating branches; trichospores obpyriform to cylindrical with about 4–6 appendages (in Simuliidae larvae) .... ***Pennella***

13(11′). Zygospores perpendicular to zygosporophore and attached medially 14

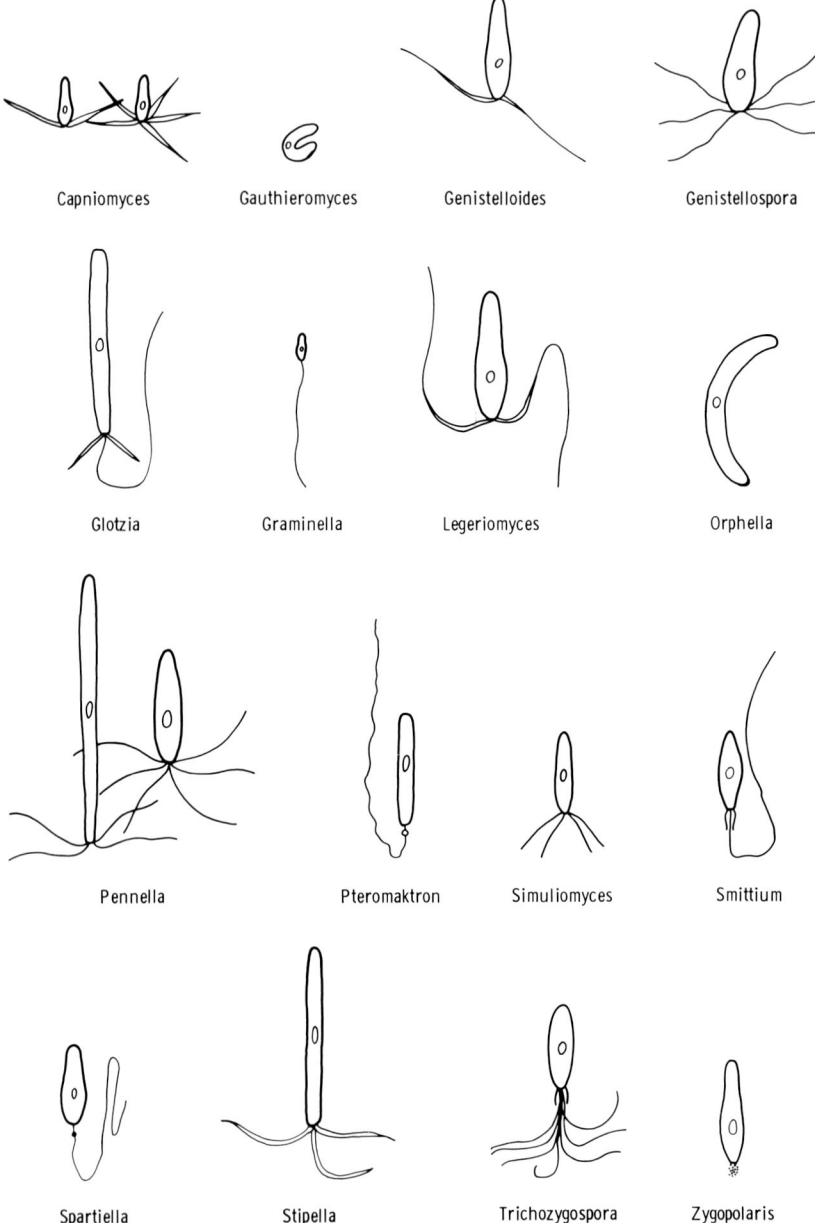

FIG. 11.7. Legeriomycetaceae trichospores: distinguishing generic characters.

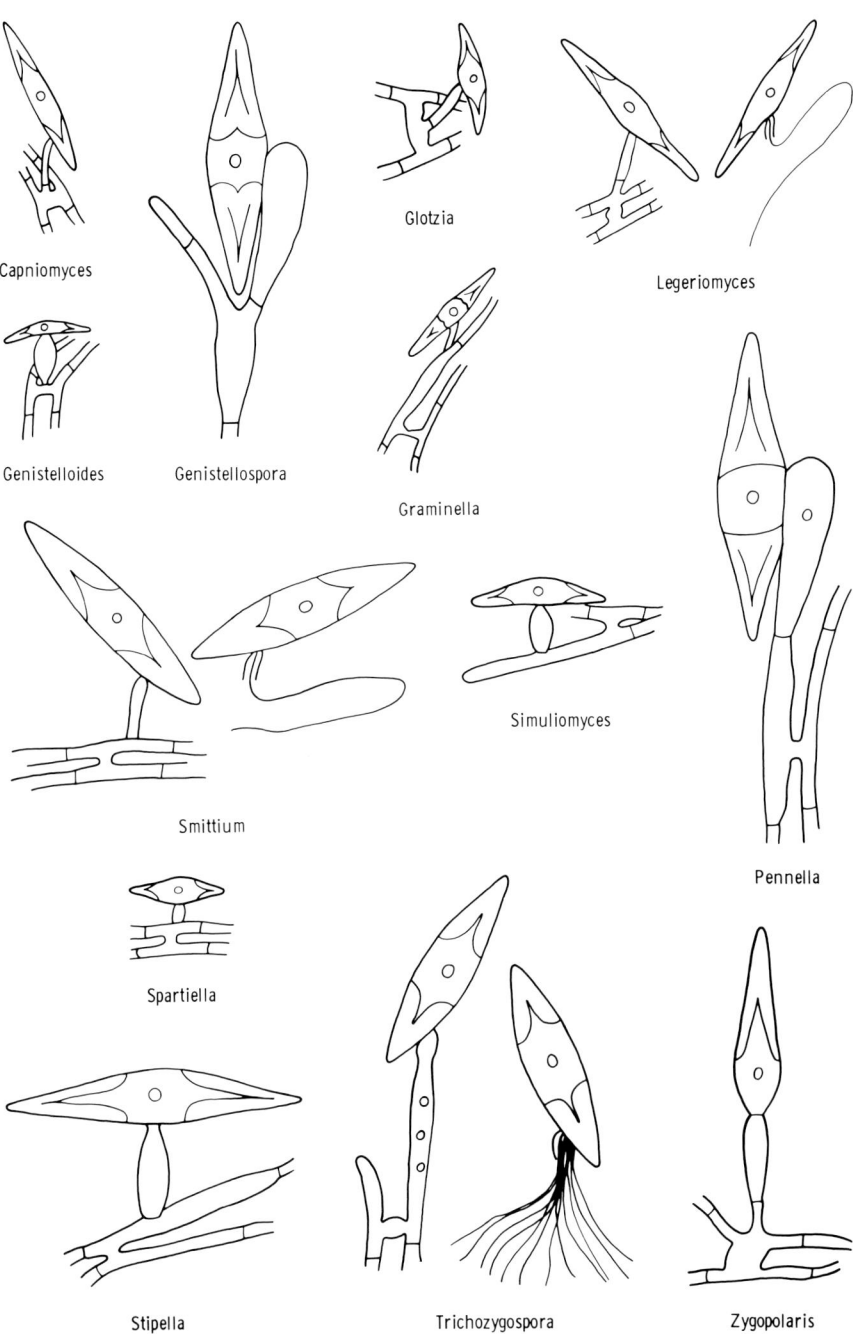

FIG. 11.8. Legeriomycetaceae zygospores: distinguishing generic characters. Zygospores unknown in *Gauthieromyces, Orphella,* and *Pteromaktron*.

13'(11'). Zygospores oblique to zygosporophore and attached
submedially 15
14(13). Trichospores cylindrical, usually with 3 appendages (in
Simuliidae larvae) .... *Stipella*
14'(13). Trichospores elongate–ellipsoidal with 2–4 appendages
(in Simuliidae or Plecoptera larvae) .... *Simuliomyces*
15(13'). Trichospores cylindrical with 2 short and 1 long appendages. In
Ephemeroptera larvae .... *Glotzia*
15'(13'). Trichospores elongate–ovoid with variable number (1–6) of
appendages of equal length. In Plecoptera larvae .. *Capniomyces*

Capniomyces

Peterson & Lichtwardt, 1983

Trichospores elongate–ovoid with 1–6 very broad appendages, collar lacking. Zygospores biconical, point of attachment to the zygosporophore submedian and the angle oblique. In hindgut of larval Capniidae (Plecoptera). Monotypic.

Type species: *Capniomyces stellatus* Peterson & Lichtwardt.

☐ *Capniomyces stellatus* Peterson & Lichtwardt, 1983

Attached to the hindgut cuticle by a discrete pad of secreted holdfast material. Trichospores elongate–ovoid, (10–)15(–19) × 4–6 μm, with 1–6 very broad basal appendages 40–150 × 2–6 μm. Generative cells 10–20 × 4–7 μm, forming a series of 3–4 trichospores on terminal branches. Zygospores biconical, (42–)52(–64) × 7–9 μm, point of attachment to the zygosporophore submedian and the angle oblique. Zygosporophores 10–15 × 2–3 μm, arising from the conjugation tube between two hyphae, or from a swollen hemispherical region of a fertile hypha distant from any conjugations of branches. In hindgut of larval Capniidae (Plecoptera). Type species.

Illustrations: Figs. 11.7, 11.8.

Host: In hindgut of larval *Allocapnia* spp., including *A. granulata* (Claassen) and *A. vivipara* (Claassen) (Plecoptera, Capniidae), in leaf packs, on sticks, and under stones in flowing streams.

Distribution: Kansas, Arkansas, Missouri, Indiana, U.S.A.

The variable number (1–6; median 4) of very broad (2–6 μm) and gelatinous–like appendages make this species easily identifiable. Another unique feature is the sexual apparatus: Zygospores arise by hyphal conjugations, as in most other harpellids, but as well from the juncture of two adjacent cells on a hypha. The species has been cultured axenically by Peterson.

References: Peterson and Lichtwardt, 1983; Peterson, 1984.

Gauthieromyces

Lichtwardt, 1983

Thallus arbusculate, bearing horseshoe–shaped trichospores attached eccentrically (basolaterally) to their generative cells. In hindgut of Baetidae (Ephemeroptera) nymphs. Monotypic.

Type species: *Gauthieromyces microsporus* Lichtwardt.

☐ *Gauthieromyces microsporus* Lichtwardt, 1983a
 [= *Genistella microspora* Gauthier, 1960, *nom. nud.*]

Arbusculate thallus up to 300 μm long, main axis and primary branches stout (15–20 μm diam), secondary branches narrower (3–6 μm diam), with a prominent basal holdfast. Trichospores horseshoe–shaped, overall shape ~10–12 × 9 μm, attached laterally near base, up to 4 per fertile branch. Trichospore appendages and zygospores unknown. Type species.

Illustrations: Fig. 11.7; and see Lichtwardt, 1983a, Figs. 1, 2.

Host: Larval hindgut of *Baetis pumilus* (Burm.) (Ephemeroptera, Baetidae).

Distribution: Streams (Bresson, Tavernolles) located near Grenoble, France.

The species is apparently quite rare, for it is reported only once in the literature. I was not able to find it in 1968 in mountain streams in the general vicinity where Gauthier originally collected it. She reported that *G. microsporus* occurred together with *Legeriomyces ramosus* in the same host hindguts. Gauthier incorrectly assigned this fungus to *Genistella* (now *Legeriomyces*); the trichospores of the two genera are not similar. *Gauthieromyces* and *Orphella* are the only Legeriomycetaceae with curved trichospores, but the reproductive branches of the latter are distinctly different and the two genera are readily separable on this basis.

References: Gauthier, 1960; Lichtwardt, 1983a.

Genistelloides

Peterson, Lichtwardt & Horn, 1981

Trichospores obpyriform, without a collar, bearing 2 appendages. Zygospores attached medially and perpendicularly to the zygosporophore. In hindgut of nymphal Capniidae (Plecoptera). Monotypic.

Type species: *Genistelloides hibernus* Peterson, Lichtwardt & Horn.

The trichospores of this genus resemble those of *Legeriomyces*. However, the zygospores of *Genistelloides* are Type I, rather than Type II as in *Legeriomyces*.

☐ *Genistelloides hibernus* Peterson, Lichtwardt & Horn, 1981

Young thalli attached to hindgut cuticle by a more or less pointed basal holdfast, often with a unilateral bulbous swelling just above the point of attachment. Long fertile branches producing 3(–7) generative cells that become distally swollen in the region where trichospores are produced. Trichospores obpyriform, (23–)29(–34) × (6–)8(–10) μm, with two distinct long appendages. Zygospores (22–)26(–30) × (4–)4.5(–6) μm, produced in clusters of conjugated branches, each attached perpendicularly midway on one side of the spore axis to a rotund zygosporophore measuring (11–)13(–15) × (6–)8(–10) μm, which develops in most cases directly from the conjugation tube. Upon release, the zygosporophore usually remains attached to the zygospore. Type species.

Illustrations: Fig. 11.7, 11.8, 11.9.

Hosts: In hindgut of *Allocapnia vivipara* (Claassen), *A. granulata* (Claassen), *A. rickeri* (Frison), *Mesocapnia* sp. (Plecoptera, Capniidae) nymphs.

Distribution: Streams in Kansas, Missouri, Arkansas, Indiana, U.S.A.

This species from winter–emerging stonefly nymphs resembles *Legeriomyces* spp. from mayfly nymphs in the basic features of its trichospores, but produces a different type of zygospore. Manier (1969b) showed a photomicrograph (her Fig. 1, Plate IX) of a "*Legeriomyces* sp." from *Capnia bifrons*, which conceivably is a species of *Genistelloides* on the

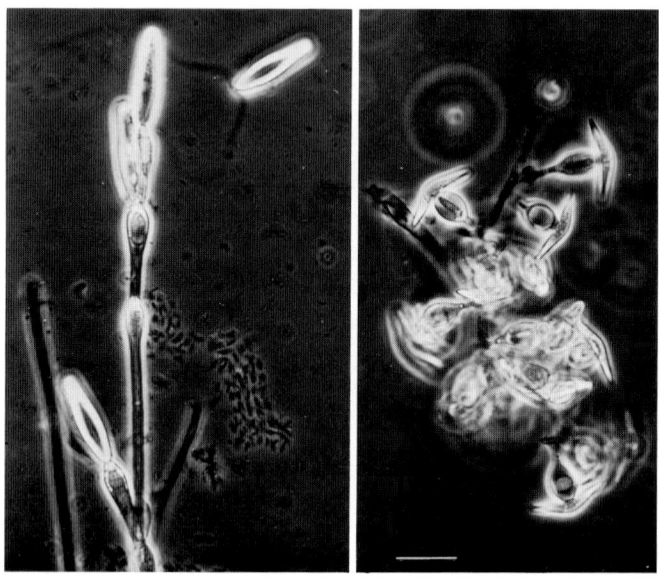

FIG. 11.9. *Genistelloides hibernus*. A. Trichospores. B. Zygospores. Scale bar = 20 μm for both figures.

basis of the Capniidae host; however, zygospores would have to be available in order to determine its generic placement. Several axenic isolates of *Genistelloides hibernus* have been obtained by both Peterson and Lichtwardt.

References: Peterson et al., 1981; Peterson, 1984.

Genistellospora

Lichtwardt, 1972

Trichospores ovoid, without a collar, bearing about 6 appendages. Zygospores produced without hyphal conjugations, borne parallel to axis of zygosporophore, to which they are attached medially. In hindgut of larval Simuliidae (Diptera). Monotypic.

Type species: *Genistellospora homothallica* Lichtwardt.

☐ *Genistellospora homothallica* Lichtwardt, 1972

Thalli generally less than 1 mm long, with moderate to prolific branching, usually with a main axis attached by a small but prominent secreted holdfast. Trichospores ovoid, (24–)30(–41) × (8–)10(–13) μm, bearing (5–)6(–7) very fine appendages, without a collar. Generative cells typically several times longer than the trichospores; lateral trichospores borne on short branchlike extensions of the generative cells. Zygospores (77–)100(–113) × (15–)20(–22) μm, produced without conjugation, attached medially and oriented parallel to the axis of the zygosporophore. Zygosporophores about (39–)40(–48) × (10–)15(–17) μm. Supporting cell (below zygosporophore) with a straight or reflexed, thumblike branch terminating in a cell. Type species.

Illustrations: Figs. 7.7, 7.16, 7.18, 11.7, 11.8, 11.10.

Hosts: In hindgut of many species of larval *Simulium* and *Prosimulium* (Diptera, Simuliidae).

Distribution: Widespread in lotic habitats in the U.S.A. Common in (but not restricted to) many blackfly populations in the Rocky Mountains and other mountain streams. One unpublished report (Moss and Lichtwardt) from southern England in Bere Stream, East Stoke, Wareham.

This species is easily identified on the basis of its blackfly host, prominent secreted holdfast, growth habit, and especially the distinctive homothallic zygospore apparatus. The trichospores most closely resemble those of several species of *Pennella*, also encountered in blackfly guts. Like *G. homothallica*, *Pennella* spp. have Type III zygospores, but conjugations of thallial branches are clearly seen in *Pennella*. Furthermore, *Pennella* thalli attach to the gut cuticle by a general secretion of mucilage along a tapered, forked, or lobed basal cell, and consequently do not present a clearly defined holdfast structure as in *G. homothallica*. Thalli of *G. homothallica* not infrequently are found with attached thalli of *Simulio-*

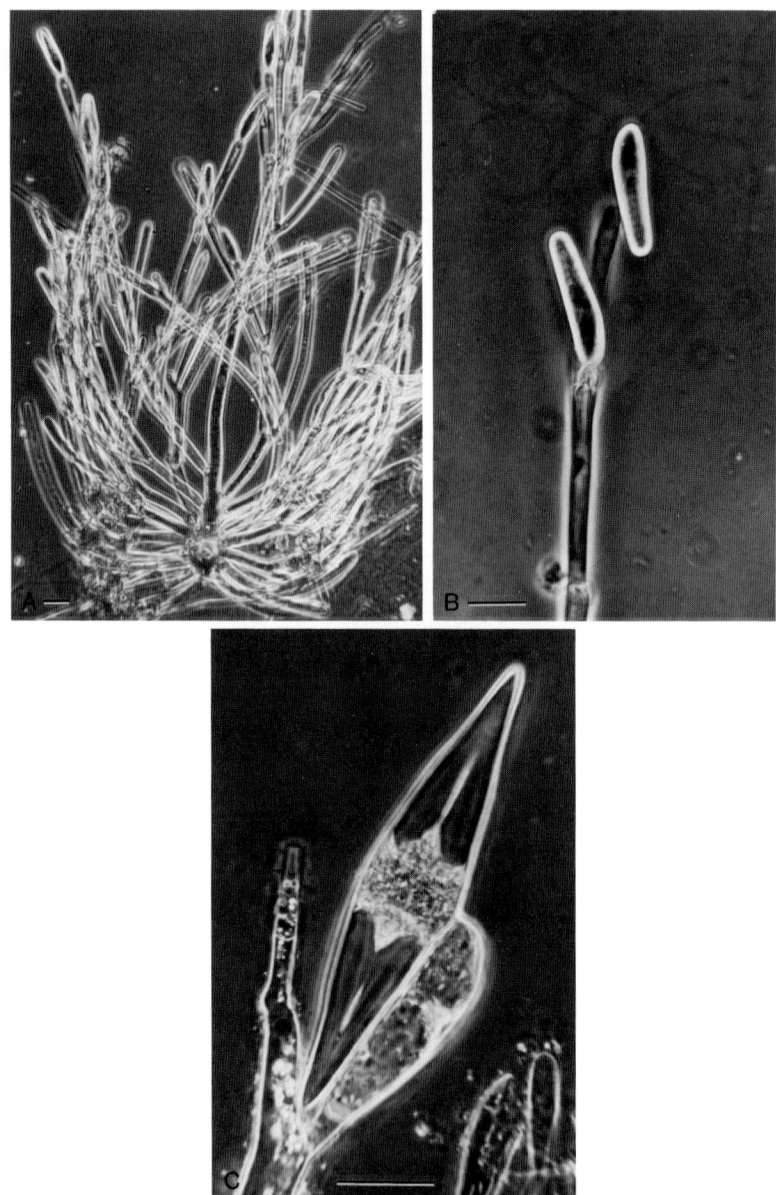

Fig. 11.10. *Genistellospora homothallica*. A. Sporulating thallus. B. Trichospores. C. Zygospore. Scale bars = 20 μm.

myces microsporus. Trichospore formation and holdfast structure of *G. homothallica* have been studied with the electron microscope (Preisner, 1973; Moss and Lichtwardt, 1974; Mayfield and Lichtwardt, 1980).

References: Lichtwardt, 1972; Preisner, 1973; Mayfield and Lichtwardt, 1980; Moss, 1975; Moss and Lichtwardt, 1976.

Glotzia

Gauthier ex Manier & Lichtwardt, 1969 (1968)

[= *Glotzia* Gauthier, 1936, *nom. nud.*]

Thalli with one or more holdfasts attached laterally to a prominent basal cell. Trichospores cylindrical, collarless, with 2 short divergent appendages and 1 long, finer appendage. Zygospores attached obliquely and submedially to a zygosporophore arising from the area of conjugation between two cells. In hindgut of Baetidae (Ephemeroptera) nymphs. Two species.

Type species: *Glotzia centroptili* Gauthier ex Manier & Lichtwardt.

Gauthier's (1936) illegitimate genus was monotypic when Manier and Lichtwardt validated it by providing a Latin diagnosis; Manier (1969b) later "validated" the type species with a Latin diagnosis, which was unnecessary according to Art. 42 of the Code. Tuzet and Manier in 1955(a) described *Glotzia grassei (nom. nud.)* from larvae of a species of blackfly, but it was later transferred to the genus *Pennella;* see *Pennella grassei*.

In 1971 the author collected in northern Sweden what is apparently an undescribed species of *Glotzia* with long–obpyriform trichospores and zygospores with 2 appendages. Further studies will probably require a slight modification of the above generic description.

—Key to *Glotzia* species—

1. Trichospores 40×4 μm; zygospores $50-60 \times 15$ μm .. **G. centroptili**
1'. Trichospores $45-70 \times 4.5-6$ μm; zygospores $28-37 \times 7.5-10$ μm .... **G. ephemeridarum**

☐ *Glotzia centroptili* Gauthier ex Manier & Lichtwardt, 1969 (1968)

[= *Gotzia centroptili* Gauthier, 1936, *nom. nud.*]

Thalli often clustered, 1–2 mm long, the slightly swollen and branched basal cell of each thallus with one lateral holdfast. Trichospores cylindrical, 6–7 per fertile branch, 40×4 μm, with a terminal refractive cap, and bearing 1 long fine appendage initially spiralled around 2 short (10 μm), broader, divergent, and rigid appendages. Zygospores $50-60 \times 15$ μm. Type species.

Illustrations: See Gauthier, 1936, Figs. 1–4.

Host: In hindgut of *Centroptilum luteolum* Müll. nymphs (Ephemeroptera, Baetidae).

Distribution: Cold pools and streams of the Dauphiné French Alps.

References: Gauthier, 1936; Manier and Lichtwardt, 1968 (1969); Manier, 1969b (1970b).

☐ *Glotzia ephemeridarum* Lichtwardt, 1972

Basal cell of thalli unbranched or irregularly branched, when mature having a series of peglike secreted holdfasts. Trichospores cylindrical, $45-70 \times$

FIG. 11.11. *Glotzia ephemeridarum.* A. Trichospore with longer appendage coiled around two shorter ones. B. Longer appendage expanded. C. Zygospores. Scale bars = 20 μm.

4.5–7 μm, tip sometimes refractive, usually with three appendages, the two outer ones short, broad and often divergent, the central one at an angle to the two others, broad at the base but tapering to a long fine filament. Zygospores 28–37 × 7.5–10 μm, flattened or incurved on the side of attachment; attachment to zygosporophore submedial and oblique; upon detachment bearing a small collar. Zygosporophore about 16–19 × 3–5 μm.

Illustrations: Figs. 7.10A, 11.11.

Hosts: In hindgut of *Baetis tricaudatus* Dodds nymphs, and possibly other Baetidae (Ephemeroptera).

Distribution: Streams in Wasatch Co., Utah; Mission Creek and tributary near National Bison Range, Wolf Creek, and N. Fork Bond Creek, Lake, Co., Montana, U.S.A.

Reference: Lichtwardt, 1972.

Graminella

Léger & Gauthier ex Manier, 1962
[= *Graminella* Léger & Gauthier, 1937, *nom. nud.*]

Thalli with a bulbous base and sparsely branched main axis, producing long series of small ovoid trichospores with a single appendage and no collar. Zygospores with an oblique, median collar. May reproduce endogenously within the gut by means of specialized vegetative propagules. In hindgut of nymphal Baetidae (Ephemeroptera). Two species.

Type species: *Graminella bulbosa* Léger & Gauthier ex Manier.

Both species of *Graminella* are unusual Harpellales because of their production of vegetative propagules; these are ontogenetically different in the two species.

—Key to *Graminella* species—

1. Trichospores 10–17 μm long .... ***G. bulbosa***
1'. Trichospores 6–8.5 μm long .... ***G. microspora***

☐ *Graminella bulbosa* Léger & Gauthier ex Manier, 1962a
 [= *Graminella bulbosa* Léger & Gauthier, 1937, *nom. nud.*]

Thalli up to 500–600 μm long, lower branches few and short, terminal ones fertile and producing a unilateral series of up to 40 ovoid trichospores 10–17 × 2–3 μm having 1 appendage and no collar. Bulbous base of thallus 8–10 μm wide, terminal branches 2–3 μm wide. Thalli often formed side by side and arranged in parallel rows. Vegetative production of new thalli occurs by means of a small bulbous detachable outgrowth from a thallus base. Zygospores unknown. Type species.

Illustrations: See Manier, 1962a, Fig. VII.

Hosts: In hindgut of *Baetis* sp. (Ephemeroptera, Baetidae) nymphs.

Distribution: Streams near Grenoble (Department of Isère) and in the Néouvieille Mountains (Department of Hautes–Pyrénées), France.

References: Léger and Gauthier, 1937; Manier, 1962a.

☐ *Graminella microspora* Moss & Lichtwardt, 1981 (*in* Lichtwardt and Moss, 1981)

Thalli sparsely branched, up to 300 μm long with a bulbous base up to 11–18 μm diam. Terminal branches 2.5–3 μm diam, some producing long series of 10–20 lateral, ovoid trichospores 6–8.5 μm long by 2–2.5 μm diam, with a single appendage and no collar upon detachment. Conjugating branches producing from one of the conjugated cells biconical zygospores 32–43 μm × 6–8 μm, each with an oblique, median collar 5–8 μm long upon detachment. Reproduction is also by single clavate to ovoid cells, which emerge from the bulbous basal cells or other swollen cells of the thallus, attach to the gut cuticle, and develop into new thalli.

Illustrations: Fig. 11.12.

Hosts: In hindgut of *Baetis tricaudatus* Dodds nymphs, and unidentified nymphs of another Baetidae species (Ephemeroptera).

Collections: N. Fork Bond Creek, Lake Co., Montana, U.S.A.; and at base of Mürrenbach Falls near Stechelberg, Switzerland.

Reference: Lichtwardt and Moss, 1981.

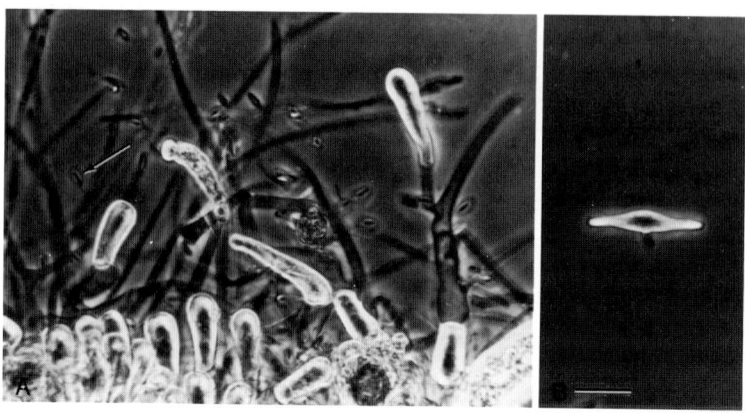

FIG. 11.12. *Graminella microspora*. A. Minute trichospores (arrow) and large propagative cells. B. Zygospore. Scale bar = 20 μm for both figures. Fig. 11.12A from Lichtwardt and Moss, 1981, by permission of Trans. Brit. Mycol. Soc.

Legeriomyces

Pouzar, 1972
= *Genistella* Léger & Gauthier, 1932

Trichospores obpyriform, without a collar, bearing 2 appendages. Zygospores biconical, attached submedially and obliquely to the zygosporophore, upon detachment retaining a collar and a single appendage. In hindgut of larval Ephemeroptera. Two species.

Type species: *Légeriomyces ramosus* Pouzar.

Five illegitimate species have been described under the name *Genistella*: *G. chironomi* Tuzet & Manier, 1953 (now *Smittium chironomi*); *G. choanifera* Tuzet & Manier, 1953 (probably a *Smittium* sp.); *G. mailleti* Tuzet & Manier, 1955 (probably *L. ramosus*); *G. microspora* Gauthier, 1960 (now *Gauthieromyces microsporus*); and *G. rhitrogenae* Tuzet & Manier, 1955 (insufficiently described).

—Key to *Legeriomyces* species—

1. Trichospores in one size range, 31–49 × 6–10 μm .... *L. ramosus*
1'. Trichospores in three size ranges: 14–23 × 3.5–5 μm, 30–36 × 6–8 μm, and 40–49 × 7–9 μm .... *L. aenigmaticus*

☐ *Legeriomyces aenigmaticus* Lichtwardt & Williams, 1983b

Thalli highly branched, sometimes in dense tufts or clumps with a mass of contorted cells in the holdfast region, producing one or more of three trichospore sizes: (1) 14–23 × 3.5–5 μm; (2) 30–36 × 6–8 μm; (3) 40–49 × 7–9 μm. Trichospores long–obpyriform, with two appendages; appendages unusually broad (3–7 μm) in the two larger spore sizes, progressively narrowing towards their free ends. Smallest trichospores most common, borne in series of up to 15 per fertile branch. The two larger size trichospores borne 2–4 per fertile branch, with one or both larger types on the same thalli that produce the smallest spores. Zygospores unknown.

Illustrations: See Lichtwardt and Williams, 1983b, Figs. 1–10.

Host: In hindgut of *Drunella spinifera* Needham (Ephemeroptera, Ephemerellidae) nymphs.

Distribution: Known from one small stream, Johnson Creek, draining into Swan Lake in nothwestern Montana, U.S.A.

The unusual feature of this species is the three class sizes into which the trichospores fall. The only known host species was found in fair abundance, mostly on silted substrates, in the otherwise clear mountain stream. At least nine other trichomycetes have been collected in the lotic insect fauna of Johnson Creek, which is also the type locality for *Harpella leptosa*.

Reference: Lichtwardt and Williams, 1983b.

☐ *Legeriomyces ramosus* Pouzar, 1972
 = *Genistella ramosa* Léger & Gauthier, 1932

Trichospores long–obpyriform, 31–49 × 6–10 μm, with 2 long appendages. Biconical zygospores 41–70 × 7–10 μm, upon detachment with a collar and a single appendage. Type species.

Illustrations: Fig. 11.13.

Hosts: In hindgut of Ephemeroptera nymphs, including *Baetis rhodani* Pict. and *B. bioculatus* L. (Baetidae), and *Ephemerella infrequens* McDunnough (Ephemerellidae).

Distribution: Streams in the French and Swiss Alps, French Pyrenees, and Montana, U.S.A.; possibly also in the foothills of the Sierra Nevada Mountains, California, U.S.A. (Whisler, 1963) and England (Moss, 1979).

This appears to be primarily a European species, and Baetidae nymphs are the usual host. The appendages of *L. ramosus* can be seen forming within the generative cell before trichospore outgrowth commences. Upon initial release of the mature trichospore the 2 appendages are intricately wound around each other, then they separate, straighten out, and become somewhat divergent. The appendages are often broad near the trichospore body; they were unusually broad in one collection in Montana (Lichtwardt and Williams, 1983b). Infrequently, 1 appendage may be noticeably shorter than the other (Moss, 1979; Lichtwardt, unpublished). Manier (1973a) has published an electronmicrographic study of the species.

References: Léger and Gauthier, 1932, 1935; Manier, 1962, 1973a; Whisler, 1961, 1963; Moss, 1979; Lichtwardt and Williams, 1983b.

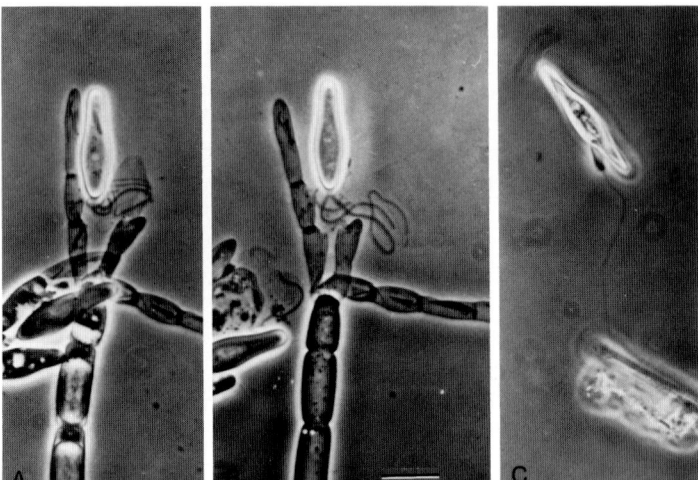

FIG. 11.13. *Legeriomyces ramosus*. A. Trichospore with its two appendages just released from the generative cell and coiled like a ball of twine. B. Seconds later the appendages are unfurling. C. Zygospore with its single appendage stuck to a piece of debris. Scale bar = 20 μm for all figures. Fig. 11.13A reprinted by permission from Mycologia, vol. 73: 482, Copyright 1981, R.W. Lichtwardt and The New York Botanical Garden.

Orphella

Léger & Gauthier, 1931

Sporulating branches in the form of a cyme. Generative cells subterminal, produced singly on separate branches, each generative cell giving rise to one curved trichospore. In hindgut of Plecoptera nymphs. Monotypic.

Type species: *Orphella coronata* Léger & Gauthier.

Orphella was the first genus of Legeriomycetaceae described. It was again described as a "new genus" by Léger and Gauthier the following year (1932), along with six other new genera of Harpellales, at which time the family Genistellaceae (now Legeriomycetaceae) was established. A second but illegitimate species, *Orphella culicis (nom. nud.)* from mosquito larvae, was described by Tuzet and Manier (1947a); it was later renamed *Smittium culicis* by Manier (1969b).

☐ *Orphella coronata* Léger & Gauthier, 1931

Thallus up to 1 mm long, attached to host cuticle by a cupulate swelling with short radiating branches. Fertile branch tips cymelike, commonly ending in three pairs of tapered branches, each of the six branches bearing one curved trichospore, 45–50 µm long, from a subapical generative cell. Trichospore appendages and zygospores unknown. Type species.

Illustrations: Fig. 11.7; also see Léger and Gauthier, 1931, Figs. 1, 2.

Host: In hindgut of *Protonemura humeralis* Pict. and possibly *Nemura cinerea* (Retz.) (= *N. variegata* Oliv.) nymphs (Plecoptera, Nemouridae).

Distribution: Streams of the Dauphiné French Alps, France.

This unusual species has not been reported since Léger and Gauthier's descriptions were published. The cymose trichospore arrangement is unique, as is also the presence of a sterile terminal cell above the single generative cell of each fertile branch. Léger and Gauthier (1931) described the trichospores as detaching at maturity, but they observed no appendages. In fact, no appendages of trichospores were described until 1935(b), by Léger and Gauthier; in that publication their illustrations of trichospore appendages of several previously described genera of Harpellales did not include *Orphella*.

References: Léger and Gauthier, 1931, 1932.

Pennella

Manier ex Manier, 1968
[= *Pennella* Manier, 1963b, *nom. nud.*]

Base of principal thallial cell simple, lobed or bifurcated, cemented to host cuticle by a mucilaginous secretion. Trichospores obpyriform to cylindrical, without a collar, bearing about 4–6 appendages. Biconical zygospores attached medially and oriented parallel to axis of zygosporophore. In hindgut of larval Simuliidae (Diptera). Five species.

Type species: *Pennella hovassi* Manier ex Manier.

—Key to *Pennella* species—

1.	Trichospores cylindrical to almost filiform	2
1'.	Trichospores long obpyriform to long–ovoid	3
2(1).	Trichospores (50–)80(–104) µm long, with 4–6 fine appendages	*P. angustispora*
2'(1).	Trichospores 49–59 µm long, with 4 petaloid appendages	*P. grassei*
3(1').	Trichospores long–ovoid, zygospores more than 80 µm long	*P. simulii*
3'(1').	Trichospores long–obyriform; zygospores less than 80 µm long	4
4('').	Trichospores more than 35 µm long; zygospores more than 14 µm diam	*P. arctica*
4'(3').	Trichospores less than 35 µm long; zygospores less than 14 µm diam	*P. hovassi*

☐ *Pennella angustispora* Lichtwardt, 1972

Thalli with little secondary branching, consisting of a main axis up to 1 mm long. Base of primary cell tapered to a rounded point, rarely slightly bifurcate, cemented to the hindgut lining by a mucilaginous substance. Trichospores cylindrical–clavate, almost filiform, straight to slightly curved or bent, (50–)80(–104) × (2.5–)4(–6) μm, without a collar, bearing 4–6(–7) fine appendages. Heterothallic. Biconical zygospores 82–92 × 15–17 μm, attached medially and oriented parallel to the zygosporophore axis, formed above the area of conjugation from the end of one of the conjugant cells. Zygosporophores 32–36 × 10–12 μm.

Illustrations: See Lichtwardt, 1972, Figs. 37–44.

Hosts: In hindgut of many species of larval *Simulium* (Diptera, Simuliidae), including *S. arcticum* Mall., *S. argus* Will., *S. virgatum* Coq., and *S. vittatum* Zett.

Distribution: Montana, Wyoming, Colorado, Utah, and California, U.S.A.; Aomori Prefecture, Honshu, Japan.

Pennella angustispora is relatively common in blackfly larvae in the montane regions of western U.S.A. It often occurs along with other blackfly trichomycetes. The species has been found once in Japan, in 1967 (unpublished). *Pennella simulii*, a rarer species, has been found in some of the same blackfly species, but not in the same host specimens; it is readily distinguishable from *P. angustispora* by the long–ovoid shape of its trichospores. The fine structure of the holdfast cell of *P. angustispora* has been studied by Mayfield and Lichtwardt, 1980.

References: Lichtwardt, 1972; Dang, 1978; Mayfield and Lichtwardt, 1980.

☐ *Pennella arctica* Lichtwardt & Williams, 1984 (*in* Lichtwardt, 1984a)

Mature thalli 500 μm or more in length, consisting of a large principal coenocytic cell whose base is simple to deeply and repeatedly bifurcate and embedded in a mucilaginous cementing substance. Trichospores long–obpyriform, (40–)48(–58) × (8–)9(–11) μm, with (2–)6(–7) fine appendages and no collar. Zygospores (65–)70(–72) × 15–18 μm, attached medially and parallel to the zygosporophores which measure 35–40 × 13–20 μm.

Illustrations: Figs. 7.10, 11.14.

Hosts: In hindgut of larval *Prosimulium ferrugineus* Wahlberg, *P. exigens* Dyar & Shannon, and *Simulium arcticum* Malloch (Diptera, Simuliidae).

Distribution: In river at Abisko, Sweden, and streams in northwestern Montana, U.S.A.

The amount of bifurcation in the basal cell of mature thalli of *Pennella arctica* is quite variable, and in some cases it may resemble *P. hovassi*. However, the dimensions and shape of the trichospores alone are sufficient to distinguish *P. arctica* from all other described species.

Reference: Lichtwardt, 1984a.

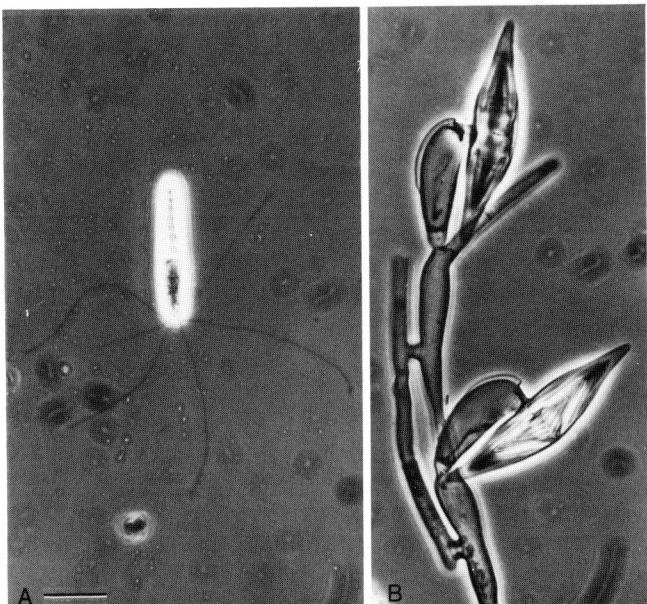

FIG. 11.14. *Pennella arctica*. A. Trichospore. B. Zygospores. Scale bar = 20 μm.

□ *Pennella grassei* Manier, 1968
[= *Glotzia grassei* Tuzet & Manier, 1955a, *nom. nud.*]

Thallus up to 1 mm long with a verrucous tapered basal cell which may be slightly digitate. Principal axis consisting of cells 7–9 μm diam and branches 5–7 μm diam. Trichospores cylindrical, 49–59 × 4.5–5.5 μm, with about 4 petaloid appendages that can subdivide at their extremity, and with no collar. Homothallic. Biconical zygospores about 90 × 15 μm, attached medially and oriented parallel to their zygosporophores which measure about 35 x 12 μm.

Illustrations: See Tuzet and Manier, 1955a, Figs. 8–10; Manier, 1963b, Fig. 4.

Host: In hindgut of *Simulium equinum* L. larvae (Diptera, Simuliidae).

Distribution: Streams of the Gorge d'Enfer, Les Eyzies, Dordogne, France.

The description of this apparently rare species was originally given under the name *Glotzia grassei* Tuzet & Manier, 1955a *(nom. nud.)*. In 1963b Manier corrected the description and illegitimately transferred the fungus to the genus *Pennella*, which she established in that paper (she provided no Latin diagnosis nor did she cite a nomenclatural type for *P. grassei*). The name *P. grassei* was validly published by Manier in 1968, but was incorrectly cited as though it were a new combination (the "basionym" was not legitimate).

There is no mention of a mucilaginous secretion around the basal cell typically seen on most thalli of other *Pennella* spp. The trichospores of *P. grassei* have unusual appendages in that they are described and illustrated (in Manier, 1963b) as being petaloid and sometimes subdivided into finer filaments; appendages of other *Pennella* spp. are usually very fine throughout their length. Despite these features, *P. grassei* seems to be placed in the correct genus.

References: Tuzet and Manier, 1955a; Manier, 1963b, 1968.

☐ *Pennella hovassi* Manier ex Manier, 1968
  [= *Pennella hovassi* Manier, 1963b, *nom. nud.*]

Mature thalli up to 1 mm long; basal cell of main axis 7–9 μm diam, simple or more often deeply and repeatedly bifucated, with numerous branches 5–7 μm diam arising laterally; basal cell partially surrounded by a secreted mucilaginous cementing substance. Trichospores long–obpyriform, 22–33 × 4.5–7 μm, lacking a collar and with 6 appendages. Heterothallic. Zygospores 68–79 × 7.5–13 μm, attached medially and parallel to the zygosporophore, which measures about 25–30 × 8–12 μm. Type species.

Illustrations: See Manier, 1963b, Figs. 5–7.

Hosts: In hindgut of larval Simuliidae (Diptera), including *Simulium monticola* Fried, *S. vittatum* Zett., and *Prosimulium* sp.

Distribution: Department of Puy–de–Dôme, France; Newfoundland, Canada.

References: Manier, 1963b, 1968; Frost and Manier, 1971.

☐ *Pennella simulii* Williams & Lichtwardt

Branched thalli up to 1 mm long, with a simple and tapered, occasionally digitate, basal cell surrounded by a mucilaginous cementing substance. Trichospores long–ovoid, (30–)33(–41) × (6.5–)8(–10.5) μm, with about 6 fine appendages and no collar. Biconical zygospores (84–)90(–96) × (19–)22(–24) μm, attached medially and oriented parallel to the zygosporophore, formed above the area of conjugation from the end of one of the conjugant cells. Zygosporophore (32–)36(–40) × (12–)14.5(–16) μm.

Illustrations: See Williams and Lichtwardt, 1971, Figs. 1–5.

Host: In hindgut of larval *Simulium* spp. (Diptera, Simuliidae), including *S. vittatum* Zett. and *S. venustum* Say.

Distribution: Wyoming and Colorado, U.S.A.

The known distribution of *Pennella simulii* is limited to a few streams in the Rocky Mountains, but infestation may be high in some populations of blackfly larvae. It is distinguishable from the more widespread species, *P. angustispora,* by the long–ovoid shape of its trichospores, slightly greater branching of thalli, and its wider zygospores.

When conjugations occur in *P. simulii* between branches of two thalli, the zygosporophore and zygospore develop at the end of one of the conjugants, the receptor cell. Donor branches may conjugate with more than one receptor branch, and may terminate their development by producing a terminal trichospore.

Reference: Williams and Lichtwardt, 1971.

Pteromaktron

Whisler, 1963

Thallus consisting of an arbusculate coenocytic branched main cell with a compact cluster of short fertile branches at its distal end. Trichospores cylindrical, collarless, with a single appendage, each trichospore produced from a separate lateral generative (subsidiary) cell arranged in a unilateral series on a terminal fertile branch. In hindgut of Baetidae (Ephemeroptera) nymphs. Monotypic.

Type species: *Pteromaktron protrudens* Whisler.

□ *Pteromaktron protrudens* Whisler, 1963

Arbusculate, mature coenocytic thalli 942–1730 μm long with a main axis 7–47 μm diam and consisting of lateral and basal branches and a compact, brushlike terminal tuft of sporulating branches that project from the anus, attached to the host cuticle by a digitate to coralloid basal outgrowth. Each fertile branch with a unilateral series of oval to cylindrical generative (subsidiary) cells 21–31 × 4–6 μm, each producing terminally a cylindrical, collarless trichospore 85–97 × 4–6 μm bearing a single long (~1.8 mm) fine appendage with a knob near the trichospore base. Zygospores unknown. In Baetidae (Ephemeroptera) nymphs. Type species.

Illustrations: Fig. 11.7, also see Whisler, 1963, Figs. 4–6.

Hosts: In hindgut of *Callibaetis pacificus* Seemann and *Cloeon dipterum* (L.) (Ephemeroptera, Baetidae).

Distribution: St. Helena Creek, Lake Co., California, U.S.A.; and Pierette Pond, near Saint–Gély–du–Fesc, Department of Hérault, France.

Pteromaktron protrudens, known only from one site in the U.S.A. and one site in France (Manier, 1969b), is perhaps the most unusual species of Harpellales with respect to its thallus structure and sporulation. The thallus consists of essentially one large coenocytic arbusculate branching cell which, according to Whisler's description and drawings, is nonseptate except to delimit the generative cells and trichospores. Whereas most other Harpellales produce trichospores directly and basipetally from terminal branches divided into a row of end–to–end generative cells, the trichospores of *Pteromaktron protrudens* are produced from individual generative cells arranged unilaterally on terminal, nonseptate branches, and their maturation on each terminal branch is acropetalous. Apparently the brushlike fertile tip of the thallus matures entirely outside of the gut. Some thalli of *Zygopolaris* spp. (also in Ephemeroptera) may likewise sporulate only after protruding from the anus, but internal sporulation is generally more common. The trichospores of *Pteromaktron protrudens* are structurally similar to those of *Spartiella barbata* (in Baetidae), but the latter are produced from typical harpellid generative cells.

References: Whisler, 1961, 1963; Manier, 1969b (1970b).

Simuliomyces

Lichtwardt, 1972

Trichospores elongate–ellipsoidal, small, collarless, bearing 2–4 appendages that are usually no longer than the trichospore body. Zygospores attached perpendicularly and medially to the zygosporophore. In hindgut of Simuliidae (Diptera) or Capniidae (Plecoptera) larvae. Two species.

Type species: *Simuliomyces microsporus* Lichtwardt.

—Key to *Simuliomyces* species—

1. Trichospores 20–30 × 4–6 µm; in Simuliidae .... **S. microsporus**
1'. Trichospores 12–16 × 2 µm; in Capniidae .... **S. spica**

□ *Simuliomyces microsporus* Lichtwardt, 1972

Branched thalli without a main axis, basal cell sometimes swollen or lobed, producing usually 4–7 generative cells per fertile branch. Trichospores collarless, long–ellipsoidal, 20–30 × 4–6 µm, tip often appearing thick–walled, with 2–4 appendages approximately as long as the trichospore; appendages usually fine, but sometimes slightly broadened near the base and tapering toward the tip. Lateral trichospores arising from a short branchlike extension of the generative cell. Generative cells usually longer than trichospores. Zygospores biconical, somewhat flattened on side of attachment, 34–45 × 7–9 µm, attached medially and perpendicularly to zygosporophore. Zygosporophore about 12–17 × 7–11 µm, arising laterally from one of the conjugating cells. Type species.

Illustrations: Fig. 7.12A, 11.15.

Hosts: In hindgut of many different species of Simuliidae (Diptera) larvae.

Distribution: Widespread in the U.S.A.: relatively common in the Rocky Mountains of Colorado, Wyoming and Montana; collections include Kansas, Utah, and California. Also known from England, France, and Sweden.

Simuliomyces microsporus is most apt to be confused with species of *Smittium*, as apparently happened in two published reports of *S. microsporus* attached to thalli of *Paramoebidium* sp. (Ingold, 1967; Moss, 1970) before *S. microsporus* was described as a new genus and species. What likely was also *S. microsporus* attached to a thallus of *Paramoebidium* sp. was identified by Manier (1955a) as *Stipella vigilans*. The released collarless trichospores with 2–4 short appendages and the Type I zygospores makes *S. microsporus* readily distinguishable. Attachment of a

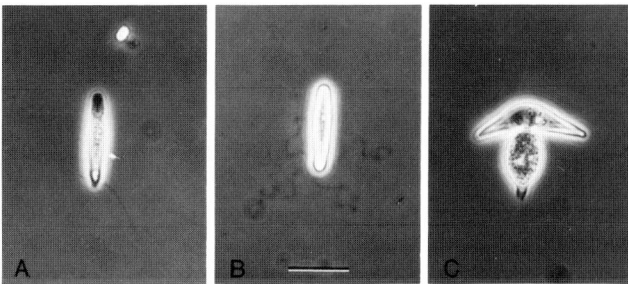

FIG. 11.15. *Simuliomyces microsporus*. A, B. Two- and four-appendaged trichospores, respectively. C. Zygospore. Scale bar = 20 μm for all figures.

thallus to other trichomycetes coinhabiting blackfly hindguts in itself suggests that it may be *S. microsporus*. If thalli are sporulating, but released trichospores (or zygospores) are not available, one can tentatively identify this species on the basis of the size and shape of the trichospores, the usually long generative cells, and the small outgrowths from the generative cells subtending the lateral trichospores. The trichospore tips, especially those mounted in lactophenol–cotton blue, often have a characteristic thick–walled appearance.

References: Lichtwardt, 1972; Manier, 1955a, Ingold, 1967; Moss, 1970.

□ *Simuliomyces spica* Peterson & Lichtwardt, 1983

Mature thalli usually with a main axis, 7–10 μm diam near the base, from which arises a lateral complex of sporulating branches. Generative cells 4–9 × 3 μm, 10–30 per fertile branch. Trichospores long–ellipsoidal, 12–14(–16) × 2 μm, with 2 fine, short basal appendages; borne on a short apical–lateral projection of the generative cell. Zygospores 36–43 × 7–10 μm, attached to the zygosporophore medially and at right angles. Zygosporophores (21–)29(–34) × 5 μm, arising from one of the conjugating hyphae, often forming a unilateral row of zygospores.

Illustrations: See Peterson and Lichtwardt, 1983, Figs. 10–18.

Host: In hindgut of *Allocapnia* sp. (Plecoptera, Capniidae) nymphs from leaf packs and stones in streams.

Distribution: Missouri, Arkansas, U.S.A.

Simuliomyces spica at present has a restricted known distribution. It is readily distinguishable from *S. microsporus* in blackfly larvae on the basis of morphology and its winter–emerging stonefly host. It somewhat resembles *Genistelloides hibernus,* also from *Allocapnia* spp. The long series of very small trichospores on fertile branches has the general appearance of *Graminella* spp. from mayfly nymphs.

References: Peterson and Lichtwardt, 1983; Peterson, 1984.

Smittium

Poisson, 1936

Trichospores ellipsoidal (or subellipsoidal) to almost cylindrical, with a short or long collar and a single appendage. Zygospores biconical to fusiform, attached to the zygosporophore obliquely and submedially, upon detachment having a collar and single appendage. In hindgut of larval Nematocera (Diptera). Twenty–three species.

Type species: *Smittium arvernense* Poisson.

In 1932(b) Poisson named a fungus, *Dixidium dixae*, from *Dixa* sp. larvae (Dixidae, Diptera), which probably is a species of *Smittium*. The description was so incomplete, however, that it has to be rejected as an acceptable name (Manier and Lichtwardt, 1968). Several illegitimate taxa, now recognized as *Smittium* species, are to be found in the literature. These include species of the genera *Orphella* and *Rubetella* (see *Smittium culicis*), and *Genistella* and *Typhella* (see *Smittium chironomi*).

—Key to *Smittium* species—

1.		Base of thallus tapered to a point and encased in a mucilaginous substance, without a secreted holdfast structure (in Simuliidae) .... ***S. pennelli***
1'.		Base of thallus not as above 2
	2(1').	Trichospores 12–14 μm long with a thick, finely punctate wall (in Chironomidae) .... ***S. incrassatum***
	2'(1').	Trichospores (except small oval spores of dimorphic species) averaging more than 14 μm long; wall not thick and finely punctate 3
3(2').		Trichospores dimorphic, some long and narrow, others short and oval 4
3'(2').		All trichospores essentially similar in shape, with one range of sizes 5
	4(3).	Larger trichospores 38–50 × 5.5–6.5 μm; zygospores 13–14 μm diam (in Chironomidae) .... ***S. dimorphum***
	4'(3).	Larger trichospores 25–40 × 6–8 μm; zygospores 9–11 μm diam (in Chironomidae) .... ***S. orthocladii***
5(3').		Collar usually flared outward and less than 2.5 μm long; usually not in Chironomidae 6
5'(3').		Collar not flared outward; if shorter than 2.5 μm, then in Chironomidae 7
	6(5).	Trichospores typically long–ovoid with greatest width below midregion of spore; not pathogenic (usually in Culicidae, less commonly in Simuliidae, Chironomidae, or Ceratopogonidae) .... ***S. culisetae***
	6'(5).	Trichospores narrowly ellipsoidal with greatest width usually near midregion of spore; often pathogenic in Culicidae .... ***S. morbosum***
7(5').		Collar usually longer than 10 μm (10–19 μm) 8

7'(5').	Collar usually shorter than 10 μm 11	
	8(7). Average length of trichospores more than 40 μm 9	
	8'(7). Average length of trichospores less than 40 μm 10	
9(8).	Trichospores (40–)46(–55) μm long, with a zigzag appendage; zygospores produced (in Chironomidae) .... *S. longisporum*	
9'(8).	Trichospores 50–60 μm long, appendage not zigzag; zygospores unknown (in Chironomidae) .... *S. gigasporus*	
	10(8'). Trichospores 6–7.5 μm diam, with a verrucose surface; zygospores unknown (in Chironomidae) .... *S. macrosporum*	
	10'(8). Trichospores 10–14 μm diam, with smooth surface; zygospores produced (in Chironomidae) .... *S. alpinum*	
11(7').	Trichospores nearly cylindrical but with a median swelling ... 12	
11'(7').	Trichospores ellipsoidal to oval 16	
	12(11). Trichospores more than 32 μm long 13	
	12'(11). Trichospores less than 32 μm long 14	
13(12).	Trichospores 6–7 μm diam; zygospores 95–115 μm long (in Chironomidae) .... *S. bisporum*	
13'(12).	Trichospores 3.5–5 μm diam; zygospores 110–150 μm long (in Chironomidae) .... *S. megazygosporum*	
	14(12'). Thallus consisting of one long branched main axis with numerous shorter basal branches radiating from the holdfast region; trichospore collar 1–2 μm long (in Chironomidae) .... *S. chironomi*	
	14'(12'). Thallus consisting of several main branches arising from the holdfast region; trichospore collar usually longer than 2 μm 15	
15(14').	Basal cell often swollen; verticillate branching uncommon; trichospores 16–28 μm long (in Simuliidae and Chironomidae, less commonly in Culicidae and Tipulidae) .... *S. simulii*	
15'(14').	Basal cell not swollen; branching often verticillate; trichospores 25–30 μm long (in Chironomidae) .... *S. typhellum*	
	16(11'). Trichospores 2.5–3 μm diam (zygospores produced) (in Chironomidae) .... *S. pusillum*	
	16'(11'). Trichospores usually more than 3 μm diam 17	
17(16').	Trichospores with a minute apical nipple (zygospores produced) (in Chironomidae) .... *S. mucronatum*	
17'(16').	Trichospores rounded at the tip 18	
	18(17'). Trichospores oval 19	
	18'(17'). Trichospores ellipsoidal 20	
19(18).	Trichospores 4–8 μm diam, collar (3–)5–9 μm long (in Culicidae, rarely in Simuliidae and Chironomidae) .... *S. culicis*	
19'(18).	Trichospores 7–9 μm diam, collar 1.5–6.5 μm long (in Chironomidae) .... *S. arcticum*	
	20(18'). Almost all thallial cells becoming fertile; trichospore collar 5–10 μm long (in Chironomidae) .... *S. cellaspora*	
	20'(18'). Only terminal branches becoming fertile; trichospore collar 2–4 μm long 21	

21(20'). Average trichospore length more than 25 μm, appendage sometimes spiraled; zygospores unknown (in Chironomidae) .... *S. elongatum*
21'(20'). Average trichospore length less than 25 μm, appendage not spiraled; zygospores produced 22
22(21'). Trichospores 6.5–7.5 μm diam, collar 5–9 μm long; zygospores 77–88 × 12–15 μm (in Chironomidae) .... *S. ouseli*
22'(21'). Trichospores 5 μm diam, collar about 2 μm long; zygospores 30–35 × 8–10 μm (in Chironomidae) .... *S. arvernense*

☐ **Smittium alpinum** Lichtwardt, 1984a

Thalli with 1–3 long generative cells per fertile branch. Trichospores oval to biconical, (23–)33(–44) × (10–)12(–14) μm, with a cylindrical collar (10–)14(–19) × (2–)3(–4) μm. Zygospores biconical, (63–)75(–83) × (14–)16(–18) μm with a single appendage and a collar (15–)22(–28) × (4–)5(–7) μm attached laterally near one end of the zygospore.

Illustrations: See Lichtwardt, 1984a, Figs. 15–25.

Hosts: In hindgut of larval *Diamesa* spp. (Diptera, Chironomidae).

Distribution: Small streams, near Continental Divide in Glacier National Park, Montana, U.S.A.; Abisko, Sweden; and Alps in France and Switzerland.

Reference: Lichtwardt, 1984a.

☐ **Smittium arcticum** Y. Kobayasi, 1969 (*in* Kobayasi et al., 1969)

Thalli 300–400 μm or more long. Axial hyphae 6.5–12 μm diam near base, arising irregularly from a short cylindrical basal cell 8–10 μm diam and producing nondivergent monopodial branches 3.5–5 μm diam. Generative cells 1–5 per fertile branch, producing terminolateral outgrowths 6.5–12 × 2–2.5 μm bearing oval trichospores (15–)20(–24) × (7–)8(–9) μm; upon detachment trichospores have a short–campanulate or cylindrical collar 1.5–6.5 μm long and bear a single appendage. Zygospores unknown.

Illustrations: See Kobayasi et al., 1969, Fig. 5 and Plate 1, Figs. A, B.

Hosts: In hindgut of Chironomidae (Diptera) larvae.

Distribution: Peters Lake, Alaska.

Reference: Kobayasi et al., 1969.

☐ **Smittium arvernense** Poisson, 1936

Thalli up to 1.2 mm long, not highly branched, attached to host cuticle by means of a disklike holdfast. Trichospores ellipsoidal, 20–25 × 5 μm, with a short collar and a single short fine appendage. Zygospores 30–35 × 8–10 μm, with a small oblique submedian collar ~2 μm long. Type species.

Illustrations: See Poisson, 1936, Figs. 3–7.

Hosts: In hindgut of *Smittia* sp. (Diptera, Chironomidae) larvae.

Distribution: Flowing waters at Besse–en–Chandesse, Department of Puy–de–Dôme, France.

Smittium arvernense apparently has not been found since Poisson's 1936 publication. He referred to what we now call trichospores as "azygospores," and described and illustrated zygospores resulting from the fusion of two hyphae. His sole illustration of a zygospore, a drawing (erroneously called an azygospore in the legend), shows a small submedian collar attached at an angle. It appears that the angle of this collar is opposite to what it should be, on the basis of zygospores of other species subsequently placed in this genus as well as other genera with Type II zygospores. Nor did Poisson mention the single appendage now known to be attached to zygospores of some other *Smittium* species. Despite these differences, it seems clear that *S. arvernense* is the correct type for this, the largest, genus of Harpellales.

Reference: Poisson, 1936.

☐ *Smittium bisporum* Manier & Coste, 1971

Thalli 500–600 μm long, with a small holdfast and 2 or 3 main branches that rebranch in a bi– or trifurcate manner; branches about 5 μm diam. Trichospores (1–)2 per fertile branch, subcylindrical with a slight median swelling, 35–40 × 6–7 μm, collar 6–7 × 3–4 μm. Zygospores fusiform, 95–115 × 10–12 μm, with a collar 6–10 × 6 μm located on lower third of zygospore, and with a single long appendage.

Illustrations: See Manier and Coste, 1971, Figs. 1–7.

Host: In hindgut of *Psectrotanypus varius* Fabr. (Diptera, Chironomidae) larvae.

Distribution: In still waters of the ancient quarries of Vendargues, Department of Hérault, France.

The trichospores of *S. bisporum* are most similar in size and shape to those of *S. mucronatum,* but lack the apical nipple of the latter species. Their respective zygospores make the two species quite distinct, however.

In an unpublished dissertation, Coste (Coste–Mathiez, 1970) used the name *S. megazygosporum* for a new species, and illustrated it with four photomicrographs (her Figs. 54, 55, 57, 58); three of these photomicrographs were later published by Manier and Coste (1971) under the new name *S. bisporum* (their Figs. 3, 5, 6). *Smittium megazygosporum* was then used by Manier and Coste (1971) as the validly published name for another species of *Smittium*.

References: Manier and Coste, 1971; Coste–Mathiez, 1970.

☐ *Smittium cellaspora* Williams, 1982

Thalli short, up to about 300 μm in length, sparsely branched, almost all cells becoming fertile. Trichospores ellipsoidal, (20–)29(–36) × (7–)8.5(–10) μm, single long appendage often coiled; collar (5–)9(–10) × 2.5 μm. Zygospores unknown.

Illustrations: See Williams, 1982, Figs. 1–4.

Hosts: In hindgut of *Sympotthastia* sp. and possibly other Chironomidae (Diptera) larvae.

Distribution: Missouri, U.S.A.

The unusual feature of this small species is that virtually all cells can produce trichospores.

Reference: Williams, 1982.

☐ *Smittium chironomi* Manier, 1970b (1969b)
[= *Typhella chironomi* (Tuzet & Manier) Manier & Mathiez, 1965, *nom. nud.*]
[= *Genistella chironomi* Tuzet & Manier, 1953, *nom. nud.*]

Mature thallus consisting of a main axis up to 800 μm long with branching along the upper part and producing numerous shorter basal branches radiating from above the disklike holdfast. Main axis and larger basal branches somewhat swollen at their base, 2.5–5 μm diam, tapering abruptly to 1.5–2 μm diam. Fertile upper branches of main axis each with 4–8 generative cells producing subcylindrical trichospores with a slight median swelling, 20–23 × 2.5–3 μm, collar 1–2 μm long × 1.5–2.5 μm wide. Zygospores unknown.

Illustrations: See Tuzet and Manier, 1953, Figs. 4, 5 (as *Genistella chironomi*). Note: the shape of the trichospores does not conform to the published description.

Hosts: In anterior hindgut of lotic Orthocladiinae larvae (Diptera, Chironomidae).

Distribution: Balaruc–le–Vieux and Sète, Department of Hérault, France; possibly in England.

Smittium chironomi was incorrectly published by Manier (1969b) as a new combination based on an illegitimate basionym, *Genistella chironomi* Tuzet & Manier, 1953 *(nom. nud.);* her 1970b (1969b) publication of the name, however, is valid. *Genistella chironomi* was transferred improperly to another illegitimate genus *Typhella* (as *T. chironomi*), by Manier (1962a). *Smittium chironomi* is distinct from other *Smittium* spp. on the basis of the habit of the thallus and the dimensions of the trichospores as described, although no accurate illustrations are available for the latter.

Moss (1972) did an electron microscopic study of a *Smittium* sp., which he called *S. chironomi*, obtained from the hindgut of *Tanytarsus* sp. (Tanytarsini, Chironomidae) larvae collected in England, but there was no accompanying description or other indication to assure that it was correctly identified.

References: Manier, 1962a, 1969b (1970b); Tuzet and Manier, 1953; Moss, 1972.

□ *Smittium culicis* Manier, 1970b (1969b)
  [= *Orphella culici* Tuzet & Manier, 1947a, *nom. nud.*]
  [= *Rubetella culicis* Tuzet, Rioux & Manier, 1961, *nom. nud.*]

Thalli with divergent, often monopodial, branches, attached to host cuticle by an inconspicuous holdfast; base sometimes pseudorhizoidal. Fertile terminal branches often arched, with 4–6 generative cells. Trichospores oval (15–)20(–32) × (4–)6(–8) μm, with a more or less campanulate collar (3–)5–9 μm long. Zygospores unknown.

Illustration: Fig. 11.16A.

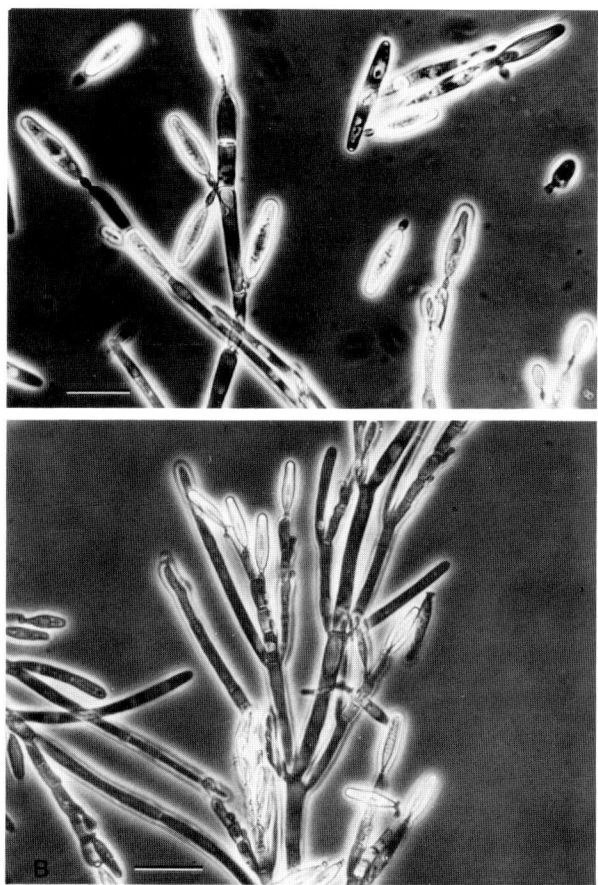

FIG. 11.16. The two most common *Smittium* species in mosquito larvae, both in axenic culture: A. *S. culicis;* B. *S. culisetae.* Scale bars = 20 μm.

Hosts: In hindgut of many genera and species of larval Culicidae (Diptera); rarely in Simuliidae and Chironomidae larvae.

Distribution: France, Tunisia, Canada, New Zealand, and U.S.A. (California, Wyoming, Nebraksa, Kansas).

In her 1969 monograph, Manier gave only average trichospore measurements of 16.5 × 4.8 μm, with a collar length of 4.7 μm. The description provided above shows a considerable range in trichospore sizes. These were determined from average measurements kindly provided by Manier in 1969 from sample collections from different mosquito larvae in France and Tunisia, and from measurements obtained by the author from his own collections and from the many available axenic cultures. Because of the somewhat different, but overlapping, ranges of trichospore size from collection to collection, in addition to the size variation within single collections and cultures, it is possible that *S. culicis* consists of a species complex of several trichospore size variants. Nevertheless, the species is fairly easy to identify based on the oval shape and size of the trichospores, the relatively long and campanulate collar, and the mosquito host. *Smittium culicis* is readily distinguishable from the other common mosquito species, *S. culisetae*, on the basis of trichospore morphology alone. Sangar et al. (1972) found a distinct immunoelectrophoretic relationship among six isolates of *S. culicis*, four obtained from mosquito larvae and one each from a blackfly and a chironomid larva, thus adding serological confirmation to the morphological data.

Even though the mosquito is its usual host, *S. culicis* was found and isolated several times by the author from *Chironomus* sp. larvae (Chironomidae) in one site in southern France, and it was cultured several times from a population of *Simulium vittatum* Zett. larvae in Leavenworth Co., Kansas, U.S.A. by Lichtwardt and Peterson (unpublished). (This blackfly population was also infested with *S. culisetae*, which was also isolated axenically.) Whisler provided the author with a culture (listed in Appendix C as CAN–X–1) that he isolated from a *Simulium* sp. larva in Canada. In southern France *S. culicis* is common and widespread in many species of mosquito larvae, whereas it has been found to occur more sporadically in the U.S.A. Its geographic distribution is undoubtedly much greater than present records indicate. It was found by the author in *Culex pervigilans* Bergroth in New Zealand.

Smittium culicis was one of the two species of Harpellales first cultured axenically (Clark et al., 1963). The species has been used in studies of host specificity (see Chapter 6) and a number of physiological experiments. The fine structure of its trichospore appendage, consisting of concentric electron–opaque rings in cross section (Moss and Lichtwardt, 1976), resembles that of *S. mucronatum* (Manier and Coste–Mathiez, 1968), a species to which *S. culicis* is serologically related (Sangar et al., 1972).

References: Manier, 1969b (1970b); Tuzet and Manier, 1947a; Tuzet et al., 1961; Manier et al., 1964; Clark et al., 1963; Chapman, 1966; Sangar,

1969; Coste–Mathiez, 1970; Müller–Kögler, 1971; Williams and Lichtwardt, 1972; Preisner, 1973; Moss and Lichtwardt, 1976; El–Buni and Lichtwardt, 1976a, 1976b; Cerniglia et al., 1978; Williams and Nagel, 1980.

☐ *Smittium culisetae* Lichtwardt, 1964
 = *Smittium inopinatum* Manier, 1970b (1969b)
 [= *Rubetella inopinata* Manier, Rioux & Whisler, 1961, *nom. nud.*]

Mature thalli large, attached to host cuticle by an inconspicuous holdfast, commonly verticillately branched, sporulating prolifically. Trichospores usually 4–10 per fertile branch, long–ovoid, (11–)16(–30) × (3–)4(–7) μm, greatest width below midregion, with a short collar 1–2.5 μm long often flared outward; appendage fine and relatively short. Zygospores rare, biconical, (46–)52(–58) × (5.5–)6(–8) μm, with a collar (3.5–)4(–4.5) μm attached almost medially and perpendicularly to the zygospore body.

Illustration: Fig. 7.14A, 9.6, 9.12, 11.16B.

Hosts: In hindgut of many species of Culicidae (Diptera) larvae; more rarely in larval Simuliidae, Chironomidae and Ceratopogonidae (Diptera) hindguts. Found once in an Ephemeroptera larva.

Distribution: Widespread in mosquito larvae in the U.S.A. (Colorado, Nebraska, Kansas, Wyoming, California, Hawaii) and Japan; also in Australia, New Zealand, and France.

Many axenic isolates of *S. culisetae* have been obtained (see Appendix C). The type culture, COL–18–3, was isolated by the author from a larva of *Culiseta impatiens* (Wlk.) in August of 1963, but the first isolate was obtained by Clark et al. (1963) from *Culiseta incidens* (Thompson), who named it *Rubetella* sp. (= *Smittium* sp.), later suggesting it was *Smittium inopinatum*. *Smittium culisetae* has been the most intensively studied trichomycete in the laboratory, as elaborated in Chapter 9. The unusual harpellid septum was first demonstrated in ultrastructural studies of this species (Farr, 1965; Farr and Lichtwardt, 1967).

Smittium culisetae occurs predominantly in mosquito larvae of several genera, including *Culiseta, Aedes, Culex,* and *Anopheles*. The smaller, long–ovoid trichospores with their short, flared collars readily distinguish *S. culisetae* from the other common inhabitant of mosquito hindguts, *S. culicis*. The latter species has been found most often in Europe. Some strains of *S. culisetae* produce trichospores not only from the usual terminal generative cells, but as well from normally vegetative cells that may spuriously become reproductive, such that all of the cells of entire sets of branches become fertile. In such cases the long vegetative cells may produce additional septa before trichospores commence to form. Although the range of trichospore lengths is great (11–30 μm), the size distribution is bimodal, with most spore sizes falling within the shorter end of the curve. The less common larger trichospores appear to develop from longer generative cells. Cultured colonies of *S. culisetae* tend to be more loose and floccose than those of other cultured *Smittium* spp. Colonies of some strains easily break apart on handling because of the weaknesses resulting from the spuriously reproducing thallial cells (generative cells, after sporulation, are devoid of cytoplasm and break readily).

Peterson and Lichtwardt in 1981 (unpublished) found a population of *Simulium vittatum* larvae (Simuliidae) along the dam spillway of Leavenworth County State Lake, Kansas, infested with both *S. culisetae* and *S. culicis* (some hindguts with both species intermixed), and were able to verify the identifications by axenically culturing both species. Other unusual hosts of *S. culisetae* include larvae of the ceratopogonid genus *Dasyhelea* (see next paragraph) and Chironomidae larvae. The latter hosts were populations of bloodworms (*Chironomus* sp.) studied by the author in rock pools of the Georges River near Cambelltown, N.S.W., Australia. These same pools contained an unidentified mayfly (Ephemeroptera) nymph and larvae of *Dasyhelea* sp., both infested with *S. culisetae*.

Smittium inopinatum is considered to be a synonym of *S. culisetae* for the following reasons. Manier et al. (1961) described *S. inopinatum* (as *Rubetella inopinata*, nom. nud.) from the hindgut of *Dasyhelea lithotelmatica* larvae found in southern France. [This ceratopogonid species is also a host of *Carouxella scalaris* (Harpellaceae).] The single character that differentiates *S. inopinatum* from *S. culisetae*, other than the host, is the reported irregular, often dark, pseudorhizoidal branches at the base of thalli. This general form of growth, although not identical to that illustrated by Manier et al., may be present when dense clumps of *S. culisetae* grow in some hosts. More importantly, the author has isolated *S. culisetae* from *Dasyhelea* sp. in southern France, and, in addition to morphological identity, this isolate (FRA-7-1) has been shown by Sangar et al. (1972) to have a strong serological affinity to other isolates of *S. culisetae*, including the type species. The author also has found and cultured *S. culisetae* in a species of *Dasyhelea* (not *D. lithotelmatica*) taken from the same rock pools in Australia referred to in the preceeding paragraph. Thus, it is clear that *Dasyhelea* spp. are occasionally hosts of *S. culisetae*, and neither host differences nor morphology can be the basis for maintaining the species *S. inopinatum*.

Smittium culisetae has trichospores that resemble the rarer mosquito pathogen, *S. morbosum*. Trichospores of *S. morbosum* tend to be more ellipsoidal than *S. culisetae*, but reference to other descriptive characters must be used to insure proper identification.

Zygospores of *S. culisetae* have been found in only two populations of mosquito larvae (*Aedes vexans* Meigen) collected by Williams (1983) and his assistant near Kearney, Nebraska in 1979. Williams was able to isolate the fungus, but no zygospores formed *in vitro;* nor was he able to find zygospores in larvae of *Aedes aegypti* (L.) infested in the laboratory, or when grown *in vitro* with two other *S. culisetae* isolates. The near median and perpendicular position of the collar on the zygospore body is not typical of other zygospore-producing *Smittium* spp., but is close to that described for the type species, *S. arvernense*.

References: Lichtwardt, 1964; Manier et al., 1961; Manier et al., 1964; Manier, 1969b (1970b); Clark et al., 1963; Farr, 1965; Farr and Lichtwardt, 1967; Chapman, 1966; Müller–Kögler, 1971; Williams and Lichtwardt, 1972; Sangar et al., 1972; Sangar and Dugan, 1973; Patrick et al., 1973; Preisner, 1973; El–Buni and Lichtwardt, 1976a, 1976b; Moss and Young, 1978; Cerniglia et al., 1978; Starr et al., 1979; Sweeney, 1981; Williams, 1983.

□ *Smittium dimorphum* Lichtwardt & Williams, 1983a

Thalli producing two trichospore types: (1) long–ellipsoidal, 38–50 × 5.5–6.5 μm with a collar (2.5–)5 μm long, narrowed at the posterior end; (2) oval, 10–12 × 5.5–6.5 μm with a collar 2.5–4 μm long. Zygospores biconical, 71–96 × 13–14 μm, with an obliquely angled collar 12–17 μm long by 5–7 μm diam connnected to the zygospore wall one–third to one–quarter the length from one end, with a single appendage.

Illustration: Fig. 11.17.

Host: In hindgut of *Boreoheptagyia lurida* (Garrett) larvae (Diptera, Chironomidae).

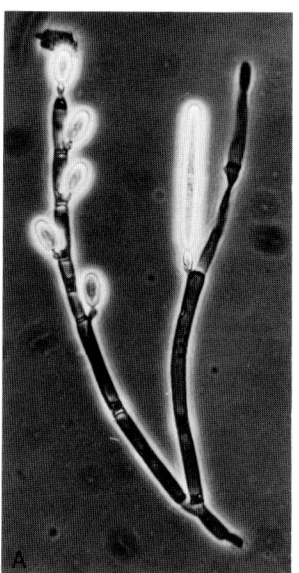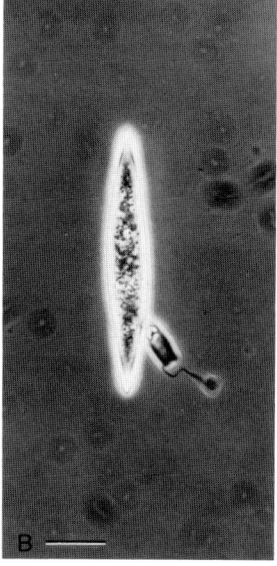

FIG. 11.17. *Smittium dimorphum*. A. Trichospores of two sizes and shapes borne on interconnecting branchlets. B. Zygospore. Scale bar = 20 μm for both figures. Fig. 11.17A reprinted by permission from Mycologia, vol. 75: 731, Copyright 1983, R.W. Lichtwardt and The New York Botanical Garden.

Distribution: Several fast–flowing streams in and near Glacier National Park, Montana, U.S.A.

This distinctive species and *Stachylina pedifer* (in the peritrophic membrane) are the only Harpellales known in larval Boreoheptagyiini. Both species often grow simultaneously in the same host specimens. Dimorphic trichospores also occur in *S. alpinum,* but the two species are easily distinguished by their zygospores and by the morphology of the larger trichospores, those of *S. dimorphum* being longer and somewhat narrower.

Reference: Lichtwardt and Williams, 1983a.

☐ *Smittium elongatum* Lichtwardt, 1972

Thalli profusely branched near base, wider hyphae often verticillately branched; holdfast inconspicuous. Fertile branches about 3 µm diam bearing long–ellipsoidal trichospores (20–)34(–44) × 3–6 µm, with a collar 2–4 µm long and a single, sometimes spiraled, appendage. Zygospores unknown.

Illustrations: See Lichtwardt, 1972, Figs. 56–58.

Hosts: In hindgut of larval *Diamesa* near *nivoriunda* Fitch, and possibly also *Cricotopus* sp. (Diptera, Chironomidae).

Distribution: Stream draining Washington Gulch near Crested Butte, Colorado, U.S.A.

Reference: Lichtwardt, 1972.

☐ *Smittium gigasporus* Williams & Lichtwardt, 1984

Thallus base often branching prolifically with a number of lateral branches that rebranch once. Trichospores 1–2, rarely 3, per fertile branch, elongate–ellipsoidal to subcylindrical, 50–60 × 7–8.5 µm, collar 10–14 × 5 µm narrowing toward the end proximal to the generative cell, with a helical fine appendage up to 10 times the spore length. Zygospores unknown.

Illustrations: See Williams and Lichtwardt, 1984, Figs. 9–15.

Host: In hindgut of larval *Pagastia* sp. (Diptera, Chironomidae).

Distribution: In algal growth on rocks and from wood in streams flowing into Hungry Horse Reservoir, Montana, U.S.A.

Reference: Williams and Lichtwardt, 1984.

☐ *Smittium incrassatum* Y. Kobayasi, 1971 (*in* Kobayasi et al., 1971)

Thalli 100–130 µm long, prolifically branched near the base, branches 4–7 µm diam, often constricted at septa. Trichospores 1–2 per fertile branch, ellipsoidal, 12–14 × 7–9 µm with a thick and finely punctate wall, developing from a 7–10 × 3–4 µm outgrowth of the generative cell; upon detachment trichospores bearing a short collar and a single long appendage. Zygospores unknown.

Illustrations: See Kobayasi et al., 1971, Fig. 15.

Host: In hindgut of Chironomidae (Diptera) larvae.

Distribution: On surface of submerged pebbles in a cold stream, Angmagssalik, Greenland.
Reference: Kobayasi et al., 1971.

☐ *Smittium longisporum* Williams, Lichtwardt & Peterson, 1982

Trichospores long fusiform–ellipsoidal, (40–)46(–55) × (6–)8(–10) μm, with a long, well-defined appendage that often has a zigzag appearance; collar (10–)13(–17) × 4 μm, slightly bulged centrally. Usually 1–2 trichospores per fertile branch, produced on long generative cells. Zygospores fusiform–biconical, (102–)110(–113) × (13–)15(–17) μm, collar 28–32 × 5 μm located near lower tip of the zygospore; single appendage several times longer than the zygospore.
Illustration: Fig. 11.18.
Hosts: In hindgut of *Cricotopus* sp. and other Chironomidae (Diptera) larvae.
Distribution: Flowing streams in Kansas, Missouri, and Nebraska, U.S.A.; small stream draining into north shore of Lake Torneträsk, Sweden.
The presently known, disjunct distribution of *S. longisporum* suggests that it remains to be found in other sites as well.
Reference: Williams et al., 1982.

☐ *Smittium macrosporum* Y. Kobayasi, 1969 (*in* Kobayasi et al., 1969)

Axial hyphae several, 6.5–13 μm diam at base, arising from a short-cylindrical basal cell 6.5–9 μm diam, branching variously. Fertile branches

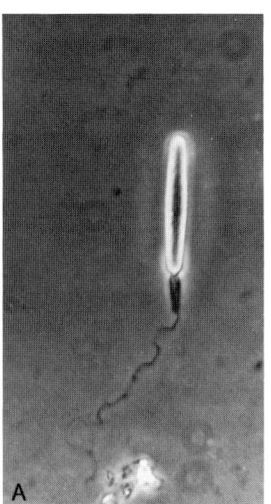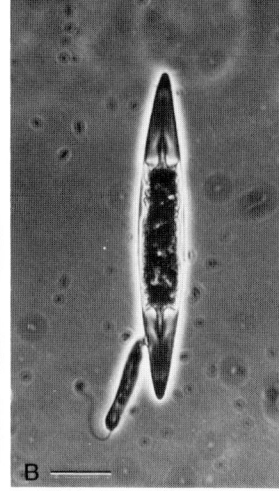

FIG. 11.18. *Smittium longisporum*. A. Trichospore. B. Zygospore. Scale bar = 20 μm for both figures.

3–6.5 μm diam, with 1, rarely 2, long generative cells producing outgrowths 13–20 × 2.5–4 μm bearing fusiform–ellipsoidal trichospores (26–)31(–42) × (6–)6.5(–7.5) μm with a finely or coarsely verrucose surface; upon detachment trichospores have a cylindrical collar 10–18 μm long and bear a single appendage. Zygospores unknown.

Illustrations: See Kobayasi et al., 1969, Fig. 6 and Plate 2, Figs. A–D.
Host: In hindgut of Chironomidae (Diptera) larvae.
Distribution: Alaska, U.S.A.
Reference: Kobayasi et al., 1969.

☐ *Smittium megazygosporum* Manier & Coste, 1971

Thalli in tufts, verticillately branched, branches 3–6 μm diam. Trichospores nearly cylindrical with a slight median swelling, 36–49 × 3.5–5 μm, with a campanulate collar 3.5–4 × 5–6 μm and a prominent single appendage. Zygospores fusiform, 110–150 × 10–12 μm, with a collar 9–10 × 6 μm located on the lower fourth of the zygospore.

Illustrations: See Manier and Coste, 1971, Figs. 15–17.
Host: Larval hindgut of *Syncricotopus rufiventris* Meig. (Diptera, Chironomidae).
Distribution: Ephemeral stream of Lirou near Roman bridge bordering route D 112, Department of Hérault, France.

The name *S. megazygosporum* was first used by Coste in an unpublished dissertation (Coste–Mathiez, 1970). The description in that dissertation apparently included collections of both *S. megazygosporum* and *S. bisporum;* the two species were separated and validly published as two species by Manier and Coste in 1971.

References: Manier and Coste, 1971; Coste–Mathiez, 1970.

☐ *Smittium morbosum* Sweeney, 1981a

Thalli forming a dense clump 0.5–2 mm diam, branches 2.5–4.5 μm diam. Generative cells 4–8 μm long, producing narrowly ellipsoidal trichospores (10–)15(–18) × (3.5)–4(–4.5) μm, collar slightly flared, 0.5–2.5 × 1–2 μm. Zygospores unknown.

Illustration: Fig. 11.19.
Host: In anterior hindgut of larval *Anopheles hilli* Woodhill & Lee, and possibly *Anopheles annulipes* Walker (Diptera, Culicidae). May be pathogenic to larvae, penetrating into anterior midgut and occasionally the Malpighian tubules. May occur in adults.
Distribution: Australia.

The unusual pathogenicity of *S. morbosum* has been reviewed in Chapter 8. The fungus has been cultured axenically by Sweeney. *Smittium morbosum* is most likely to be confused with *S. culisetae,* for the morphological differences in the trichospores of these two mosquito–inhabiting species are subtle: those of *S. culisetae* have a wider range in size and are more

consistently wider below the middle of the spore body. The branching of *S. culisetae* is more often verticillate, and colonies in culture tend to be more floccose than those of *S. morbosum*. Refer to the description of *S. culisetae* for other differences.
Reference: Sweeney, 1981a, 1981b.

☐ *Smittium mucronatum* Manier & Mathiez ex Manier, 1970b (1969b)
[= *Smittium mucronatum* Manier & Mathiez, 1965, nom. nud.]

Trichospores elongate–ellipsoidal, 33–37 × 6.5–7 μm, with a minute apical nipple, collar cylindrical to campanulate, 7.5–9 × 3.3–4 μm; single appendage long and fine. Zygospores biconical, 44–60 × 11–13.5 μm, collar 11–13.5 × 3.4–5.5 μm attached to lower third of zygospore, bearing a single fine appendage.
   Illustrations: See Manier and Mathiez, 1965, Figs. IV–VI; Lichtwardt, 1976b, Figs. 24.2, 24.3.
   Host: In hindgut of larval *Psectrocladius sordidellus* (Zett.) Edw. (Diptera, Chironomidae).
   Distribution: In waters of ancient quarries of Vendargues, and on rocks in large shallow pond north of St. Gély–du–Fesc, Department of Hérault, France.
   Smittium mucronatum has been cultured axenically by Lichtwardt (in 1968) and Coste–Mathiez (1970). It was indistinguishable serologically from *S. culicis* (Sangar et al., 1972), a species common in mosquito larvae in the general region where *S. mucronatum* has been found. The trichospores of *S. mucronatum* are more elongate than those of *S. culicis,* and characteristically have a minute nipple at the tip of the trichospore, but that structure is not readily seen on all trichospores in cultured material. An electron microscopic study revealed that the nipple ("mucron") is a thickening of the outer trichospore wall, and that the long trichospore appendage is coiled and twisted within the region of the collar before spore release. In cross section, the appendage shows concentric electron–opaque rings (Manier and Coste–Mathiez, 1968; Coste–Mathiez, 1970), in this respect resembling *S. culicis* (Moss and Lichtwardt, 1976).
   Smittium mucronatum is capable of infesting mosquito larvae under laboratory conditions (see Chapter 6), but the infestation is not persistent, and trichospores and zygospores formed within mosquito guts may be abnormal (Manier and Mathiez, 1965; Coste–Mathiez, 1970; Williams and Lichtwardt, 1972a). The presumption is that the fungus is not a natural commensal in the gut of mosquito larvae. No zygospores have been found or induced in axenic cultures. El–Buni (1975) and El–Buni and Lichtwardt (1976a, 1976b) have studied some facets of the physiology of this species in axenic culture (see Chapter 9). The sole sterol produced by *S. mucronatum* was desmosterol (Starr et al., 1979).
   References: Manier and Mathiez, 1965; Manier and Coste–Mathiez, 1968; Manier, 1969b (1970b); Coste–Mathiez, 1970; Williams and Licht-

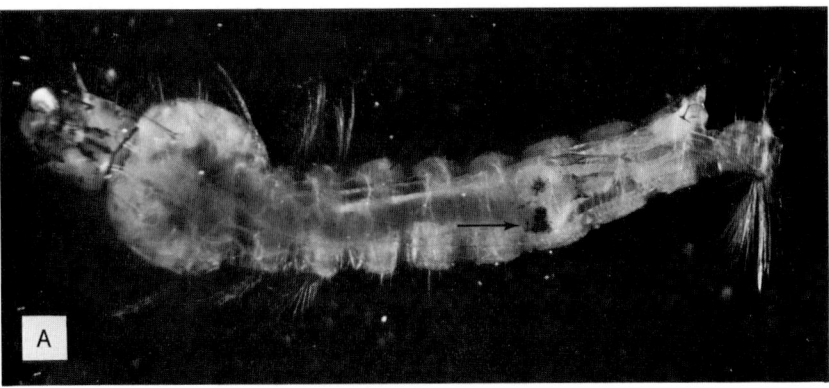
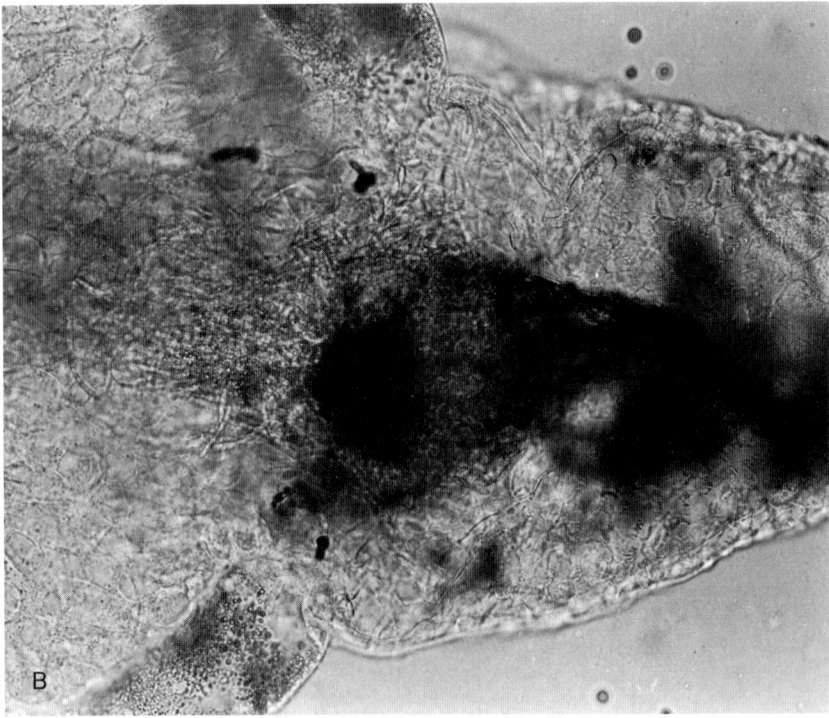

Fig. 11.19. *Smittium morbosum.* A. Dead *Anopheles hilli* larva with a blackened region in the sixth abdominal segment (arrow), an indication of infection by *S. morbosum*. B. Fungus growing in the pyloric chamber and posterior midgut of an *A. hilli* female. C. Axenic culture of the fungus; scale bar = 20 μm. From Sweeney, 1981a, by permission of the author, who kindly provided the photographs, and Trans. Brit. Mycol. Soc.

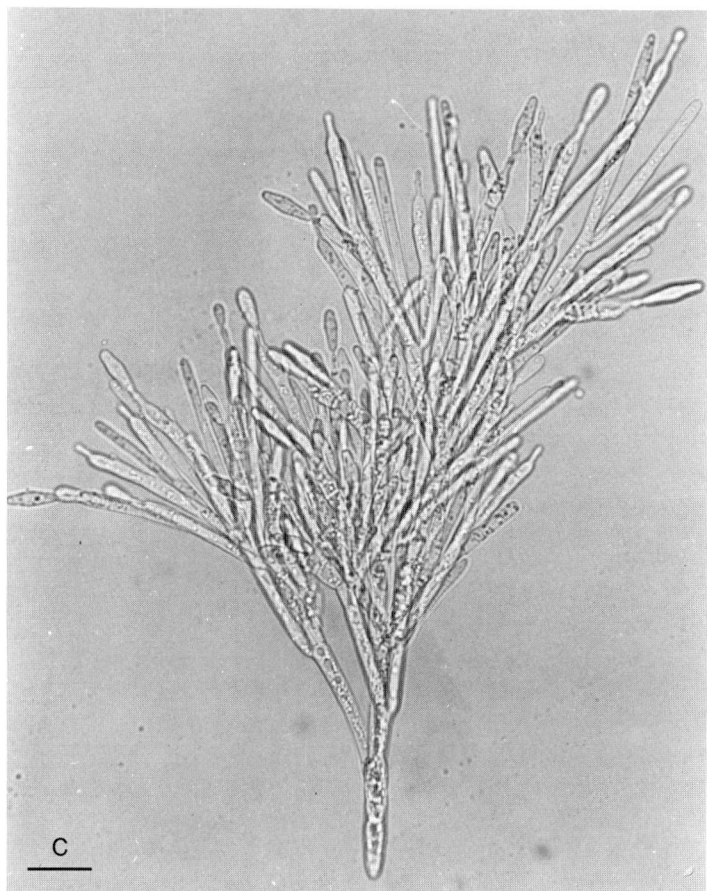

wardt, 1972a; Sangar et al., 1972; Preisner, 1973; El–Buni, 1975; El–Buni and Lichtwardt, 1976a, 1976b; Starr et al., 1979.

☐ *Smittium orthocladii* Manier, 1970b (1969b), *emend.* Lichtwardt, 1984a

Thalli compactly branched, with basal branches of smaller thalli often fascicled and sometimes enveloped in a brownish gelatinous sheath. Generative cells 2–10 per fertile branch. Trichospores either long–ellipsoidal, (25–)30(–40) × (6–)7(–8) μm with a cylindrical collar 5–10 μm long, or small and oval, 8–10 × 5–6 μm with a cylindrical collar 5–10 μm long. Zygospores biconical, subcylindrical in the middle, (81–)87(–98) × (9–)10(–11) μm, with a collar 12–18 μm long attached close to one end of the zygospore, and bearing a single appendage.

Illustration: Fig. 11.20.

Hosts: In hindgut of larval *Orthocladius* spp., *Diamesa* sp., and other lotic Chironomidae (Diptera).

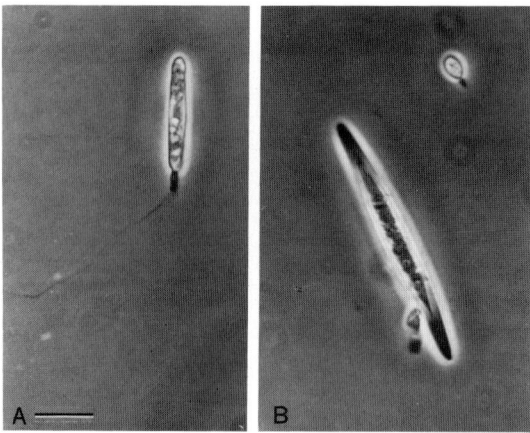

FIG. 11.20. *Smittium orthocladii*. A. Elongate trichospore. B. Zygospore and small, oval trichospore. Scale bar = 20 μm for both figures.

Distribution: Department of Hérault, France; northwestern Montana, U.S.A.

Manier (1969b) validly published the name *S. orthocladii*, but erroneously as a new combination based on the illegitimate species, *Rubetella orthocladii* Manier & Mathiez, 1965 *(nom. nud.)*. It was discovered in larvae of *Orthocladius* gr. *rubicundus* Meigen from an ephemeral stream. Lichtwardt's collections of this species were from a small alpine stream near Logan Pass, Glacier National Park (in 1975), and Doris Creek draining into Hungry Horse Reservoir (in 1977) in Montana, from midge larvae belonging to at least two subfamilies. The emended description provided information on zygospores and the small oval trichospores seen by Manier and Mathiez but interpreted as younger stages of development. The morphology of the regular larger trichospores are virtually identical in the European and North American collections. Like *Smittium dimorphum*, both trichospore types may occur on the same thallus. The fascicled appearance of the thallial base may be evident only in smaller thalli. Larger thalli form a dense, radiating growth such that the basal portion cannot be readily seen. Lichtwardt saw no brownish sheath around the basal branches in his specimens. The dimensions and shape of the longer trichospores with their long collar distinguish this *Smittium* species from others even in the absence of the distinctive zygospores.

References: Manier, 1969b (1970b); Manier and Mathiez, 1965; Coste–Mathiez, 1970; Lichtwardt, 1984a.

☐ *Smittium ouseli* Williams & Lichtwardt, 1984

Thalli producing 1–2(–3) trichospores per fertile branch. Trichospores long–ellipsoidal, 21–25 × 6.5–7.5 μm with a collar 5–9 μm long. Zygospores biconical, 77–88 × 12–15 μm, with an obliquely angled collar connected to the zygospore wall one-third to one-fourth the length from one end, bearing a single appendage upon release.

Illustrations: See Williams and Lichtwardt, 1984, Figs. 16–21.

Host: In hindgut of larval *Eukiefferiella* sp. (Diptera, Chironomidae).

Distribution: Mountain stream just south of Glacier National Park, Montana, U.S.A.

Reference: Williams and Lichtwardt, 1984.

☐ *Smittium pennelli* Lichtwardt, 1984a

Thalli up to 300 μm long, sparsely branched, mucilaginous basal cell tapered to a point. Trichospores in series of 2–4, long fusiform–ellipsoidal, (31–)41(–55) × (6–)7(–8) μm, with a short rounded collar 3.4–4 μm long. Terminal trichospores often produced subapically. Zygospores unknown.

Illustration: Fig. 11.21.

Hosts: In hindgut of larval *Prosimulium exigens* Dyar & Shannon, *P. onychodactylum* Dyar & Shannon, *Simulium defoliarti* Stone & Peterson, and *Simulium* sp. (Diptera, Simuliidae).

Distribution: Lotic habitats in Rocky Mountain National Park, Colorado, and northwestern Montana, U.S.A.

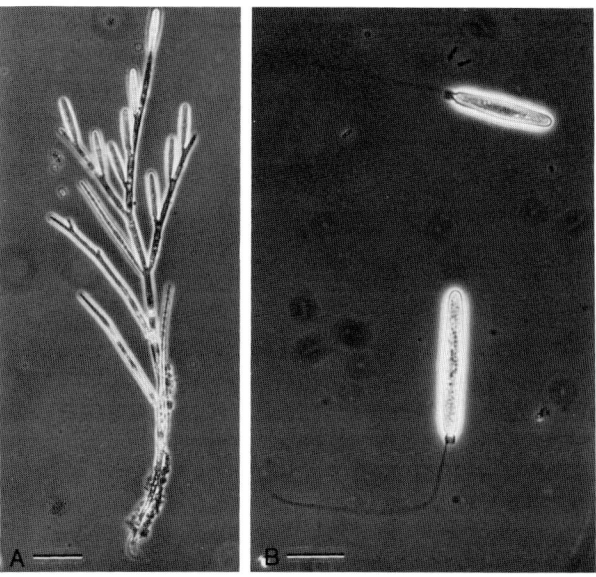

FIG. 11.21. *Smittium pennelli*. A. Entire thallus with a typical mucilaginous base. B. Trichospores. Scale bars: A, 40 μm; B, 20 μm. Fig. 11.21A from Lichtwardt, 1984a, by permission of Mycotaxon.

The thallus of *Smittium pennelli* somewhat resembles that of species of *Pennella*, which also grow in Simuliidae larvae, but the trichospores are clearly a *Smittium*. Thalli of *S. pennelli* sometimes are found growing together in bundles.
Reference: Lichtwardt, 1984a.

□ *Smittium pusillum* Manier & Coste, 1971

Thalli usually 200–300 μm long, attached to host cuticle by a pseudorhizal base; branches 4–6 μm diam. Trichospores 4–6 per fertile branch, ellipsoidal, 14–22 × 2.5–3 μm, collar 2–3.5 × 1.2–1.5 μm. zygospores fusiform, 50–60 × 6–7 μm, collar 5–6 × 2–3 μm.
  Illustrations: Fig. 7.14B; also see Manier and Coste, 1971, Figs. 9–11.
  Host: In hindgut of *Procladius* sp. (Diptera, Chironomidae) larvae.
  Distribution: In still waters of the ancient quarries of Vendargues, Department of Hérault, France.
  References: Coste–Mathiez, 1970; Manier and Coste, 1971.

□ *Smittium simulii* Lichtwardt, 1964a
[= *Rubetella simulii* Manier, 1963b, *nom. nud.*]

Basal cell often swollen around a refractive holdfast and giving rise to several main branches usually without verticillate branching. Terminal fertile branches producing usually 4–6 generative cells. Trichospores cylindrical but swollen in the middle, (16–)23(–30) × (3–)5(–7) μm, collar cylindrical, (1.5–)3.5(–4.2) μm long; single appendage relatively short. Zygospores unknown.
  Illustrations: Figs. 3.1, 7.19, 9.7, 11.22.

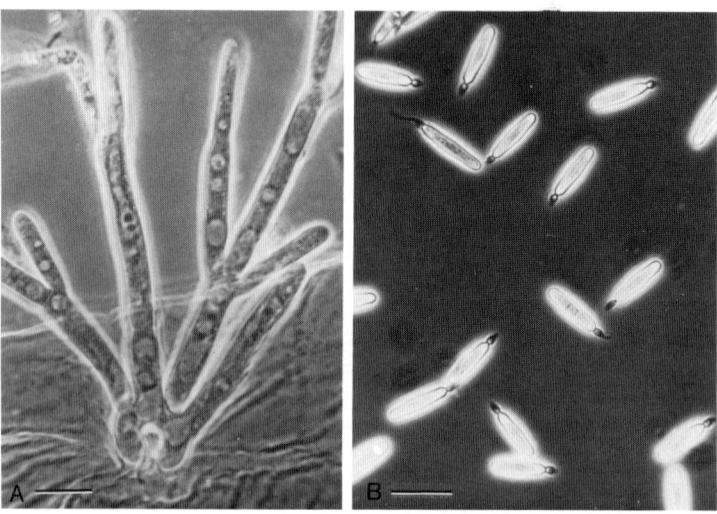

FIG. 11.22. *Smittium simulii*. A. Characteristic base of a young thallus. B. Trichospores. Scale bars = 20 μm.

Hosts: In hindgut of many species of larval Simuliidae and Chironomidae (Diptera); more rarely in Culicidae and Tipulidae larvae.

Distribution: Streams, ditches, and drainpipes with running water [or still water with Culicidae larvae in Japan] in many states of the U.S.A., in Japan, Australia, New Zealand, England, Sweden, and France.

Smittium simulii is relatively common in lotic midge and blackfly larvae in the U.S.A. (California, Montana, Wyoming, Minnesota, Kansas). Manier in 1963(b), unknown to the author at the time, described *Rubetella simulii (nom. nud.)* from several species of *Simulium* larvae collected from sites north of Montpellier, France; her species appears to be *S. simulii*. The author, with Drs. Y. Kobayasi and H. Indoh, collected *S. simulii* in several regions of Japan, the most unusual find coming from mosquito larvae. Williams and Lichtwardt in 1982 (unpublished) found what appeared to be *S. simulii* in larvae of *Elliptra* sp. (probably *E. astigmatica* Alexander) (Tipulidae) from Step Falls, part of a stream near route US 2 draining from Tranquil Basin, south of Glacier National Park, Montana, U.S.A. A number of axenic isolates have been obtained from blackflies, midges, and (two) mosquitoes from the U.S.A. and Japan. The author has found *S. simulii* in the endemic blackfly *Austrosimulium australense* (Shiner) in New Zealand; and in Australia in various endemic Simuliidae, in *Chironomus* sp., and in at least two genera of Culicidae. In Queensland, Australia, *S. simulii* was found in a predaceous mosquito, *Culex halifaxii*. An immunoelectrophoretic study by Sangar et al. (1972) using seven cultures representing three host families showed that they are serologically related, thus confirming the morphological basis for speciation. Williams and Lichtwardt (1972a) found that only one of four isolates obtained from blackfly and midge larvae would infest mosquito larvae *(Aedes aegypti)*, whereas the other three infested 1st instars slightly and the infestation did not persist through moltings to the 4th instar. It would seem, therefore, that mosquitoes are infestable only by some strains. *Smittium simulii* cultures have been used in several physiological and experimental studies as well (see Chapter 9). The type culture is CAL-8-1, isolated from a larva of *Simulium argus* (Will.).

References: Lichtwardt, 1964a; Manier, 1963b, 1969b (1970b); Chapman, 1966; Williams and Lichtwardt, 1972a; Sangar et al., 1972; Preisner, 1973; El-Buni and Lichtwardt, 1976a, 1976b; Cerniglia et al., 1978; Starr et al., 1979.

□ *Smittium typhellum* Manier & Coste, 1971

Thalli up to 1.2 mm long, from the basal region producing several main branches about 6 µm diam from which arise verticillate secondary branches about 4 µm diam. Generative cells short (7–10 µm), 5–6 per fertile branch; producing nearly cylindrical trichospores with a slight median swelling, 25–30 × 3–3.5 µm, with a narrow collar 3–3.5 × 1.5–2 µm. Zygospores unknown.

Illustrations: See Manier and Coste, 1971, Figs. 12–14.

Host: Larval hindgut of *Chironomus* sp. (*plumosus* complex) (Diptera, Chironomidae).

Distribution: Small pond at junction of routes N 109 and D 5 E, Department of Hérault, France.

Manier and Mathiez' (1965) Fig. 7, published under the name of *Typhella choanifera* Manier, 1962a *(nom. nud.)* [= *Genistella choanifera* Tuzet & Manier, 1953 *(nom. nud.)*], is identical to Manier and Coste's (1971) Fig. 14. Manier (1969, p. 622) stated that *Typhella (Genistella) choanifera*, which came from *Chironomus plumosus* L. larvae, was probably a *Smittium* sp., but she did not validate that species due to lack of a sufficient description.

References: Manier and Coste, 1971; Manier and Mathiez, 1965.

Spartiella

Tuzet & Manier ex Manier, 1968
[= *Spartiella* Tuzet & Manier, 1950a, *nom. nud.*]

Trichospores obpyriform, without a collar, with a single appendage. Zygospores attached perpendicularly and medially to the zygosporophore, with a collar upon detachment. In hindgut of Baetidae (Ephemeroptera) nymphs. Monotypic.

Type species: *Spartiella barbata* Tuzet & Manier ex Manier.

The new genus and species were invalidly described by Tuzet and Manier (1950a), with further observations published by the two authors in 1953. Manier (1962b) later provided a generic Latin diagnosis and a more accurate description. The genus and species became validated by Manier only in 1968 upon citing a nomenclatural type (Art. 37); she also provided a Latin description of the species in that paper, which was not necessary (Art. 42).

☐ *Spartiella barbata* Tuzet & Manier ex Manier, 1968
[= *Spartiella barbata* Tuzet & Manier, 1950a, *nom. nud.*]

Thallus up to 1 mm long. Mature basal cell often branched, 10–12.5 μm diam with several bulbous swellings around the zone of attachment to the host cuticle. Terminal branches about 5 μm diam. Obpyriform trichospores 5–10 per fertile branch, 22–27 μm long by 7.5–10 μm at their widest diameter, 3.2–5 μm diam nearer the apex, without a collar. Single trichospore appendage often with a small spherical knob near the trichospore base. Zygospores 25–30 × 6–7.5 μm, attached medially and perpendicularly to the zygosporophore, upon detachment retaining a collar 3–3.6 μ long by 2.5–3.3 μm wide. Type species.

Illustrations: Figs. 11.7, 11.8; also see Manier, 1962b, Figs. I–V.

Hosts: In hindgut of *Baetis rhodani* Pictet, *B. gemellus* Eaton, and other Baetidae nymphs (Ephemeroptera).

Distribution: Streams (le Lez, la Mosson), Department of Hérault, France; England; Wales.

Moss and Lichtwardt (unpublished) collected *Spartiella barbata* in

Baetidae from the Lake District of England (stream draining from Little Langdale Tarn and River Duddon), and in Wales (River Ystwyth). The appendage of this species characteristically has a small spherical knob just below the base of the spore. The appendage often does not unfurl completely after release of the trichospore, and may form a tangled clump at the base of the spore; however, when it unravels it can be seen to consist of a very fine and extremely long filament. Trichospores of *Pteromaktron protrudens* (also from Baetidae) have a knobbed, long single appendage somewhat resembling that of *Spartiella barbata*, but the two species are otherwise quite distinct.

References: Manier, 1950 (1951), 1962b, 1968; Tuzet and Manier, 1950a, 1953.

Stipella

Léger & Gauthier, 1932

Base of thallus attached to hindgut cuticle by mucilaginous secretion. Trichospores cylindrical, without a collar, bearing 3(–4) equal appendages. Biconical zygospores attached perpendicularly to the zygosporophore. In hindgut of larval Simuliidae (Diptera). Monotypic.

Type species: *Stipella vigilans* Léger & Gauthier.

The name *Stipella* is conceivably a homonym of *Stypella* Möller 1895 (Tremellales). However, the General Committee on Botanical Nomenclature, when asked for an opinion, was divided in its vote on this issue: 8 to 6 that the names should be treated as variants (Voss, 1973). As a consequence of this split opinion, I have elected to maintain the name *Stipella* for the trichomycete genus.

□ *Stipella vigilans* Léger & Gauthier, 1932

Thalli up to 1 mm long, consisting of a coenocytic principal cell 6–10 μm diam with a simple or forked verrucose base producing a mucilaginous adhesion substance, and branches 3–4 μm diam. Trichospores cylindrical, 37–80 × 3–5 μm, without a collar, bearing 3(–4) broad (petaloid) appendages, often produced from a long series of generative cells. Homothallic; zygospores 80–105 × 15–18 μm, attached perpendicularly to a zygosporophore ~27 μm long × 15 μm wide arising from one of the conjugated cells. Type species.

Illustrations: Figs. 11.7, 11.8; also see Manier, 1963b, Figs. I–III; Moss, 1970, Figs. 5–10.

Hosts: In hindgut of various larval Simuliidae (Diptera), including *Simulium equinum* L., *S. ornatum* Meig., *S. variegatum* Meig., and *S. bezzii* Corti.

Distribution: Streams in several French Departments (Dauphiné, Hérault, Puy–de–Dôme), and in England (Bere stream, Dorset).

Nonsporulating thalli of *Stipella* can be confused with the more common genus *Pennella*, because both are found in blackfly hindguts attached by a mucilaginous secretion surrounding the base of the principal cell. The genera are distinguishable by the larger number of trichospore appendages in *Pennella* spp., and especially by their different zygospore types, those of *Stipella* being Type I and those of *Pennella* Type III. The cylindrical trichospores of *Stipella* with 3 appendages somewhat resemble those of *Glotzia* (from mayfly nymphs), but one of the 3 appendages of *Glotzia* normally is much longer than the other two. The zygospore types of *Glotzia* and *Stipella* also are different. Manier (1955a) described *Stipella vigilans* as sometimes attaching to thalli of *Paramoebidium* sp., but the epiphytic fungus was no doubt a *Simuliomyces*, a genus not then described. Moss (1970) states that *S. vigilans* sometimes protrudes from the rectum of blackfly larvae.

References: Léger and Gauthier, 1932; Tuzet and Manier, 1950a; Manier, 1950 (1951), 1955a, 1963b, 1969b (1970b); Moss, 1970.

Trichozygospora

Lichtwardt, 1972

Trichospores ellipsoidal with a prominent collar from which emanate multiple fine appendages. Biconical zygospores with a prominent collar, attached obliquely to the zygosporophore, having numerous fine appendages upon release. In hindgut of larval Chironomidae (Diptera). Monotypic.

Type species: *Trichozygospora chironomidarum* Lichtwardt.

☐ *Trichozygospora chironomidarum* Lichtwardt, 1972

Branched thalli attached to hindgut cuticle by an inconspicuous holdfast. Trichospores ellipsoidal, 17–29 × 4.5–11 µm, with a bulbous collar 2–13 µm long narrowing at the free end, bearing 5–7 very fine appendages. Zygospores biconical, (58–)65(–69) × (15–)16(–19) µm, bulbous collar 13–19 × 7–9 µm narrowing at the free end, bearing 10 or more fine appendages upon release. Collar attached obliquely ⅕ to ¼ the distance from one end of the zygospore and connected to a zygosporophore that develops without septation from the free end of one of the conjugating branches. Type species.

Illustrations: Figs. 11.7, 11.8, 11.23.

Hosts: In larval hindgut of *Orthocladius* spp., *Cricotopus* sp., and *Diamesa valkanovi* Saether, and other unidentified midges (Diptera, Chironomidae).

Distribution: In fast–flowing streams in Wyoming, U.S.A., northern Sweden, Switzerland, and possibly England.

Trichozygospora chironomidarum was first recorded from a stream in the Grand Teton National Park in the Rocky Mountains of Wyoming, and several years later was found at the mouth of Abiskojåkka, a river draining into Lake Torneträsk located some 125 miles north of the Arctic Circle

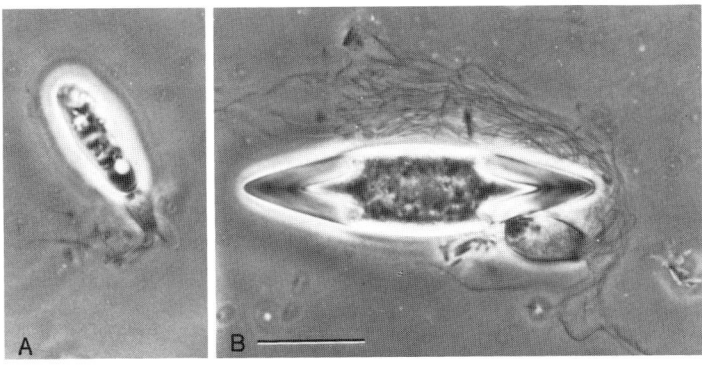

FIG. 11.23. *Trichozygospora chironomidarum*. A. Trichospore. B. Zygospore. Scale bar = 20 μm for both figures. Reprinted by permission from Mycologia, vol. 64: 176, 177, Copyright 1972, R.W. Lichtwardt and The New York Botanical Garden.

in Sweden. Such disjunct habitats suggest that it exists in other Holarctic sites as well. Moss and Lichtwardt, working in the Lake District of northern England in 1980, collected what appeared to be this species (based on trichospore–producing thalli only) in an unidentified species of midge larva. The Switzerland collection (from a stream draining Steingletscher, below Col de Susten) had zygospores as well as trichospores, so its identity was confirmed.

The midge larvae in Sweden, *Orthocladius* sp. and *Diamesa valkanovi*, were abundantly infested, and about a dozen axenic isolates of *T. chironomidarum* were obtained by the author. However, they grew poorly, and the final isolate was lost some 18 months later despite attempts to satisfy its requirements for growth.

Moss and Lichtwardt (1976, 1977) studied the ultrastructure of the trichospores and zygospores of *T. chironomidarum*. The results are discussed in Chapter 7.

References: Lichtwardt, 1972; Moss and Lichtwardt, 1976, 1977.

Zygopolaris

Moss, Lichtwardt, & Manier, 1975

Trichospores elongate–obpyriform, with no collar or appendages. Modified biconical zygospores attached at one pole with axis in line with that of the zygosporophore. In hindgut of Ephemeroptera nymphs. Two species.

Type species: *Zygopolaris ephemeridarum* Moss, Lichtwardt, & Manier.

—Key to *Zygopolaris* species—

1. Trichospores 25–38 μm long .... *Z. ephemeridarum*
1'. Trichospores 44–65 μm long .... *Z. borealis*

□ *Zygopolaris borealis* Lichtwardt & Williams, 1984

Branched thalli up to 1.5 mm long, rarely longer, attached to host cuticle by a well-defined basal holdfast, often showing dichotomous (or trichotomous) branching below the sporulating branchlets. Trichospores elongate–ovoid, (44–)50(–65) × 6–9 μm, borne on sporulating tufts of branchlets each consisting of 3–12 or more generative cells 10–20 μm long, sometimes projecting from the host's anus. Upon release, trichospores without a collar, or with a very short outwardly flared collar, and bearing no appendages. Modified biconical zygospores (74–)79(–84) × 18–22 μm, attached at one pole to a zygosporophore, 29–38 × 13–21 μm.

Illustrations: See Lichtwardt and Williams, 1984b, Figs. 1–8.
Host: In hindgut of larval *Epeorus longimanus* (Eaton) (Ephemeroptera, Heptaginiidae).
Distribution: In streams, Montana, U.S.A.
Reference: Lichtwardt and Williams, 1984.

□ *Zygopolaris ephemeridarum* Moss, Lichtwardt, & Manier, 1975

Branched thalli up to 2 mm in length attached to hindgut cuticle by a short secreted holdfast formed laterally near the base of a slightly swollen basal cell. Main axis of the thallus producing infrequently septate branches 7–10 μm diam, which may project out of the host anus. Sporulating branchlets forming a more or less compact head, each branchlet composed of 3–15 generative cells 8–15 μm long. Trichospore elongate–obpyriform (25–)32(–38) × 5–8 μm, upon release having no collar or appendages (rarely 3 appendages). Modified biconical zygospores (40–)55(–86) × (13–)15(–26) μm, attached at one pole to a zygosporophore 17–31 μm in length. Type species.

Illustration: Fig. 11.24.
Hosts: In hindgut of lotic Ephemeroptera nymphs, especially in Baetidae such as *Baetis bicaudatus* Dodds, *B. tricaudatus* Dodds, and *B. parvus* Dodds; also in Ephemerellidae such as *Ephemerella inermis* Eaton and possibly other species of *Ephemerella*.
Distribution: Mountain streams of Colorado, Wyoming, and Montana, and stream in Seattle, Washington, U.S.A.

Zygopolaris ephemeridarum is relatively common in Rocky Mountain streams in mayfly nymphs of the Baetidae, less commonly in Ephemerellidae. The species is easily identified, but could be confused with the less common and geographically more restricted species, *Z. borealis*, if spore measurements are not made. However, *Z. borealis* lives in Heptageniidae. The two species of *Zygopolaris* have occasionally been found in the same stream but in different families of mayflies.

Both trichospores and zygospores are devoid of well-developed appendages. Trichospores rarely may have a small blob of material at their

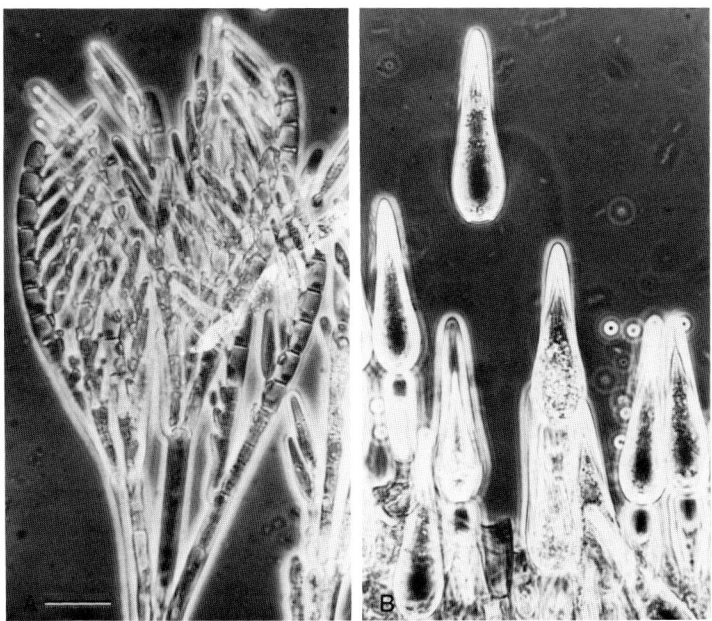

FIG. 11.24. *Zygopolaris ephemeridarum*. A. Trichospores. B. Zygospores. Scale bar = 20 μm for both figures.

base upon release (Moss et al., 1975). Lichtwardt and Williams (1984) reported finding 3 short appendages on a released trichospore of *Z. ephemeridarum* and discrete appendages within the generative cells of unreleased spores. Zygospores have a fibrous substance at their base within the zygosporophore (Moss and Lichtwardt, 1977) which, upon release of the zygospore, may expand into a large, almost invisible skirt of clear material surrounding the base of the zygospore (Lichtwardt and Williams, 1984).

References: Moss et al., 1975; Moss and Lichtwardt, 1977; Lichtwardt and Williams, 1984.

Asellariales

Manier ex Manier & Lichtwardt, 1978 (*in* Lichtwardt and Manier, 1978)
[= Asellariales Manier, 1951 (1950), *nom. nud.*]

Thalli branched, septate, reproducing by fragmentation of the branches into uninucleate arthrospores. Sexual reproduction unknown. Attached by a specialized basal holdfast cell or holdfast apparatus to the hindgut cuticle of Isopoda or Insecta (Collembola).

Consisting of the single family Asellariaceae.

Asellariaceae

Manier ex Manier & Lichtwardt, 1969 (1968)
[= Asellariaceae Manier, 1951 (1950), *nom. nud.*]

Characters same as those of the order. Two genera.

Type genus: *Asellaria* Poisson.

Formerly included in this family, the genus *Trichoceridium* Poisson (1932b), with the single species *T. ramosum,* is not recognized by the author as a good taxon, and is consequently excluded from this treatment. Although recognizing it was incompletely described, the monotypic genus was placed in the Asellariaceae by Manier in 1951 (1950) (also see Manier, 1963a, 1969b, and Manier and Lichtwardt, 1968) solely on the basis of Poisson's report of some disarticulating cells in the branched thallus, a feature of reproduction in the Asellariaceae. The "arthrospores" he described do not appear to have the ontogeny and structure of typical Asellariaceae; furthermore, on occasions the author has seen similar cells of Legeriomycetaceae that break away [e.g., *Genistellospora homothallica* (Lichtwardt, 1972, Fig. 12)]. *Trichoceridium ramosum* has the type of branching and holdfast structure found in many Legeriomycetaceae, and, significantly, its larval host is a species of *Trichocera* (Diptera, Tipulidae). All other Asellariaceae occur in Isopoda or Collembola. Lichtwardt and Williams (unpublished) in 1982 found two species of *Smittium* (Legeriomycetaceae) in cranefly larvae (Tipulidae) in the Rocky Mountains of Montana, U.S.A. Possibly Poisson's fungus was, in fact, a species of *Smittium* in a nonsporulating stage.

—Key to Genera of Asellariaceae—

1. Attached to hindgut cuticle of Isopoda (Crustacea); basal cell usually not branched .... **Asellaria**
1'. Attached to hindgut cuticle of Collembola (Insecta); basal cell usually branched .... **Orchesellaria**

Asellaria

Poisson, 1932a

Branched thalli consisting of mostly coenocytic cells prior to development of uninucleate arthrospores. Basal cell morphologically distinct. Attached to hindgut cuticle of aquatic, marine and terrestrial Isopoda (Crustacea). Six species.

Type species: *Asellaria caulleryi* Poisson.

—Key to *Asellaria* species—

1. Basal cell more or less cylindrical, producing a whorl of distal branches, sometimes branched on lateral wall, attached to host cuticle by many short basal rhizoidal projections ... **A. armadillidii**

1'. Basal cell not more or less cylindrical, unbranched, without basal
   rhizoidal projections 2
2(1'). Basal cell long and tapered to one or more points, with
   longitudinal rows of pectinate outgrowths 3
2'(1'). Basal cell rounded at the base and invaginated to form a
   furrow, without rows of pectinate outgrowths 5
3(2). Basal cell not divided longitudinally, with 2 parallel rows of
   pectinate outgrowths .... **A. caulleryi**
3'(2). Basal cell divided longitudinally, each prong with rows of pectinate
   outgrowths 4
4(3'). The two prongs of basal cell appressed, each bearing one or
   more rows of short pectinate outgrowths; basal cell without
   accessory outgrowths .... **A. aselli**
4'(3'). The two prongs of basal cell somewhat divergent and talon-
   like, each bearing one row of short to long pectinate
   outgrowths; basal cell with accessory talon-like or bulbous
   outgrowths .... **A. unguiformis**
5(2'). Basal cell campanulate, 18–27 μm diam .... **A. gramenei**
5'(2'). Basal cell with a bulbous expansion, 40–50 μm diam ..... **A. ligiae**

□ *Asellaria armadillidii* Tuzet & Mainer ex Manier, 1968
[= *Asellaria armadillidii* Tuzet & Manier, 1953, *nom. nud.*]

Thalli up to 1.1 mm long, consisting of a main axis with lateral branches 7–8 μm diam and secondary branches. Principal (basal) cell more or less cylindrical, 60–165 μm long by 8–17 μm diam, attached at its base to host cuticle by many short rhizoidal projections embedded in a mucilaginous substance. Principal cell producing a whorl of branches at its distal end, sometimes branching at its base above the rhizoidal outgrowths and at other parts of the side wall. Arthrospores 6–23 μm long.

Illustrations: Fig. 11.25; also see Manier, 1963a, Fig. 5.

Host: In hindgut of *Armadillidium simoni* Dollfus, *A. vulgare* (Latreille), and *A. nasatum* Budde–Lund (Isopoda, Armadillidiidae). In laboratory experiments, infestation also established in the following Isopoda: *Chaetophiloscia elongata* (Dollf.) (Oniscidae); *Acaeroplastes melanurus* B.–L., *Porcellio laevis* (Latr.), *Porcellio lamellatus* Verh., *Protracheoniscus occidentalis* Vand. (Porcellionidae).

Distribution: Terrestrial habitats in Department of Hérault, France, and in Kansas, U.S.A.

All of the naturally infested hosts of *Asellaria armadillidii* are species of *Armadillidium*, but Manier (1963a) was able in the laboratory to transmit the fungus successfully to five additional species that were uninfested in their natural habitats, by keeping them in containers with infested *A. simoni* (see Chapter 6). The other two naturally infested isopods, *Armadillidium vulgare* and *A. nasatum*, were collected several times in different sites in Kansas by the author. Those two isopods are also hosts of *Parataeniella*

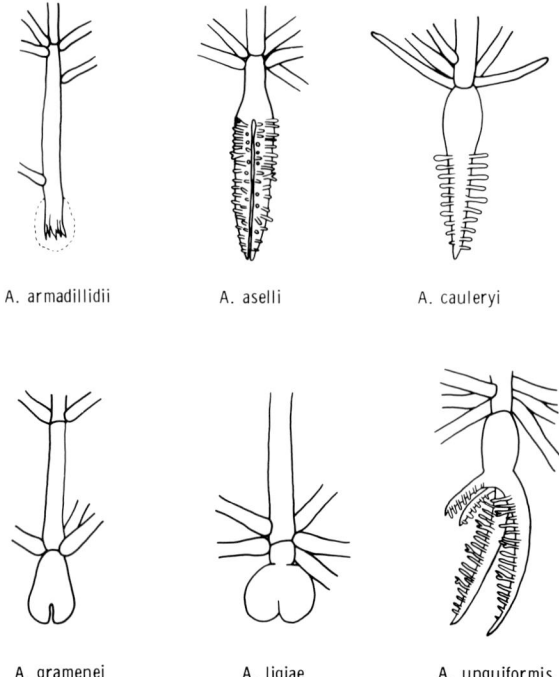

Fig. 11.25. Species of *Asellaria* as distinguished by their basal cells.

armadillidii (Eccrinales). It is likely that *Asellaria armadillidii* is much more widely distributed than present records indicate.

References: Tuzet and Manier, 1953; Manier, 1963a, 1968, 1969b (1970b); Chapman, 1966.

☐ *Asellaria aselli* Scheer ex Moss & Lichtwardt, 1984 (*in* Lichtwardt and Moss, 1984a)
[= *Asellaria aselli* (Sheer) Sheer, 1972a, *nom. nud.*]
[= *Recticharella aselli* Sheer 1944, *nom. nud.*].

Mature thalli up to 2.8 mm long. First cell of the main axis (above the holdfast cell) bearing a whorl of branches that may rebranch; other cells of the main axis often with one or more branches. Basal holdfast cell up to 170 μm long by 37 μm diam, major lower part obovoid and divided into two appressed tapered prongs, each with pectinate rows of short to long (10 μm), sometimes branched, lateral outgrowths. Arthrospores usually (15–)50–93 × (6–)14–20 μm.

Illustrations: Figs. 7.10D, 11.25; also see Lichtwardt and Moss, 1984a, Figs. 10–13.

Host: In hindgut of *Asellus aquaticus* L. (Isopoda, Asellidae).

Distribution: Freshwater ponds, lakes, and streams in north–central Europe (Poland, East Germany, West Germany), and in England.

Scheer (1944) first described this species, under the new generic name

Recticharella (nom. nud.), and later (1972a, 1972b) he reassigned it correctly, but invalidly, to the genus *Asellaria*. The species is widely distributed in north–central Europe and England, and one can assume it is also present in other, unreported localities because of the wide distribution of its host species.

Asellus aquaticus also harbors the morphologically similar (type) species, *A. caulleryi*. The most obvious way the two can be differentiated is by the structure of the basal holdfast cell, that of *A. aselli* being divided into two appressed prongs, each prong with one row of pectinate outgrowths; *A. caulleryi*'s basal cell has no bifurcation, and has two series of pectinate outgrowths. Lichtwardt and Moss (1984) reported that some thalli of *A. aselli*, which they collected in England, had a mucilaginous secretion around the basal cell.

References: Lichtwardt and Moss, 1984a; Moss, 1979; Sheer, 1944, 1972a, 1972b.

□ *Asellaria caulleryi* Poisson, 1932a

Thalli 200–1500 μm long, main axis (above the holdfast cell) ~10 μm diam, from whose basal region arise many branches without secondary branches. Basal holdfast cell coenocytic, 30–140 μm long, up to 20–25 μm diam, tapering basally and bearing on its incurved surface two parallel, pectinate rows of short outgrowths. Arthrospores 40–45 μm long. Type species.

Illustrations: Fig. 11.25; also see Poisson, 1932a, Figs. I–IX.

Hosts: In hindgut of *Asellus aquaticus* L. and *A. meridianus* Rac. (Isopoda, Asellidae).

Distribution: Ponds, and occasionally in streams, in Departments of Pas–de–Calais and Ille–et–Vilaine, France.

This freshwater species does not appear to have been collected by investigators other than Poisson. He provided a lengthy description of its morphology, cytology and development. The isopods were most highly infested when feeding on *Volvox*. When he fed them plant materials such as dead leaves, thalli broke loose and were voided with the excreta, presumably due to abrasive action of those materials. Thalli were not maintained in isopods fed a carnivorous diet such as earthworm fragments.

The holdfast cell of *Asellaria caulleryi* has some resemblance to that of *A. aselli*, which also lives in *Asellus aquaticus*, but the holdfast cell of the latter species is divided into two appressed prongs. *Asellus meridianus*, the other host of *A. caulleryi*, has been found infested also with *A. gramenei*, in southern France.

Reference: Poisson, 1932a.

□ *Asellaria gramenei* Tuzet & Mainer ex Manier, 1968
 [= *Asellaria gramenei* Tuzet & Manier, 1950a, *nom. nud.*]

Thalli up to 1.8 mm long, main axis 15–23 μm diam, consisting of a row of several cells from whose bases arise one or more branches without secondary branches. Basal holdfast cell 40–50 μm long by 18–27 μm diam, campanulate, with the base invaginated to form a furrow. Arthrospores 40–65 μm long.

Illustration: Fig. 11.25.

Host: In arterior hindgut of *Asellus meridianus* Rac. (Isopoda, Asellidae).

Distribution: In streams draining a salt marsh, Department of Hérault, France.

The original descriptions of *Asellaria gramenei* [Tuzet and Manier, 1950a; Manier, 1950 (1951)] identified the host as *Asellus aquaticus*, but subsequent publications called it *A. meridianus*. This isopod is a freshwater species, but was collected from streams with unspecified salinity draining the Salins de Gramenet near Montpellier.

Asellaria gramenei is very similar morphologically to *A. ligiae* from marine isopods (*Ligia* spp.), one species of which, *L. italica*, can be found infested with *A. ligiae* at Palavas on the Mediterranean coast not far from the single reported inland locality for *A. gramenei*. *Asellaria ligiae*, however, is considerably more dispersed geographically than *A. gramenei*. Other than their different host preferences, *A. gramenei* can be distinguished from *A. ligiae* by the shape and dimensions of the basal holdfast cell (it is campanulate and longer than wide), by lesser branching in mature thalli, and by having generally longer arthrospores.

References: Tuzet and Manier, 1950a; Manier, 1950 (1951), 1963a, 1968, 1969b (1970b), 1978.

□ *Asellaria ligiae* Tuzet & Manier ex Manier, 1968
[= *Asellaria ligiae* Tuzet & Manier, 1950a, *nom. nud.*]

Mature thalli much branched, sometimes more than 1 mm long by almost 2 mm wide. Basal branches 13–25 μm diam. Basal holdfast cell 30–50 μm long by 40–50 μm diam, with a spherical, bulbous expansion invaginated at the base to form a furrow. Arthrospores 23–47(–70) μm long.

Illustrations: Frontispiece; Figs. 7.11C, 11.25, 11.26.

Hosts: In hindgut of *Ligia italica* Fab., *L. exotica* Roux, and other *Ligia* spp. (Isopoda, Ligiidae).

Distribution: Probably worldwide. On marine shorelines in Europe (France, Yugoslavia), U.S.A. (North Carolina, Florida, California, Hawaii), and Japan. Also from an undescribed *Ligia* sp. living in a freshwater stream and pool in Hawaii.

Asellaria ligiae is the most widespread member of the Asellariales, and the only marine species. In the author's experience, infestation of different populations varies from very low to very high. Infestation rates may depend upon the density of the isopods as well as the season and the molting frequencies. *Ligia exotica* is found in many oceans and seas, and this may account for the geographic dispersal of the fungus. These isopods, often called the rock louse, scamper about on rocks, pilings, seawalls, etc.,

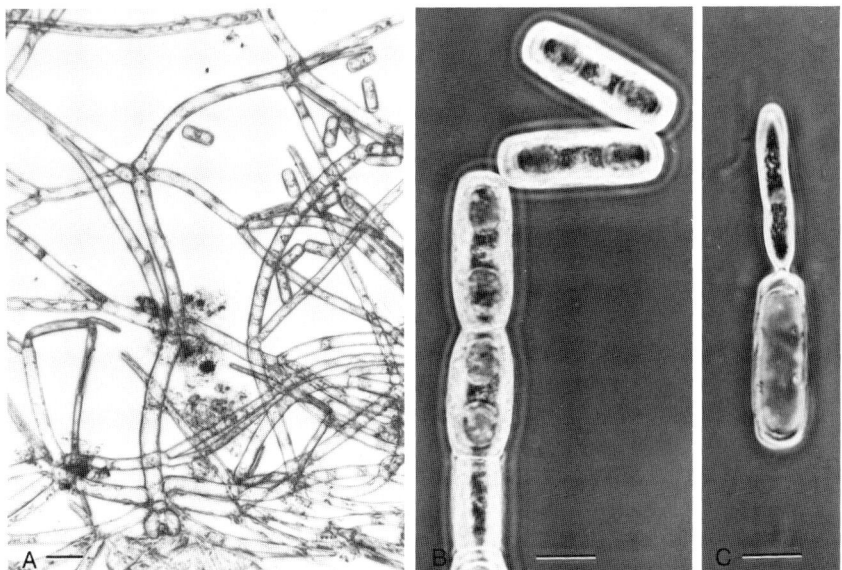

FIG. 11.26. *Asellaria ligiae*. A. Thallus attached to a piece of hindgut cuticle, with some loose arthrospores. B. Arthrospores disarticulating. C. Germinated arthrospore. Scale bars: A, 50 μm; B, C, 20 μm. Fig. 11.26A from Lichtwardt, 1976, by permission of Granada Publishing Ltd.

near the water's edge, but can be considered marine in the sense that some species (e.g., *L. exotica*) can remain submerged in seawater for prolonged periods, if aerated. In Hawaii there exists an apparently undescribed and endemic species of *Ligia* that lives upstream in and near perfectly fresh water; this population was found to be highly infested with *A. ligiae* (see Chapter 4).

Manier (1973b) did an electron microscopic study of *A. ligiae*, and found the septal structure to resemble that found in the Harpellales. Manier (1963a) and Lichtwardt (1973a) germinated arthrospores of *A. ligiae in vitro* and found that they produced a limited exogenous growth that somewhat resembled a trichospore, but these did not detach. Such germinated arthrospores closely resemble the disarticulated generative cell–trichospore unit of *Carouxella scalaris* (Harpellaceae). Conjugating thallial branches and released arthrospores have been observed in *A. ligiae*, but there was no zygospore development (Lichtwardt, 1973a).

The author has observed in heavily infested isopods on several occasions thalli of *A. ligiae* without the usual bulbous basal cell. Such thalli were held to the hindgut by a branching rhizoidal outgrowth that penetrated through the cuticle (Fig. 7.11C), and were interpreted as being branches that had broken away from other normal thalli in the gut and had attached by means of this unusual adventitious growth (see Chapter 7). *Asellaria ligiae* is capable of complete conversion of the thallus to arthrospores, leaving only the basal holdfast cell.

Asellaria ligiae morphologically is most similar to *A. gramenei* from the aquatic isopod *Asellus meridianus*. *Asellaria ligiae* has a basal cell that is more bulbous and usually at least as wide as it is long, and has a more developed branching system.

References: Tuzet and Manier, 1950a; Manier, 1950 (1951), 1963a, 1968, 1973b; Lichtwardt, 1973a, 1976.

☐ *Asellaria unguiformis* Lichtwardt, 1984 (*in* Lichtwardt and Moss, 1984a)

Mature thalli highly branched. Basal holdfast cell consisting of two principal, curved talon–like cell extensions 70–120 μm long and tapered to a point, each with a pectinate series of short to long irregular projections and with accessory talon–like or bulbous outgrowths with irregular projections. Arthrospores usually 60–100 x 10–20 μm.

Illustrations: Fig. 11.25; also see Lichtwardt and Moss, 1984a, Figs. 1–9.

Hosts: In hindgut of *Lirceus hoppinae* (Faxon) and *Caecidotea laticauda* (Williams) (Isopoda, Asellidae).

Distribution: Streams in Arkansas and Florida, U.S.A.

All branches of *A. unguiformis* may disarticulate into arthrospores, leaving only the strikingly complex basal holdfast cell attached to the host cuticle as evidence of a former thallus.

Reference: Lichtwardt and Moss, 1984a.

Orchesellaria

Manier ex Manier & Lichtwardt, 1969 (1968)
[= *Orchesellaria* Manier, 1958, *nom. nud.*]

Branched thalli with a proliferation of cells or branches arising from the base of the original cell, attached to host cuticle by a holdfast structure secreted by one or more basal cells, or by basal rhizoidal cells. Arthrospores arising from disarticulated branches, or terminally. Attached to hindgut cuticle of Collembola (Insecta). Four species.

Type species: *Orchesellaria lattesi* Manier ex Manier & Lichtwardt.

The generic diagnosis provided above is emended so as to encompass species described subsequent to Manier and Lichtwardt's diagnosis.

—Key to *Orchesellaria* species—

1. Attached to host cuticle by several small swollen cells with rhizoidal bases .... *O. podurae*
1'. Attached to cuticle by some other structure 2
   2(1'). Attached to cuticle by a secreted flat shieldlike holdfast; reproducing by single terminal filiform deciduous cells .... *O. pelta*

2'(1'). No flat shieldlike holdfast secretion produced; reproducing
 by disarticulation of many thallial cells 3
3(2'). Base of thallus a cluster of radiating branches; thallial cells slightly
 wider at their distal end .... *O. mauguioi*
3'(2'). Base of thallus with a lateral growth from which arise a series of
 parallel short and long branches; thallial cells more or less
 cylindrical .... *O. lattesi*

□ *Orchesellaria lattesi* Manier ex Manier & Lichtwardt, 1969 (1968)
 [= *Orchesellaria lattesi* Manier, 1958, *nom. nud.*]

Thalli 300–800 μm long by 150–400 μm wide, branches 2–5 μm diam. Basal holdfast complex consisting of a curved filamentous lateral growth from the initiating spore, this lateral growth secreting a holdfast substance and giving rise to a series of parallel short and long thallial branches more or less perpendicular to the initiating spore cell. Arthrospores 20–60 × 2–5 μm, bearing 3–4 serrations on the lateral wall near their apical end. Type species.

Illustrations: Fig. 11.27; also see Manier, 1958, Figs. I–III.
Hosts: In hindgut of *Orchesella villosa* L. (Collembola, Entomobryidae).
Distribution: On water and damp vegetation by water's edge, Department of Hérault, France.

The fungus can be found throughout the hindgut, but in less infested individuals thalli may be present only in the posterior rectum. The unusual serrations that develop on the arthrospores even before they release later produce the characteristic lateral growth described above. This original cell (the arthrospore) may persist as part of the developing thallus, but in a degenerated form.

References: Manier, 1958, 1964c, 1979b; Manier and Lichtwardt, 1968 (1969).

□ *Orchesellaria mauguioi* Manier ex Manier, 1970b (1969b)
 [= *Orchesellaria mauguioi* Manier, 1964c, *nom. nud.*]

Thalli 200–400 μm long by 100–150 μm or more wide, attached to host cuticle by a somewhat bulbous basal region. Initiating spore remains per-

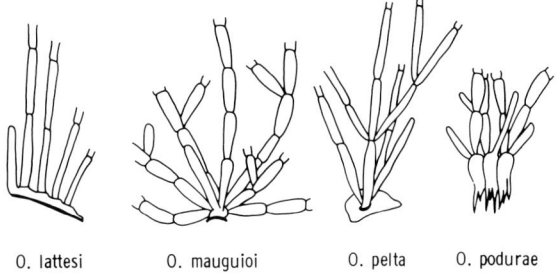

O. lattesi   O. mauguioi   O. pelta   O. podurae

FIG. 11.27. Species of *Orchesellaria* as distinguished by their basal cells.

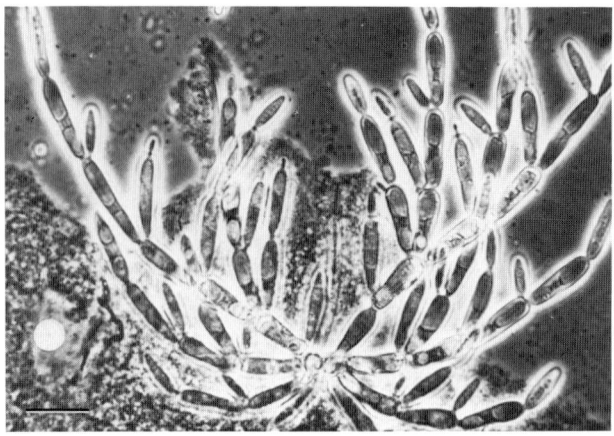

FIG. 11.28. *Orchesellaria mauguioi* thallus. Scale bar = 20 μm.

sistent, producing at its base, directly and indirectly, a cluster of radiating branches that may in turn rebranch. Most thallial cells slightly wider at their distal ends, commonly 20–35 × 4–9 μm, disarticulating to form arthrospores.

Illustrations: Figs. 11.27, 11.28.

Hosts: In hindgut of *Isotomurus palustris* (Müller), *Isotoma* sp., and *Agrenia bidenticulata* (Tullberg) (Collembola, Isotomidae).

Distribution: On surface of lotic and lentic waters, and adjacent vegetation and other damp substrates, in Department of Hérault, France, and in U.S.A. (Washington, Montana).

Orchesellaria mauguioi has been found in the widespread springtail species *Isotomurus palustris* in France and the U.S.A. The fungus was reported by Lichtwardt and Moss (1984a) to be present in Montana in all three of the presently known hosts: *Isotoma* sp. and *I. palustris* from a cattail swamp, and *Agrenia bidenticulata* from rocks at the base of a waterfall.

Moss (1975) found the septal structure of *O. mauguioi* to be of the harpellid type, as it is also in *O. podurae* and *Asellaria ligiae*. Lichtwardt and Moss (1984a) reported finding numerous spherical intercalary, or more rarely terminal, chlamydospores in one springtail.

References: Manier, 1964c, 1969b (1970b), 1979b; Moss, 1975, 1979; Lichtwardt and Moss, 1984a.

☐ *Orchesellaria pelta* Lichtwardt, 1984 (*in* Lichtwardt and Moss, 1984a)

Branched thalli growing singly or in multiple tufts up to 200 μm long, attached to host cuticle by a flat shieldlike holdfast of irregular shape.

Branches 2–6 μm diam, producing terminally single filiform deciduous cells 40–60 × 1.2–2.2 μm.

Illustrations: Fig. 11.27; also see Lichtwardt and Moss, 1984a, Figs. 14–22.

Host: In hindgut of *Hydroisotoma schaefferi* (Krausbauer) (Collembola, Isotomidae).

Distribution: On water and marginal leaf packs in streams in Missouri and Arkansas, U.S.A.

This species of *Orchesellaria* is unusual in two respects: It produces arthrospores that are filiform and develop singly at the ends of branches, and it is attached to the host's cuticle by a secreted holdfast structure that is flat and of irregular shape. Adjacent tufts of thalli may share a common shieldlike holdfast structure. Spherical chlamydospores have been found in some springtails.

Reference: Lichtwardt and Moss, 1984a.

□ *Orchesellaria podurae* Manier, 1979b

Thalli 120–170 μm long by 170–250 μm wide. Basal region attached to host cuticle by several small swollen cells with rhizoidal bases. Initiating spore persistent during thallus development. Thalli much branched, consisting of cells slightly wider at their apical ends, 23–25 × 4–5 μm, which disarticulate basipetally to form arthrospores.

Illustrations: Fig. 11.27; also see Manier, 1979b, Figs. 1, 2 and Plate I.

Host: In hindgut of *Podura aquatica* L. (Collembola, Poduridae).

Distribution: On bordering vegetation and water of two ephemeral streams, Department of Hérault, France.

Manier's 1979(b) description of *O. podurae* included an electron micrographic study that showed the septal structure to resemble the harpellid type, found also in *O. mauguioi* and *Asellaria ligiae* (Asellariales). The fungus has a thallus similar to *O. mauguioi*'s, but has a distinctively different holdfast apparatus.

Reference: Manier, 1979b.

Eccrinales

Léger & Duboscq, 1929a
sensu Manier & Lichtwardt, 1969 (1968)

Thalli coenocytic, unbranched or (rarely) branched at the base. Producing basipetally one or more types of sporangiospores. Attached by a secreted holdfast to hindgut or foregut of Diplopoda, Crustacea, or Insecta.

Consisting of the families Eccrinaceae, Palavasciaceae, and Parataeniellaceae.

Léger and Duboscq's (1929a) concept of the order included the families Eccrinaceae, Arundinulaceae, and Taeniellaceae. The latter two were incorporated into the Eccrinaceae by Manier and Lichtwardt [1968 (1969)], at which time two new families, the Palavasciaceae and Parataeniellaceae, were established.

Terms: *Primary infestation sporangiospores* are released from the gut and germinate only after ingestion by a suitable host; they are usually either thin walled and uninucleate, or thick walled and 1- to 4-nucleate (Fig. 7.20A, B). *Secondary infestation sporangiospores* germinate within the gut where they are produced, thus increase infestation endogenously; they are typically thin walled and 4- to 8-nucleate (Fig. 7.20C). *Spore mother-cells* are the spores that give rise to a new thallus by a germination process; they may degenerate soon after germinating, or they may persist at the apex of maturing thalli as a distinct structure with or without a crosswall separating them from the thallus proper (Fig. 11.34D).

Eccrinaceae

Léger & Duboscq, 1929a
emend. Manier & Lichtwardt, 1969 (1968)

Thalli unbranched, or (in a few species) branched near the base. Two basic types of sporangiospores formed singly in basipetal series of terminal sporangia, usually on different thalli: (1) primary infestation spores that are uninucleate and thin walled, or (usually upon molting of the host) thick walled, with or without appendages, and 1- to 4-nucleate; and (2) secondary infestation spores that are multinucleate and thin walled. Other cell types may be produced. In hindgut or foregut of Diplopoda, Crustacea and Insecta. Twelve genera.

Type genus: *Enterobryus* Leidy.

—Key to Genera of Eccrinaceae—

1.	Thalli growing in amphipods	2
1'.	Thalli not growing in amphipods	5
2(1).	Thalli not dimorphic	***Taeniellopsis***
2'(1).	Thalli clearly dimorphic	3
3(2').	Macrothalli with the spore mother-cell persisting apically; producing thick-walled cells	***Astreptonema***
3'(2').	Macrothalli with the spore mother-cell persisting at the base near the holdfast; producing no thick-walled spores	4
4(3').	Thalli branched at the base, giving rise to several fertile axes	***Ramacrinella***
4'(3').	Thalli unbranched, with one fertile axis	***Paramacrinella***
5(1').	Thalli producing thick-walled bilocular spores, each chamber with a 4-nucleate cell. (In millipedes)	***Eccinidus***
5'(1').	Thalli either not producing thick-walled spores or, if present, they are not bilocular	6
6(5').	Thalli typically growing in tufts, attached to a multiple holdfast system. (In foregut of crustaceans)	***Enteromyces***

	6'(5'). Thalli growing singly (one exception, in beetles) 7
7(6').	Bases of thalli prominently lobed or inflated unilaterally. (In isopods) .... **Alacrinella**
7'(6').	Bases of thalli not prominently lobed or inflated unilaterally 8
	8(7'). Thalli growing in foregut (stomach), as well as in hindgut. (In decapods) .... **Arundinula**
	8'(7'). Thalli growing in hindgut only 9
9(8').	Thalli producing no thick–walled spores 10
9'(8').	Thalli producing thick–walled spores at some stage 11
	10(9). Thalli may fuse in scalariform fashion. [In mud shrimps (Anomura)] .... **Enteropogon**
	10'(9). Thalli never fusing. [Usually in millipedes; a few species in beetles or mole crabs (Anomura)] .... **Enterobryus**
11(9').	Thick–walled spores without appendages, usually with a channel penetrating each end; in millipedes and isopods ...... **Eccrinoides**
11'(9').	Thick–walled spores with 2 appendages at each pole; in decapods .... **Taeniella**

Alacrinella

Manier & Ormières ex Manier, 1968

[= *Alacrinella* Manier & Ormières, 1961b, *nom. nud.*]

Thalli dimorphic, with spore mother–cell persisting apically. Macrothalli lobed or inflated at the base, producing thick–walled, oval, 4–nucleate primary infestation sporangiospores, or multinucleate secondary infestation sporangiospores. Microthalli cleaving into a series of small cells. In hindgut of marine Isopoda (*Limnoria* spp.) (Crustacea). Two species.

Type species: *Alacrinella limnoriae* Manier & Ormières ex Manier.

This genus is found only in the small marine wood–boring isopods commonly known as gribbles. *Alacrinella* morphologically is very similar to the genus *Astreptonema* from amphipods. *Astreptonema* species do not have lobed or inflated bases (see Fig. 11.29), and the oval resistant spores in some species have two appendages, one at each pole. It should be noted that Hibbits [1978, p. 252 and Fig. 22 (d–1)] found limited material of a dimorphic fungus, which she called *Astreptonema* sp., from the isopod *Exosphaeroma amplicauda* (Stimpson). The macrothalli of this unnamed species have a swollen base and produce oval spores (20–22 × 6–7 μm) with a knob at each end and with a long (up to 50 μm) appendage attached to one of the knobs. Thus, her isopod eccrinid seems to have characters of both genera.

—Key to *Alacrinella* species—

1. Base of macrothalli consisting of several bulbous swellings or lobes; thick–walled oval spores 10–15 μm long .... ***A. limnoriae***
1'. Base of macrothalli inflated unilaterally and bent at an angle; thick–walled oval spores 15–18 μm long .... ***A. sanjuanensis***

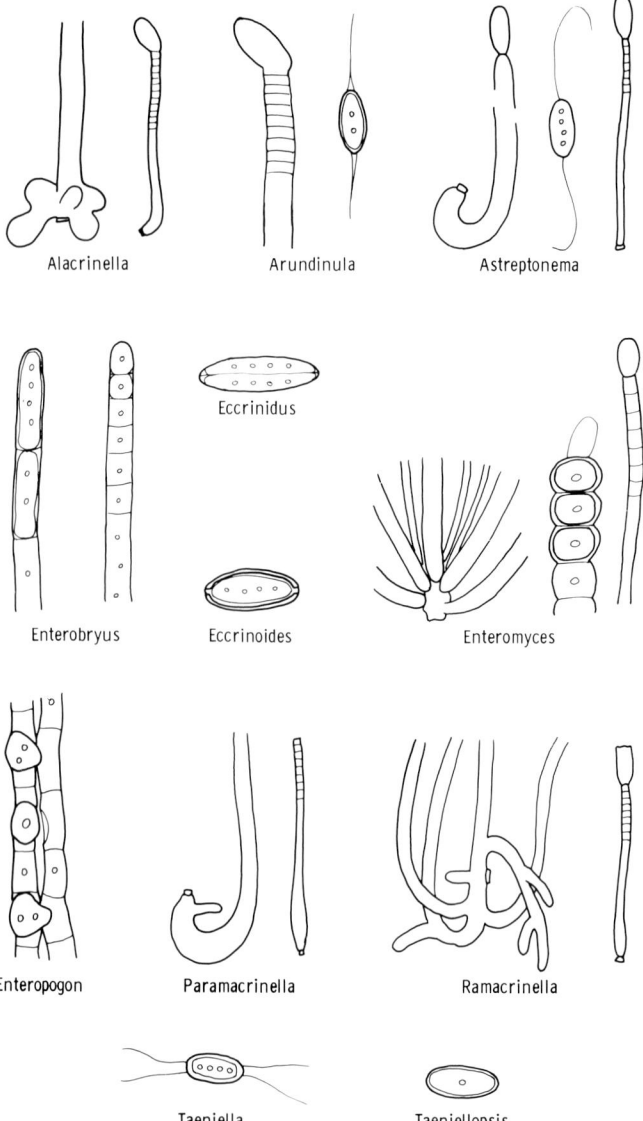

FIG. 11.29. Eccrinaceae: distinguishing generic characters.

☐ *Alacrinella limnoriae* Manier & Ormières ex Manier, 1968
 [= *Alacrinella limnoriae* Manier & Ormières, 1961b, *nom. nud.*]

Macrothalli 450–600 μm long by 7–12 μm diam distally, basal region 50–75 μm wide consisting of several bulbous swellings and lobes just above a small flat holdfast. Primary infestation sporangiospores thick–walled, oval, 10–15 × 5–7 μm, becoming 4–nucleate. Microthalli up to 250 μm by

~4.5–5 μm diam basally to 2–3 μm diam distally, distal part cleaving into more or less isodiametric uninucleate cells. Type species.

Illustrations: See Manier and Ormières, 1961b, Figs. I–IV.

Host: In rectum of adult *Limnoria tripunctata* Menzies (Isopoda, Limnoidae).

Distribution: Marine, boring in old wood, in the Mediterranean Étang de Thau near Sète, Hérault, France.

Manier and Ormières found about 80% of the adult gribbles infested. The number of thalli per isopod was not high, but usually both thallus types were present. The cells of the microthalli have no known function. The thick–walled primary infestation spores have 1, 2, or 4 nuclei, but all probably become 4–nucleate at full maturity. The secondary infestation spores may have up to 10 or 12 nuclei.

Reference: Manier, 1968, 1969b (1970b); Manier and Ormières, 1961b.

☐ *Alacrinella sanjuanensis* Hibbits, 1978

Macrothalli 750 μm long by 7–19 μm diam, basal region inflated unilaterally and bent at an angle, with a small holdfast at the interior angle of the bend. Primary infestation sporangiospores thick–walled, oval, 4–nucleate, 15–18 × 6–7 μm. Secondary infestation sporangiospores 4–nucleate (or more?), 40 × 10–25 μm. Microthalli 200 μm long by 3–5 μm diam, with a slightly inflated base, cleaving into cells 2 × 5 μm diam.

Illustrations: See Hibbits, 1978, Fig. 23.

Host: In hindgut of *Limnoria lignorum* (Rathke) (Isopoda, Limnoidae).

Distribution: Marine, in driftwood, near San Juan Island, Washington, U.S.A.

Hibbits (1978) described the oval primary infestation spores as having a small knob at each end, but cautioned that these observations were made on preserved material and that the knobs could have been fixation artifacts. The macrothalli in some cases initially produced secondary infestation sporangia, and this was followed on the same thallus by production of the more numerous primary infestation spores.

References: Hibbits, 1978; Galt, 1971.

Arundinula

Léger & Duboscq, 1906

Thalli with spore mother–cell persisting apically, attached to foregut (stomach) and usually also to hindgut, of Decapoda (Crustacea). Secondary infestation sporangiospores usually nearly oval, not more than twice as long as broad, occasionally more elongate. Primary infestation sporangia often flat and disklike. Oval or ellipsoidal spores with appendages produced in some species. Five species.

Type species: *Arundinula capitata* Léger & Duboscq.

The genus name first used by Léger and Duboscq (1905a, 1905b) was

Arundinella, but it was changed to *Arundinula* in 1906 because the former name had priority as a genus of Gramineae: *Arundinella* Raddi 1823.

Most species of *Arundinula* are found in hermit crabs or other marine anomurids, the only exception being *A. orconectis* in freshwater crayfish. They inhabit the stomach of the crustaceans (like *Enteromyces callianassae*) as well as the hindgut in most species. Sporulation is characteristically very prolific, possibly required to ensure successful transmission of spores among individuals of these types of hosts.

—Key to *Arundinula* species—

1. In foregut only; thalli wider at distal end (50–83 μm) tapering to a narrower base (27–47 μm) .... **A. galatheae**
1'. In foregut and hindgut; thalli not narrowing toward the base 2
2(1'). In crayfish; thalli mostly thin–walled .... **A. orconectes**
2'(1'). In anomurids; thalli mostly thick–walled 3
3(2'). Producing no appendaged spores; hindgut thalli of 4 types .... **A. haplogaster**
3'(2'). Producing oval or ellipsoidal spores with 1 appendage at each pole; hindgut thalli of 2 types 4
4(3'). Oval spores binucleate at maturity, with spinelike appendages 18–25 μm long; all holdfasts essentially basal .... **A. capitata**
4'(3'). Ellipsoidal spores uninucleate, with tapering appendages 120 μm long; anteriormost thalli in foregut with holdfasts variously lateral .... **A. washingtoniensis**

☐ *Arundinula capitata* Léger & Duboscq, 1906
[= *Arundinella capitata* Léger & Duboscq, 1905a, *nom. nud.*]

Foregut thalli mostly thick–walled, 1–2(–3) mm long, spore mother–cell persisting at apex until sporulation commences; thalli 23–30 μm diam producing 5–30 4– to 6–nucleate secondary infestation sporangiospores 13–15(–30) μm long; thalli 17–22 μm diam producing series of (30–)70–80(–120) flat, disklike uninucleate primary infestation sporangiospores. Hindgut thalli 1 cm or more long, anterior ones 35–60 μm diam, posterior ones 9–17 μm, producing same kinds of spores as foregut thalli; upon molting of host larger thalli producing (rarely) numerous thick–walled, oval spores ~30 × 14 μm, becoming binucleate and bearing a spinelike appendage 18–25 μm long at each pole. Type species.

Illustrations: See Duboscq et al., 1948, Plates IV, V.

Hosts: In fore– and hindgut of *Pagurus maculatus* Hell. and *Eupagurus cuanensis* Thomps. (Decapoda, Paguridae), and possibly other hermit crabs.

Distribution: Marine, intertidal, Departments of Pyrénées–Orientales and Hérault, France.

The above description is based mostly on Manier's 1969(b) description. The original name used by Léger and Duboscq (1905a, 1905b) was *Arun-*

dinella capitata (see note under generic description). The species was not illustrated until Duboscq, Léger, and Tuzet redescribed it in 1948. The oval resistant spores appear to be rare, for they found them in only 3 out of 150 molting hermit crabs. They considered these resistant spores to arise by a fusion of intrathallial "gametes," but this was not confirmed in Hibbit's (1978) study of similar spores in *A. washingtoniensis,* nor does such development conform to resistant spore production in other Eccrinales. Manier and Ormières (1962) listed *Pagurus spinimanus* Luc as another host of *A. capitata.*

References: Léger and Duboscq, 1905a, 1905b, 1906; Duboscq et al., 1948; Manier, 1950 (1951), 1969b (1970b); Manier and Ormières, 1962.

□ *Arundinula galatheae* Manier & Ormières ex Manier, 1968
[= *Arundinula galatheae* Manier & Ormières, 1962, *nom. nud.*]

Thalli rigid and thick–walled, up to 3 mm long, 50–83 μm diam near the apex, narrowing to 27–47 μm diam near the base, with a spore mother–cell persisting apically until sporulation. Holdfasts narrow with a flared base, 25–30 μm long. Secondary infestation sporangiospores multinucleate, becoming oval, 45–65 × 30–45 μm. Upon molting of host, thalli completely converting to series of flat, multinucleate sporangia that release from the apex round (27–31 μm diam) or oval thick–walled primary infestation sporangiospores.

Illustrations: See Manier and Ormières, 1962, Figs. 1, 2, and Plates I, II.

Host: In foregut of *Galathea strigosa* L. (Decapoda, Galatheidae).

Distribution: Marine, deep water off coast of Sète, Department of Hérault, France.

Arundinula galatheae has several features that are unusual. The anomurid hosts were collected at depths of 15–25 m. The only other trichomycete collected from such depths is *Enteromyces callianassae,* but some of its hosts also live in the intertidal zone. The tapered thalli of *E. galatheae* have exceptionally heavy walls, up to 5–8 μm thick. Manier and Ormières stated that occasionally some thalli would "molt" by slipping out of the end of the outer wall layer, and reattach to the gut cuticle; the old thallus wall remained in position attached by its original holdfast. The multinucleate thick–walled primary infestation spores of *A. galatheae* apparently have no appendages, as they do in *A. capitata* and *A. washingtoniensis.*

Other species of *Arundinula* can live in both the foregut and hindgut; at least it appears that the fungi in both locations belong to the same species. The host of *A. galatheae, Galathea strigosa,* also has an eccrinid that lives in the hindgut, but Manier and Ormières came to the conclusion that those thalli were not an *Arundinula,* and assigned them to the genus *Taeniella,* as *T. galatheae,* considered here to be a synonym of *T. carcini.*

References: Manier, 1968; Manier and Ormières, 1962.

□ *Arundinula haplogaster* Hibbits, 1978

Foregut thalli thick–walled, spore mother–cell more or less persistent, of two types: (1) thalli slightly curved, 3 mm × 16–35 μm, producing 8–nucleate, doliiform secondary infestation sporangiospores 20–35 × 17–35 μm; (2) thalli usually straight, producing thin–walled uninucleate primary infestation sporangiospores 20–35(–43) × 5–7.5 μm; both thallus types may also occur in anterior part of hindgut. Hindgut thalli of four types (anterior to posterior): (1) undulate thalli 6 mm × 12–25 μm, producing 4–nucleate sporangiospores 60–95 × 10–12.5 μm, in exuviae dividing to form uninucleate spores 20 × 10 or 5–7 × 14 μm; (2) coarse thalli 15 mm long by 17–25 μm diam distally, 41–50 μm diam proximally, with holdfasts 40 μm long, producing 1–6 spores 26–70 × 17–30 μm; (3) thalli 7 mm long, 37–43 μm diam proximally, tapering to 16–21 μm distally, producing fewer than 20 spores 32–37 × 14–21 μm; (4) thick–walled thalli 8 mm long by 55–105 μm diam near the base, tapering to 9 μm diam at the tip, sometimes internally cleaved into cells.

Illustrations: See Hibbits, 1978, Figs. 12–14.

Host: In foregut and hindgut of *Haplogaster mertensii* Brandt (Decapoda, Lithodidae).

Distribution: Marine, lowest tidal zone, rocky outcroppings of San Juan Island, San Juan Archipelago, Washington, U.S.A.

The above description is adapted from Hibbits' (1978) paper, and exemplifies the complexities of growth found in this large anomurid. It also raises the question of spore functions and conspecificity of all thallus types, a matter that cannot be resolved satisfactorily without additional studies using experimental approaches.

References: Hibbits, 1978; Galt, 1971.

□ *Arundinula orconectis* Lichtwardt, 1962

Thalli up to 2 cm long by ~20 μm diam, thin–walled, spore mother–cell persistent, sporulating prolifically. Holdfast base slightly wider than thallus. Secondary infestation sporangiospores 8– or more nucleate, ~36 × 20 μm, slightly swollen at the proximal end, produced in series of up to more than 200 per thallus. Primary infestation sporangia variable in length, ~20 μm diam, producing spores that become spherical after release, 24–34 μm diam.

Illustrations: See Lichtwardt, 1962, Figs. 1–10.

Host: In foregut and hindgut of *Orconectis nais* (Faxon) (Decapoda, Astacidae).

Distribution: In freshwater streams, Kansas, U.S.A.

Arundinula orconectis is the only freshwater species of the genus. Although it occurs in a crayfish, rather than an anomurid, it appears to be, nevertheless, an *Arundinula*. The spherical spores described above were seen lying among thalli, and are only presumed to be matured primary infestation spores of this species.

Reference: Lichtwardt, 1962.

☐ *Arundinula washingtoniensis* Hibbits, 1978

Foregut thalli thick–walled, spore mother–cell more or less persistent, of three types: (1) thalli 1.8 mm × 22–45 μm, holdfasts becoming variously lateral to the thallus base, producing up to 45 4–nucleate doliiform secondary infestation sporangiospores 20–35 × 10–20 μm; (2) thalli 4.2 mm × 17–27 μm, holdfasts basal, producing series of more than 230 very flat sporangia 3–6 × 22–27 μm diam, releasing upon molting of the host uninucleate, thin–walled, ellipsoidal primary infestation sporangiospores with a long tapering appendage 120 μm long at each pole; (3) rare thalli 500–650 × 60–120 μm with basal holdfasts and very thick walls, sporulation unknown; thallus types (1) and (2) occasionally occur in anterior part of hindgut. Hindgut thalli thick–walled, of two types: (1) thalli 6.5 mm × 20–45 μm, producing 4–nucleate secondary infestation sporangiospores 40–80 × 14–19 μm; (2) thalli 13.5 mm × 11–30 μm, base with a lateral bulbous thickening in cell wall, producing 4– to 8–nucleate secondary infestation sporangiospores 135–390 × 11–20 μm, or rarely cleaving to produce many spherical uninucleate bodies.

Illustrations: See Hibbits, 1978, Figs. 8–10.

Host: In foregut and hindgut of *Paguristes turgidus* (Stimpson) (Decapoda, Paguridae).

Distribution: Marine, subtidal, shores of Waldron Island, San Juan Archipelago, Washington, U.S.A.

Arundinula washingtoniensis occurs in a large hermit crab and, like *A. haplogaster,* constitutes one of the morphologically most complex species of trichomycetes, assuming all of the thallus types are conspecific. To complicate matters, Hibbits (1978) occasionally found what she considered to be *Taeniella carcini* in the rectum of this same species of hermit crab. Despite the careful study done by Hibbits, further investigations are needed to clarify the developmental interrelationships of the thallus and spore types. For instance, she presented only indirect evidence that the appendaged ellipsoidal spores found in molts have their origin in the foregut thalli. The appendaged spores of *A. washingtoniensis* differ from those of *A. capitata* in wall thickness and number of nuclei, as well as in other morphological features. In both species they occurred in exuviae, and their origin was not established with certainty. Nevertheless, they are of interest because only a few species of freshwater and marine Eccinales produce appendaged spores (see *Astreptonema* and *Taeniella*). Also of interest is the report that some thalli in the hindgut (Type 2) rarely cleave internally to produce sporelike uninucleate cells. This is a feature found in the Parataeniellaceae, and if such cleavage is not artifactual, it could be a transitional character.

References: Hibbits, 1978; Galt, 1971.

Astreptonema

Hauptfleisch, 1895

Thalli dimorphic, spore mother–cell more or less persisting apically. Macrothalli typically curved at the base, which is not noticeably lobed or swollen, producing thick–walled, oval, usually 4–nucleate primary infestation sporangiospores, some with 1 appendage at each pole, or producing multinucleate secondary infestation sporangiospores. Microthalli cleaving into a series of small cells. In hindgut of marine or freshwater Amphipoda (Crustacea). Five species.

Type species: *Astreptonema longispora* Hauptfleisch.

The generic description above is an extension and reinterpretation of Hauptfleisch's, and is based on studies of subsequent species. See the discussion under the type species, *A. longispora*.

The spore mother–cell at the apex of thalli is the persistent cell of the spore from which the thallus evolved. In *Astreptonema* it is usually narrower than the macrothalli and wider than the microthalli (see Fig. 11.29). It disappears in the macrothalli after sporulation has commenced, but often remains at the tip of the less common microthalli even after these have produced small, usually uninucleate, cells of unknown function. The resistant oval primary infestation spores have two appendages in *A. gammari* and *A. typica*. Further studies of the other species, preferably using living material, may show that appendages are a feature of all such oval spores in this genus. It should be noted that Hibbits (1978) described and illustrated, but did not name, an "Astreptonema sp." that she found in the hindgut of an isopod, *Exosphaeroma amplicauda*. This fungus had oval spores with a knob at each end, and had an appendage projecting from only one of the knobs. Her *Astreptonema* sp. has some characters of the genus *Alacrinella*.

—Key to *Astreptonema* species—

1. Oval spores 7–10 μm long by 2–2.6 μm diam .... ***A. longispora***
1'. Oval spores longer and wider 2
    2(1'). Oval spores 10–15 μm long .... ***A. corophii***
    2'(1'). Oval spores longer than 15 μm 3
3(2'). Mature macrothalli not exceeding 350 μm in length; oval spores 17–20 μm long .... ***A. typica***
3'(2'). Mature macrothalli normally exceeding 350 μm in length; oval spores more than 20 μm long 4
    4(3'). Oval spores about 7–10 μm diam; microthalli about 2 μm wide; in freshwater streams .... ***A. gammari***

4'(3'). Oval spores about 10–13 μm diam; microthalli about 5 μm
wide; littoral .... *A. pacificum*

□ *Astreptonema corophii* (Manier) Manier ex Manier, 1968
[= *Astreptonema corophii* (Manier) Manier, 1964b, *nom. nud.*]
[= *Eccrinella corophii* Manier, 1961b, *nom. nud.*]

Macrothalli up to 700 μm long by 7–12 μm diam near the curved base and 5–10 μm diam distally, spore mother–cell persisting. Primary infestation sporangiospores oval, 4–nucleate, 10–15 × 5–7 μm. Secondary infestation sporangiospores 1–8 per thallus, multinucleate, 14–30 μm long. Microthalli up to 200 μm long by 1.5–2.2 μm diam.

Illustrations: See Manier, 1961b, Fig. 1 (drawing) and Figs. 1–7 (photomicrographs).

Host: In hindgut of *Corophium volutator* (Pall.) (Amphipoda, Corophiidae).

Distribution: Marine, littoral muddy sand, Department of Finistère, France.

No appendages have been reported on the oval spores, possibly because they were not sought in living material (Manier, 1964b).

References: Manier, 1961b, 1964b, 1968.

□ *Astreptonema gammari* (Léger & Duboscq) Manier, 1964b
= *Eccrinella gammari* Léger & Duboscq, 1933

Macrothalli 1–2 mm long by 8–18 μm diam, curved at the base, spore mother–cell persisting. Primary infestation sporangiospores 4–nucleate, oval, thick–walled, bearing 1 long appendage at each pole, of two types: (1) more common, 25–30 × 7–10 μm, lying obliquely in series of up to 40–50 per thallus; (2) more rare, ~54–55 × 10 μm, lying longitudinally in series of 8–14 per thallus. Secondary infestation sporangiospores 2– to 8–nucleate, cylindrical with rounded ends, 25–50 × 8–10 μm. Microthalli uncommon, ~300 μm long by 2 μm diam, dividing into many small uninucleate cells.

Illustrations: Fig. 11.30; also see Léger and Duboscq, 1933, Figs. 1–15.

Hosts: In hindgut of *Gammarus pulex* (L.), *G. roeselii* Gervais, and *Echinogammarus berilloni* (Catta) (Amphipoda, Gammaridae), and possibly other freshwater Gammaridae.

Distribution: Freshwater streams in the Dauphiné (French Alps) and Rhône valley, and Departments of Calvados and Hérault, France; in East Germany; in England; possibly also in the U.S.A.

The name of the basionym, *Eccrinella gammari,* was first used by Léger and Duboscq in 1906, but the species was not described by them until 1933. Poisson (1929) had also found this species, in southeastern France, but did not describe it. In 1964(b) Manier transferred all current species

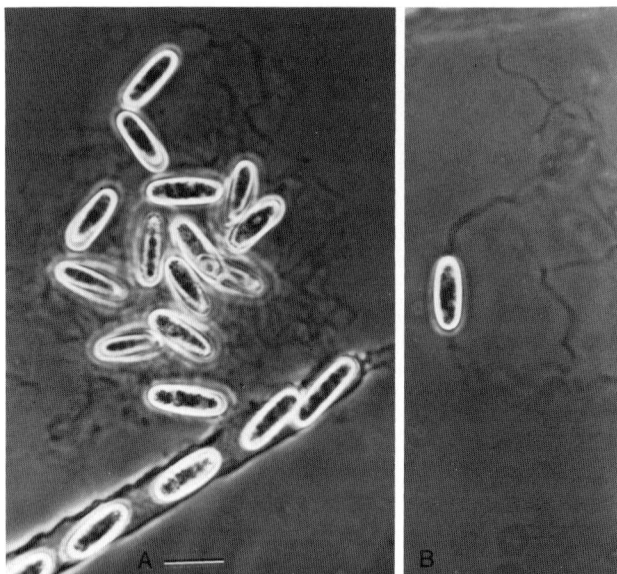

FIG. 11.30. *Astreptonema gammari*. A. A clump of primary infestation sporangiospores emerged from a thallus and entangled by their appendages. B. A sporangiospore showing a single appendage at each pole. Scale bar = 20 μm for both figures. Reprinted by permission from Mycologia, vol. 65: 8, Copyright 1973, R.W. Lichtwardt and The New York Botanical Garden.

of *Eccrinella* to *Astreptonema*. The diagnosis above is based on these sources and some collections made by the writer in France and England, and including what may be this species in the U.S.A. The amphipod host, *G. roeselii*, was recorded by Maessen (1955) in East Germany. Scheer (1976a) made *A. gammari* a synonym of *A. longispora*, the type species, but there appears to be no justification for doing this on the basis of the oval spore measurements of *A. longispora* (which see).

The peculiar, very long polar appendages of this species were apparently not observed until Manier described them in 1964(b). They are best seen in phase contrast after release from living thalli. Prior to this, the appendages are tightly folded within each end of the sporangium. Moss (1979) refers to them as mucilaginous, but in the writer's experience they appear to be ribbon–like and nonmucilaginous. Groups of spores released more or less simultaneously from a thallus may remain clumped by appendage entanglement. In the natural aquatic environment they perhaps function like the appendages of trichospores in the Harpellales. Whether the appendages of the two orders are at all homologous and have any similarities in their development have not been satisfactorily determined, and this knowledge would be of considerable interest in helping to understand the interrelationships of the two orders (Lichtwardt, 1973a). Moss (1972, 1975) has done an electron microscopic study of *A. gammari*, including the structure of the appendaged spores within their sporangia and the perforate septum of the sporangial crosswalls (see Chapter 7), and Grizel (1971) has also studied several aspects of the fine structure of this species. The present evidence is that the appendages of the Eccrinales form in a manner dissimilar to trichospore appendages.

References: Léger and Duboscq, 1906, 1933; Poisson, 1929; Manier, 1950 (1951), 1964b; Maessen, 1955; Moss, 1972, 1975, 1979; Grizel, 1971; Lichtwardt, 1973a; Scheer, 1976a.

☐ *Astreptonema longispora* Hauptfleisch, 1895

Thalli curved at the base, producing series of oval multinucleate (up to 8 nuclei) primary infestation sporangiospores 7–10 × 2–2.6 μm. Type species.

Illustrations: See Hauptfleisch, 1895, Figs. 1–10.

Host: In the posterior hindgut of *Gammarus pulex* (L.) (Amphipoda, Gammaridae). (See remarks below.)

Distribution: Freshwater habitats, Germany (Thüringen).

The diagnosis provided by Hauptfleisch (1895) was extremely brief, and was based upon limited fungal material. Nevertheless, it appears to correspond to other, more thoroughly studied related species that are now placed in this genus and which enable us to define better the generic characters. There were no appendages reported on the oval spores of this species [they were unknown until Manier (1964b) described them in the genus]. Hauptfleisch considered his fungus to belong to the Saprolegniaceae (Oomycetes), and believed the sporangia to be oogonia bearing single, multinucleate oospores. He apparently was not aware of the eccrinid literature of that time. Leidy (1920) was the first to recognize the similarity of *Astreptonema* and *Enterobryus*.

Hauptfleisch stated that the host of *A. longispora* was *Gammarus locusta* (L.). This is a seashore (marine) amphipod. Scheer (1976a) questioned the identification of the amphipod, and presented evidence that the locality where Hauptfleisch collected his specimens (Ichtershausen) is located some 350 km from any coastal areas, and that the misidentified host must have been the freshwater species, *Gammarus pulex*. The writer accepts this identification. *Gammarus pulex* is also one of the hosts of *Astreptonema gammari*, and Scheer synonymized the two species. The writer, however, does not find the argument at all convincing. *Astreptonema longispora* was stated by Hauptfleisch to have oval spores measuring 7–10 × 2–2.6 μm, whereas the smaller of the two types of oval spores of *A. gammari* are considerably larger than that [25–30(–55) × 7–10 μm]. Scheer's argument is that the size of spores can vary with the diet of the host (an unsubstantiated statement), and that the *ratio* of length to width is more important than the actual dimensions. This would not conform to the general concepts of species identification in other trichomycetes. Nevertheless, that Hauptfleisch may have incorrectly measured the spore sizes cannot be definitively discounted.

Manier (1969, Figs. 8–11) provided photographs of an "*Astreptonema* sp. from *Gammarus locusta*," and according to the size scales in her Figs. 9 and 11, the (appendaged) oval spores are approximately 24 × 8 μm and 18 × 5.5 μm, respectively. Further studies of the fungi in these amphipods from their type localities are clearly needed.

References: Hauptfleisch, 1895; Scheer, 1976a.

□ *Astreptonema pacificum* Hibbits, 1978

Macrothalli up to 750 μm long by 17–38 μm diam near the curved base and 10–15 μm diam distally, attached to the cuticle by a long (15–45 μm) striated holdfast. Producing oval, thick–walled, uninucleate primary infestation sporangiospores 22–35 × 10–13 μm, or 4–nucleate secondary infestation sporangiospores 22–43 × 10–18 μm. Microthalli 60 × 5 μm, dividing into cells ~1 μm long.

Illustrations: See Hibbits, 1978, Figs. 20 and 21.

Host: In hindgut of *Orchestia traskiana* Stimpson (Amphipoda, Talitridae).

Distribution: Marine, uppermost intertidal zone of sandy shores, San Juan Archipelago, Washington, U.S.A., and British Columbia, Canada.

The amphipod host is commonly called a beach hopper or sandflea, and was found around decaying seaweed. No appendages were seen on the oval spores, and they were uninucleate (possibly immature?).

References: Hibbits, 1978; Galt, 1971.

□ *Astreptonema typica* Manier ex Manier, 1968
[= *Astreptonema typica* Manier, 1964b, *nom. nud.*]

Macrothalli up to 350 μm long, curved at the base, 10–13 μm diam. Primary infestation sporangiospores oval, thick–walled, 4–nucleate, 17–20 × 7–9 μm, with one long appendage at each pole. Secondary infestation sporangiospores 4–nucleate, 15–25 μm long. Microthalli 100–150 × 2–2.5 μm, producing a terminal series of isodiametric cells.

Illustrations: See Manier, 1964b, Figs. 1, 2; Grizel, 1971, Plate I, Figs. 5–8.

Host: In hindgut of *Gammarus duebeni* Lilljeborg (Amphipoda, Gammaridae).

Distribution: Aquatic, in fresh and brackish water, Department of Hérault, France.

References: Manier, 1964b, 1968; Grizel, 1971.

Ecrinidus

Manier, 1970a (1969a)

Primary infestation sporangiospores cystlike, bilocular (rarely unilocular), each chamber producing and releasing a 4–nucleate cell. Attached to hindgut cuticle of Diplopoda. Monotypic.

Type species: *Eccrinidus flexilis* (Léger & Duboscq) Manier.

☐ *Eccrinidus flexilis* (Léger & Duboscq) Manier, 1970a (1969a)
 = *Eccrina flexilis* Léger & Duboscq, 1906

Thalli up to 3 or 4 mm long by 5–30 μm diam, generally spiraled near the base. Holdfast columnar with a disklike basal expansion. Upon molting of the host producing oval cystlike primary infestation sporangiospores of two types: (1) bilocular, each chamber with a 4–nucleate cell 50–65 × 11–12 μm, or (2) unilocular, containing one 4–nucleate cell 25–50 × 5–12 μm. Secondary infestation sporangia variable in length and with perpendicular to oblique end walls. Type species.

Illustrations: Fig. 11.31; also see Manier, 1969b, Plate II Fig. 2, Plate Figs. 1–3, Plate IV Figs. 5–7, Plate V Figs. 5, 6.

Hosts: In hindgut of *Glomeris marginata* Villers, *G. annulata* Brandt, *G. connexa* Koch, *G. conspersa* Koch, and *G. hexasticha* Brandt (Diplopoda, Glomeridae).

Distribution: Terrestrial, widespread in France.

Eccrinidus flexilis is a variable yet distinct species known only in glomerid millipedes. Its morphology is basically similar to that of many *Enterobryus* spp., but *E. flexilis* is distinguished by the production of cystlike primary infestation spores when the host molts. The two–chambered ones are more common than those with one chamber, and they may be found in large numbers in the exuviae. The secondary infestation spores are short to long (up to 100 μm), but more or less consistent in length in any given thallus. The crosswalls that delimit the sporangia are sometimes very oblique. Other cell types of uncertain function have also been described in this species, including uninucleate and binucleate ones. Manier

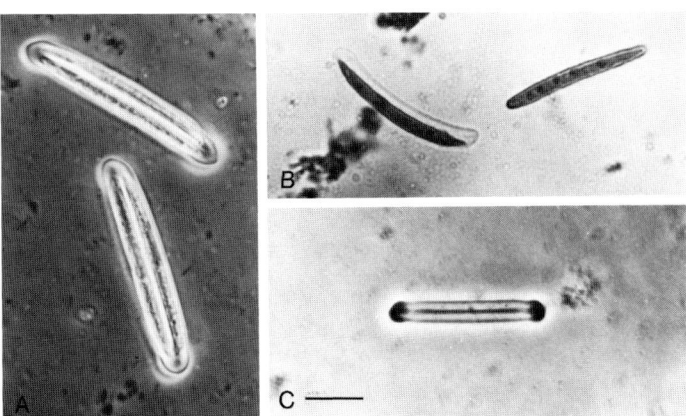

FIG. 11.31. *Eccrinidus flexilis*. A. Two mature, two–chambered primary infestation sporangiospores. B. Primary infestation sporangiospore with one of the two 4–nucleate inner cells emerged. C. Empty spore case clearly showing its bilocular structure. Scale bar = 20 μm for all figures. Fig. 11.31B reprinted by permission from Mycologia, vol. 65: 12, Copyright 1973, R.W. Lichtwardt and The New York Botanical Garden.

(1969b, p. 581) states that *E. flexilis* may also attach to the cuticle of intestinal nematodes, as do several species of *Enterobryus*. Manier and Grizel (1972) studied the fine structure of the holdfast in *E. flexilis* (see Table 7.3).

Duboscq et al. (1948) and Manier (1950) wrote extensively on *E. flexilis*, yet many of their observations on the variable spore forms need confirmation. Manier (1954) attempted to culture this species, but was able to get only limited stages of development *in vitro*. Duboscq et al. reported seeing some unusual developmental stages within the oval resistant spores involving nuclear divisions, aborting nuclei, plasmogamy, and karyogamy, suggesting a curious form of sexuality not found in other trichomycetes, or in other fungi for that matter. Manier (personal communication) has not confirmed such development, although she earlier (Manier, 1954, p. 268) claimed to have seen nuclear fusions. Nor has the author seen such stages in his study of limited material collected near Montpellier, France. The diagnosis for the species provided above is adapted from Manier (1969b).

References: Manier, 1950 (1951), 1954, 1969a (1970a), 1969b (1970b); Léger and Duboscq, 1906, 1929a; Duboscq et al., 1948; Manier and Grizel, 1972; Lichtwardt, 1973a.

Eccrinoides

Léger & Duboscq, 1929a

Primary infestation sporangiospores oval, 4-nucleate, thick-walled, usually with a channel penetrating the endwalls at each pole. Attached to hindgut cuticle of Diplopoda and Isopoda. Four species.

Type species: *Eccrinoides henneguyi* Léger & Duboscq.

—Key to *Eccrinoides* species—

1. In millipedes 2
1'. In isopods 3
    2(1). Thalli not spiraled; thick-walled spores oval; multinucleate spores produced in sporangia with oblique crosswalls .... *E. henneguyi*
    2'(1). Thalli spiraled at the base; thick-walled spores allantoid; sporangial walls transverse .... *E. broelemanni*
3(1'). Thick-walled spores of one type, 24–28 × 6–7 μm ... *E. monticola*
3'(1'). Thick-walled spores of two types: one 47–58 × 12–18 μm, the other 36–40 × 12–13 μm .... *E. helleriae*

□ *Eccrinoides broelemanni* Léger & Duboscq, 1929a

Thalli up to 5 mm long by 11 μm diam, with 3–4 spirals at the base. Secondary infestation sporangiospores 4-nucleate, up to more than 100 μm long. Primary infestation sporangiospores of variable length. Thick-walled resistant spores more or less allantoid, ~30 μm long.

Illustrations: See Duboscq et al., 1948, Figs. X(1–6).

Host: In hindgut of *Protoglomeris vasconica* Bröl. (Diplopoda, Glomeridae).

Distribution: Terrestrial, in Pyrenees mountains, France.

The name of this species, but with a minimal description, was established by Léger and Duboscq (1929a) in the same paper describing the type species of the genus, *E. henneguyi*. The description of *E. broelemanni* was amplified by Duboscq et al. in 1948.

References: Léger and Duboscq, 1929a; Duboscq et al., 1948.

□ *Eccrinoides helleriae* Manier ex Lichtwardt, 1984c

[= *Eccrinoides helleriae* (Léger & Duboscq) Manier, 1969, *nom. nud.*]

[= *Eccrinopsis helleriae* Léger & Duboscq, 1906, *nom. nud.*]

Thalli up to 4 mm long by 7–18 μm diam. Primary infestation sporangiospores uninucleate, variable in length. Secondary infestation sporangiospores 4–nucleate, 40–150 μm long. Upon molting of the host producing thick–walled, oval, 4–nucleate spores 47–50 × 12–18 μm or 36–40 × 12–13 μm.

Illustrations: See Duboscq et al., 1948, Figs. XX–XXV and Plate III; and Manier, 1963, Figs. 1, 2.

Host: In hindgut of *Helleria brevicornis* Ebner (Isopoda, Tylidae).

Distribution: Terrestrial, oak forests, Mediterranean coast, and islands east of the Rhône delta, France.

Manier made a new combination for this species in 1969(b), but it was not valid because the basionym, *Eccrinopsis helleriae*, is a *nomen nudum*. The name *Eccrinopsis helleriae* was used by Léger and Duboscq in 1906 and 1916, but in neither publication was the fungus described. The first description appeared in Duboscq, Léger, and Tuzet's 1948 publication, at which time Art. 36 would apply, but no Latin diagnosis was provided. Furthermore, the type species of *Eccrinopsis*, *E. hydrophilorum*, has been transferred to the genus *Enterobryus*. The name *Eccrinoides helleriae* was validated by Lichtwardt (1984c).

Eccrinoides helleriae is said by Duboscq et al. (1948) to produce rarely a smaller, 2–nucleate type of resistant spore, which is somewhat allantoid and measures 18 × 8 μm. This species has developmental stages in the two larger resistant spores that need reinvestigation.

References: Lichtwardt, 1984c; Manier, 1963a, 1969b (1970b); Léger and Duboscq, 1906, 1916; Duboscq et al., 1948.

□ *Eccrinoides henneguyi* Léger & Duboscq, 1929a

Thalli up to 7 mm long by 10–13 μm diam, not spiraled. Secondary infestation sporangia with oblique endwalls, producing 4–nucleate sporangiospores ~75 μm long. Uninucleate primary infestation spores produced in repeating long series of sporangia alternating with nonsporulating coenocytic segments of the thallus. Upon molting of the host producing oval, 4–nucleate, thick–walled spores 30–34 μm long with a channel at each pole. Type species.

Illustrations: See Léger and Duboscq, 1929a, Figs. I–VI.

Hosts: In hindgut of *Loboglomeris rugifera* Verh. and *L. pyrenaica* Bröl. (Diplopoda, Glomeridae).

Distribution: Terrestrial, in Pyrenees mountains, France.

This species is similar in some respects to *Eccrinidus flexilis*, which also lives in glomerid millipedes, but the oval resistant spores of *E. flexilis* are usually bilocular and do not have the polar channels in the walls that distinguish the genus *Eccrinoides*.

References: Léger and Duboscq, 1929a; Manier, 1969b (1970b).

☐ *Eccrinoides monticolae* (Poisson) Manier, 1970b (1969b)
= *Eccrinopsis monticolae* Poisson, 1931a

Thalli up to 525 μm long by 11–14 μm diam, with a small holdfast. Primary infestation sporangiospores uninucleate, 30 × 10–12 μm. Secondary infestation sporangiospores up to 8–nucleate, 50–60 × 8–9 μm. Upon molting of the host producing thick–walled, oval, 4–nucleate spores 24–28 × 6–7 μm with a channel penetrating each pole.

Illustrations: See Poisson, 1931a, Fig. VII.

Host: In hindgut of *Porcellio monticola* Lereboullet (Isopoda, Porcellionidae).

Distribution: Terrestrial, in Pyrenees mountains, France.

References: Manier, 1969b (1970b); Poisson, 1931a.

Enterobryus

Leidy, 1849a

Primary infestation sporangiospores uninucleate and thin walled, generally isodiametric, sometimes lacking. Secondary infestation spores cylindrical, usually 4– to 8–nucleate. Other cell types sometimes produced. Attached by a well–defined secreted holdfast to the hindgut cuticle of Diplopoda, less often Insecta (Coleoptera) or Crustacea (Decapoda). Twenty three species.

Type species: *Enterobryus elegans* Leidy.

Enterobryus (as *Enterobrus*) was the first genus of trichomycetes described. It is also the largest genus and, without question, the most difficult and taxonomically perplexing of all trichomycete genera. The basic difficulties lie in incomplete—sometimes misleading or erroneous—descriptions and the considerable morphological variation that exists in some of the species. In addition, a few species produce supplemental cell types whose function, if any, remains unknown. No species of *Enterobryus* has been cultured axenically in order to compare morphology under similar environments, nor have host specificity determinations been made. These problems have been outlined briefly in Chapter 10, with selected examples. Also see the note under *Enterobryus isoporostrepti*, and the description of *E. tuzetae*.

More than 20 species of *Enterobryus* have been illegitimately described, and are not included in this section. Other species have been effectively and validly published, but they are incompletely described and have not been accepted by the writer. All excluded taxa are listed at the end of this chapter and in the Index, and can be found as well in Appendix B together with the names of their hosts.

Leidy himself did not describe the reproductive features of the type species, *E. elegans*, but the species was restudied by Lichtwardt (1954b), and thus we have a reasonably good basis upon which to determine the characters of both the species and the genus. Nevertheless, the generic characters are not as exclusive and clearly defined as would be desirable, certainly less so than many other genera of Eccrinales. The minimal number of diagnostic characters used for *Enterobryus* possibly has led to the inclusion in the genus of some unrelated species by virtue of their exclusion from other genera of Eccrinales.

—Key to *Enterobryus* species—

1.	In hindgut of Insecta (Coleoptera) or Decapoda (Anomura) ... 2	
1'.	In hindgut of Diplopoda (millipedes) 5	
	2(1). In marine mole crabs (*Emerita* spp.) ..... ***E. halophilus***	
	2'(1). In aquatic or terrestrial Coleoptera (beetles) 3	
3(2').	In aquatic Hydrophilidae .... ***E. hydrophilorum***	
3'(2').	In terrestrial Passalidae 4	
	4(3'). Growing in anterior hindgut, often in clusters with a multiple holdfast system .... ***E. attenuatus***	
	4'(3'). Growing near the anus, thalli always individual .... ***E. compressus***	
5(1').	Thalli typically bifurcate, with 2 divergent fertile branches ...... ***E. bifurcatus***	
5'(1').	Thalli typically with one fertile axis 6	
	6(5'). Producing long sporangia containing a partially folded multinucleate spore which, upon release, is straight and tapered apically to a point 7	
	6'(5'). Not producing sporangiospores as above 10	
7(6).	Point of tapered multinucleate spore with a bulbous wall thickening or apical cap; in millipede family Odontopygidae ... 8	
7'(6).	Point of tapered multinucleate spore without a wall thickening or cap; in millipede family Spirostreptidae 9	
	8(7). Wall at tip of tapered multinucleate spore with an even thickening; in *Plethocrossus acutiformis* .... ***E. adjanohouni***	
	8'(7). Wall at tip of tapered multinucleate spore with a small bulbous thickening; in *Peridontopyge gasci* .... ***E. peridontopygei***	

9(7').	Thalli up to 6 mm long and 28 μm diam; in *Onychostreptus* spp. .... *E. onychostrepti*
9'(7').	Thalli up to 2.1 mm long and 14 μm diam; in *Isoporostreptus bouixi* .... *E. isoporostrepti*
10(6').	Uninucleate primary infestation spores rare or lacking 11
10'(6').	Uninucleate primary infestation spores often present 15
11(10).	Some thalli often attached to gut nematodes .... *E. elegans*
11'(10).	Thalli rarely, if ever, attached to gut nematodes 12
12(11').	Thalli long–fusiform, bluntly pointed at apex .... *E. apheloriae*
12'(11').	Thalli essentially cylindrical, not pointed at apex 13
13(12').	Holdfast up to 60 μm long, usually without a basal disk .... *E. ahlesi*
13'(12').	Holdfast not exceeding 30 μm, with a prominent basal disk or expansion 14
14(13').	Holdfast with essentially no stalk; secondary infestation spores often with more than 8 nuclei ...... *E. cherokiae*
14'(13').	Holdfast with a stalk; secondary infestation spores 4–nucleate .... *E. cingaloboli*
15(10').	Some thalli often attached to gut nematodes .... *E. borariae*
15'(10').	Thalli rarely, if ever, attached to gut nematodes 16
16(15').	Thalli frequently coiled throughout their length 17
16'(15').	Thalli never coiled throughout their length 18
17(16).	Holdfast stalk very short; secondary infestation spores 9–11 μm diam .... *E. dixidesmi*
17'(16).	Holdfast stalk 12–35 μm long; secondary infestation spores 5–8 μm diam .... *E. moniliformis*
18(16').	Thallus tips typically curved .... *E. euryuri*
18'(16').	Thallus tips not typically curved 19
19(18').	Four types of thalli produced .... *E. tuzetae*
19'(18').	One or two thallus types produced 20
20(19').	Thalli of one type; uninucleate spores with 3 size ranges, some elongated 21
20'(19').	Thalli dimorphic; uninucleate spores more or less isodiametric or like stacked disks 22
21(20).	Secondary infestation spores 8–nucleate; holdfast 9–19 μm long .... *E. leptoiuli*
21'(20).	Secondary infestation spores 4–nucleate; holdfast 5–6 μm long .... *E. cylindroiuli*
22(20').	Producing more or less isodiametric uninucleate spores; base of larger thalli 15–21 μm diam .... *E. oxidi*
22'(20').	Producing isodiametric and flat disklike uninucleate spores; base of larger thalli 21–31 μm diam .... *E. oxydesmi*

□ *Enterobryus adjanohouni* Manier, Gasc & Bouix, 1975 (1974)

Thalli up to 4.5 mm long by 8–16 μm diam, slightly swollen (10–20 μm diam) near the base, which may be recurved, often spiraled or sinuous.

Holdfast flat or isodiametric, usually located laterally near the base. Producing 1–4 terminal sporangia 250–480 × 8–12 μm, within each a partially folded secondary infestation sporangiospore 260–520 μm by ~10 μm diam with a tapered, evenly thickened apical cap. Sporangia may collapse in a pleated fashion after spore release. Some thalli producing 1–3(–6) often clavate multinucleate terminal cells 65–290 × 9–18 μm that are packed with nuclei.

Illustrations: See Manier et al., 1974, Fig. 5.

Host: In very anterior part of hindgut of *Plethocrossus acutiformis* Demange (Diplopoda, Odontopygidae).

Distribution: Terrestrial habitats, Dahomey, Africa.

According to the authors, the multinucleate cells may internally produce uninucleate cells measuring 10–15 μm diam.

See the note under *Enterobryus isoporostrepti* concerning the problem of identity of this and other species from Dahomey.

Reference: Manier et al., 1974 (1975).

☐ *Enterobryus ahlesi* Lichtwardt, 1960

Thalli straight or curved near the base, 1 cm or more long by (7–)12–24 μm diam, tip bluntly pointed. Holdfast 6–60 μm long, longitudinally striated, usually without a basal disk. Secondary infestation sporangiospores 43–70 × 7–10 μm, or ~160 × 11–14 μm, with bulbous proximal ends. Primary infestation spores rare or absent.

Illustrations: See Lichtwardt, 1960, Figs. 13–17.

Host: Throughout hindgut of *Apheloria montana* (Bollman) (Diplopoda, Xystodesmidae).

Distribution: Terrestrial habitats, Tennessee, U.S.A.

The author reported also finding in this host species two other larger spore types: one was fusiform; the other cylindrical, vacuolate, and with about 8 peripheral nuclei.

Reference: Lichtwardt, 1960.

☐ *Enterobryus apheloriae* Lichtwardt, 1954b

Thalli up to 10 mm (or slightly more) long by 7–12 μm diam, long–fusiform, tip bluntly pointed, and the base tapering gradually to the holdfast. Holdfast long (often ~60 μm) with a basal disk or enlargement. Secondary infestation sporangiospores (68–)70–100(–265) × 7–12 μm with a bulbous basal end.

Illustrations: Fig. 11.32C; aso see Lichtwardt, 1954b, Figs. 24–28.

Host: In hindgut of *Apheloria iowa* Chamberlin (Diplopoda, Xystodesmidae).

Distribution: Terrestrial habitats in Illinois, U.S.A.

Lichtwardt (1954b) also reported that this fungus produces cells, packed with nuclei, that measure ~60 × 11 μm. He erroneously thought that

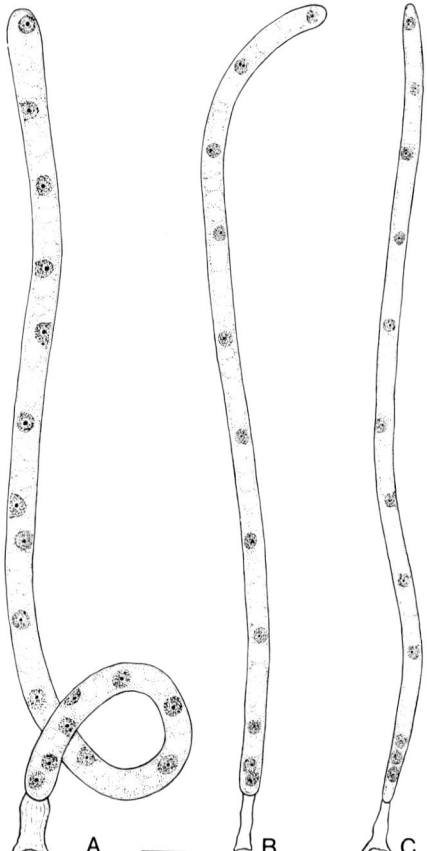

FIG. 11.32. *Enterobryus* species: morphology of young thalli. A. *E. elegans*. B. *E. euryuri*. C. *E. apheloriae*. Scale bar = 20 μm. Reprinted by permission from Mycologia, vol. 46: 566, Copyright 1954, R.W. Lichtwardt and The New York Botanical Garden.

some thalli with chains of binucleate cells that he described from the hindgut of this millipede were an unusual reproductive stage of *E. apheloriae*. However, it was undoubtedly an organism related to the non–trichomycete genus *Mononema;* see the last section of Chapter 12.

References: Lichtwardt, 1954a, 1954b.

☐ *Enterobryus attenuatus* Leidy, 1849b, *emend*. Lichtwardt, 1957a
 = *Eccrinopsis attenuatus* (Leidy) Léger & Duboscq, 1916
 [= *Trichella attenuatus* (Leidy) Poisson, 1931a, *nom. nud.*]

Thalli in anterior hindgut (ileum) growing singly but more often in whorled clusters with individual thalli attached to a common multiple holdfast sys tem, up to 1.4 mm long by 8–16 μm diam, producing uninucleate primary infestation sporangia 10–22 × 9–15 μm, or multinucleate secondary infestation sporangiospores 240–250 × 5–10 μm diam. Thalli in midregion of hindgut (colon) up to ~2 mm long, sinuous and tapering toward the apex, growing singly or less commonly in small tufts, usually nonsporulating.

Illustrations: Fig. 11.33; also see Lichtwardt, 1957a, Figs. 2–19.

Host: In hindgut of adult and larval *Popilius disjunctus* (Ill.) (= *Passalus cornutus* Fab.) and other Passalidae (Insecta, Coleoptera).

Distribution: Terrestrial, usually in decaying logs, in U.S.A., Brazil, Panama, and Trinidad.

Leidy (1849b, 1853) described only the nonsporulating thalli of *Enterobryus attenuatus* living in the colon of *Popilius disjunctus*, a large wood-inhabiting beetle that is sometimes called the Bess beetle. The fungus was more fully described by Lichtwardt (1957a). The author has found *E. attenuatus* in most populations of passalid beetles he has examined in the U.S.A. and the Neotropics. Manier and Théoridès (1965) partly described and illustrated a fungus they called *E. attenuatus* growing in the gut of a passalid beetle (*Leptaulax dentatus* F.) from Laos, but the extremely long holdfasts of the thalli, although sometimes fused together, suggest that it

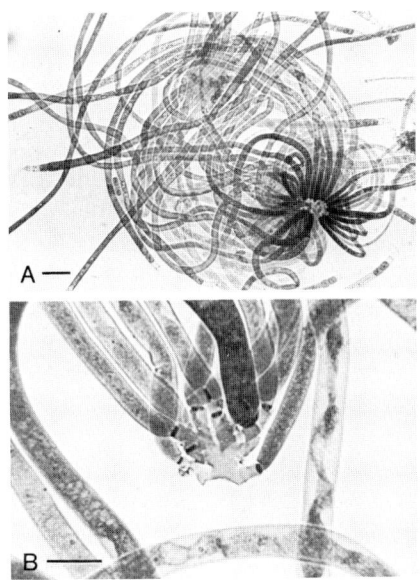

FIG. 11.33. *Enterobryus attenuatus*. A. Whorled cluster of individual thalli attached to a common multiple holdfast system. B. Holdfast system in lateral view. Scale bars: A, 20 μm; B, 20 μm. Reprinted by permission from Mycologia, vol. 49: 468, 470, Copyright 1957, R.W. Lichtwardt and The New York Botanical Garden.

may be another species. The fine structure of the multiple holdfast of *E. attenuatus* has been studied by Mayfield and Lichtwardt (1980), and is described in Chapter 7.

Passalid beetles also may have in their posterior rectum *Enterobryus compressus*, and, farther forward, an unnamed species of *Enterobryus* described by Heymons and Heymons (1934). The author has collected specimens of both, and the latter appears to be a distinct species of *Enterobryus*. More anterior in the rectum, some of these beetles have a small branched organism growing in the longitudinal fissures of the cuticle that looks like a small member of the Asellariales, but its method of reproduction and affinity to the trichomycetes have not been established.

References: Leidy, 1849b, 1853; Lichtwardt, 1954a, 1957a, 1968, 1978; Léger and Duboscq, 1916; Manier and Théoridès, 1965; Dang, 1978; Mayfield and Lichtwardt, 1980.

☐ *Enterobryus bifurcatus* Whisler, 1963

Thallus bifurcate, consisting of two divergent arms with a median lateral holdfast, often 1.2(–2.1) mm in overall length by 10(–14) μm diam. Primary infestation sporangiospores uninucleate, ~10 × 10 μm. Secondary infestation sporangiospores 4–nucleate, (43–)49(–56) × (7–)9 μm.

Illustrations: See Whisler, 1963, Figs. 1, 2.

Host: In posterior hindgut of *Californiobolus uncigerus* Wood (Diplopoda, Spirobolidae).

Distribution: In decaying logs, California, U.S.A.

This unusual thallus type develops due to the simultaneous growth of a lateral extension at the base of the thallus, while the more normal arm grows directly from the germinating spore. The terminal portions of both arms produce series of either primary or secondary infestation sporangiospores (Whisler, 1961), although the former are more common. In other respects, *E. bifurcatus* is a typical *Enterobryus*. Refer also to *Enterobryus tuzetae*, which produces a bifurcated thallus type, and to *Ramacrinella raibauti*, a member of the Eccrinaceae from amphipods, with multiple branching from the base of the thallus.

The anterior part of the hindgut of the same millipede species also contained another, unnamed *Enterobryus* sp., described by Whisler (1963) as having a large basal holdfast and averaging 7.2 mm in length by 21 μm diam, with multinucleate terminal cells measuring 91 × 21 μm. He considered such thalli likely to represent a different species.

References: Whisler, 1961, 1963.

☐ *Enterobryus borariae* Lichtwardt, 1958

Thalli up to 5 mm long, (3.5–)8–18(–50) μm diam. Holdfasts up to 100 μm long, holdfast stalk usually 6–12 μm diam with a prominent basal disk or swelling, longer stalks longitudinally striated. Primary infestation sporangiospores 9–16 μm long × 9–18 μm diam. Secondary infestation sporangiospores 45–170 × 8–32 μm, with a basal bulbous swelling. May produce other spore or cell types of unknown function or origin.

Illustrations: Figs. 7.11B, 7.12B; also see Lichtwardt, 1958, Figs. 1–20.

Host: In hindgut of *Boraria carolina* (Chamberlin) (Diplopoda, Xystodesmidae). Also attached to cuticle of oxyuroid nematodes in the hindgut, or to thalli of its own species.

Distribution: In leaf litter and under soil in North Carolina, U.S.A.

Enterobryus borariae is one of the most variable species of the genus, which makes it difficult to delineate the characters of the species (Lichtwardt, 1978a). Although known only from one millipede species, it attaches not only to the hindgut cuticle, but to the cuticle of parasitic nematodes as well. In addition, Lichtwardt (1958) illustrated spores, germinating spores, and even sporulating thalli of *E. borariae* attached to other thalli and sporangia of its own species.

In a few instances thalli have been seen producing a series of terminal secondary infestation sporangiospores, followed by a series of primary infestation sporangiospores on the same thallus, or vice versa, suggesting that the type of sporangium produced on a given thallus is not due to some innate dimorphism factor. Combinations of both sporangial types have also been seen in single thalli of *E. dixidesmi*.

Enterobryus borariae produces several other kinds of cells or spores, some which have no known function or origin. One of unknown function is the type tightly packed with nuclei, which has been seen in a few other species of *Enterobryus* as well. Also, peculiar cystlike pyriform bodies have been found beneath the hindgut lining of some millipedes (Fig. 7.11B). These were found in various stages of germination and penetration through the cuticle into the gut lumen, some having produced thalli with spores at their tips, indicating they were thalli of *E. borariae*. In the anterior region of some millipede hindguts were seen very narrow thalli producing overlapping bundles of sinuous, filiform sporangiospores, each in its own sporangium; these were intermixed with narrow thalli of *E. borariae* producing normal sporangiospores, and possibly were a stage of the same species. Finally, oval spores producing thalli as they germinated were observed in a few instances, but they were not associated with any stages of *E. borariae* producing such oval spores.

References: Lichtwardt, 1954a, 1958, 1978a.

☐ *Enterobryus cherokiae* Lichtwardt, 1960

Thalli straight or loosely spiraled, often 5–7 mm long by (8–)11–20(–26) μm diam. Mature holdfasts usually expanded into a broad disk up to 45 μm diam, without a holdfast stalk. Secondary infestation sporangiospores often with more than 8 nuclei, generally falling into one of three size ranges: (1) 100–240 × 8 μm; (2) 60–70 × 10–15 μm, with a distal bulbous end; and (3) 140–240 × 19–26 μm.

Illustrations: See Lichtwardt, 1960, Figs. 9–12.

Host: In hindgut of *Cherokia georgiana* (Bollman) (Diplopoda, Xystodesmidae).

Distribution: In leaf litter, North Carolina, U.S.A.

In addition to primary infestation sporangiospores, *E. cherokiae* produces series of terminal cells tightly packed with nuclei, as found in some other species of *Enterobryus*. Those of *E. cherokiae* are usually large, measuring 60–83 × 15–30 μm.

References: Lichtwardt, 1954a, 1960.

☐ *Enterobryus cingaloboli* Rajagopalan, 1967

Thalli up to 5 mm long, coiled or curved near the basal region, 15–22 μm diam near the base, gradually narrowing to 8–14 μm distally. Holdfast up to 30 μm long, with a prominent basal disc up to 20 μm diam. Secondary infestation sporangiospores 4–nucleate, 60–100 × 6–8 μm, borne on slender thalli. Wider thalli bearing 1–10 multinucleate cells 50–120 × 15–21 μm. Primary infestation spores rare or absent.

Illustrations: See Rajagopalan, 1967, Figs. 1–6.

Host: In hindgut of *Cingalobolus carli* Attems (Diplopoda, Pachybolidae).

Distribution: In plant debris, Kerala State, India.

Reference: Rajagopalan, 1967.

☐ *Enterobryus compressus* Thaxter, 1920

[= *Trichella compressus* (Thaxter) Poisson, 1931a, *nom. nud.*]

Thalli 0.5–3.5 mm long by 20–35 μm diam, with walls 2–4 μm thick. Holdfast like an inverted cup. Apex of thallus becoming partially flattened upon producing a series of terminal sporangia 5–8 μm long by 20–35 μm wide, each containing a thin–walled, multinucleate, more or less allantoid sporangiospore slightly smaller than the sporangium.

Illustrations: See Thaxter, 1920, Figs. 47–52.

Host: On anal plates and posterior rectum of Passalidae (Insecta, Coleoptera).

Distribution: Terrestrial, usually in rotting logs, in Dominica (West Indies), Panama, and Brazil.

This is an unusual species of *Enterobryus* in several respects. The flattened and sometimes slightly twisted ends of the thick–walled thalli produce series of short but wide sporangia, each containing a sausage–shaped multinucleate spore that lies at a right angle to the thallus axis. These spores are thin–walled, appear to be 4–nucleate at maturity, and emerge from the sidewall of the sporangium. They are morphologically secondary infestation spores. However, in this species the position of the thalli on the anal plates of the passalid beetle exposes the thalli and spores to the exterior of the gut; consequently, if such spores are functional they would function as primary infestation spores. Thaxter described thalli as "growing outside the anus," but in the author's collections of *E. compressus* (unpublished) from Panama and Brazil, very long sporulating thalli also occur within the gut, attached to the rectum. The spores produced by the long thalli often lie within the gut lumen, and are identical to the more exposed spores; possibly they can function as secondary infestation spores. These same beetles often contain *Enterobryus attenuatus*, a more common species, in the anterior hindgut.

References: Thaxter, 1920; Lichtwardt, 1978.

☐ *Enterobryus cylindroiuli* Manier, 1969
[= *Enterobryus cylindroiuli* Tuzet & Manier, 1949, *nom nud.*]
[= *Enterobryus cylindroiuli* Tuzet & Manier *in* Manier, 1950 (1951), *nom. nud.*]

Thalli essentially straight, 600–800(–1200) μm long by 4–9.5 μm diam. Holdfasts 5–6 μm long, narrowing at the top. Narrower thalli (4–6 μm) producing 4–nucleate secondary infestation sporangiospores 40–60 × 4–6 μm. Wider thalli (7–9.5 μm) producing 2–20 uninucleate cells which are either (1) swollen and isodiametric (7 × 7 μm) or elongated (17–20 × 7–9 μm), or (2) cylindrical, elongated, and vacuolate (24 × 7 μm).

Illustrations: None available.

Host: In hindgut of *Cylindroiulus londinensis* Leach (Diplopoda, Julidae).

Distribution: Terrestrial habitats, Department of Hérault, France.

The name of this species was first used and designated as a new species by Tuzet and Manier in 1949, but no description was provided. The first description (without calling it a new species or validating the name) appeared in Manier's 1950 published dissertation. In 1969, Manier validly published the species name (but as though she were validating her 1950 species), and provided a concise description upon which my diagnosis is primarily based. There is no indication that either of the uninucleate cell types that Manier described function as sporangia for primary infestation sporangiospores; they may be merely cells of unknown function similar to others that are not uncommon in *Enterobryus* spp.

References: Manier, 1950 (1951), 1969; Tuzet and Manier, 1949.

☐ *Enterobryus dixidesmi* Lichtwardt, 1960c

Thalli tightly coiled throughout their length, or curved predominantly near the base, up to 1.5 mm long (usually much shorter) by (5–)9–13(–17) μm diam. Holdfast stalk usually very short, with or without a basal disk. Secondary infestation sporangiospores 40–90 × 9–11 μm. Primary infestation sporangiospores 8–11 × 9–13 μm.

Illustrations: See Lichtwardt, 1960c, Figs. 1–8.

Host: In hindgut of *Dixidesmus tallulanus* Chamberlin (Diplopoda, Polydesmidae).

Distribution: In leaf litter, North Carolina, U.S.A.

The spiraled thalli of *E. dixidesmi* resemble those of *E. moniliformis*, but the two species differ in the sizes of their spores and holdfasts. Spiraling is most prevalent in well-infested guts. It should be noted that thalli in this species, and other spiraled species, can become straighter during dissection and preparation of slides. Sporangia producing primary infestation spores were found in series of up to 42 per thallus. Like *E. borariae*, single thalli of *E. dixidesmi* may occasionally produce in sequence both secondary and primary infestation sporangiospores. Small ellipsoid spores of unknown origin also were observed germinating in the gut of this millipede host, similar to those seen in some millipedes harboring *E. borariae*.

Reference: Lichtwardt, 1954a, 1960c.

□ *Enterobryus elegans* Leidy, 1849a, *emend*. Lichtwardt, 1954b

Thallus usually having a single loose coil near the base, up to 5 mm long, 19–23 μm diam, tapering to ~12 μm diam near the apex. Young thalli often apically clavate. Holdfast irregularly conical or campanulate, sometimes cylindrical above the flared base, usually large and prominent when fully formed but variable in size, commonly 30–85 μm long by 35–50 μm diam at the base. Secondary infestation sporangiospores typically 50–65 μm long by ~12 μm diam, and 8–nucleate. Primary infestation sporangiospores absent or rare. Some thalli producing slightly swollen terminal cells with numerous scattered nuclei, 50–100 × 15–22 μm. Type species.

Illustrations: Figs. 7.13, 11.32A, 11.34.

Hosts: *Narceus americanus* (Beauvois), *N. annularis* (Rafinesque), and probably many other *Narceus (Spirobolus)* spp. (Diplopoda, Spirobolidae). Often attached to the cuticle of oxyuroid nematodes within the millipede hindgut.

Distribution: In and around decomposing leaves and logs in eastern North America, at least as far west as Kansas, U.S.A.

Enterobryus elegans was the first species of trichomycetes to be described. The genus name was first spelled *Enterobrus*, but Leidy corrected the orthography in 1850(b). The fungus is present in many, if not most, species of *Narceus* in eastern U.S.A. The author has found it in millipedes in Kansas, Illinois, Michigan, New York, North Carolina, and Georgia. Leidy collected his material in Pennsylvania. Wright (1979) studied *E. elegans* in *N. annularis* from Ontario, Canada. The nomenclature of the original host species, *N. americanus*, has been in a state of flux, and its precise distribution remains uncertain. Leidy (1849a) called it *Julus marginatus* Say, whereas Hoffman (1951) established the name *Spirobolus americanus* (Beauvois). Dogma (1975) reported *E. elegans* from an unidentified millipede species in the Philippines, but provided no assurance that his identification was correct.

Leidy's descriptions of *E. elegans* (1849a, 1853) were very incomplete,

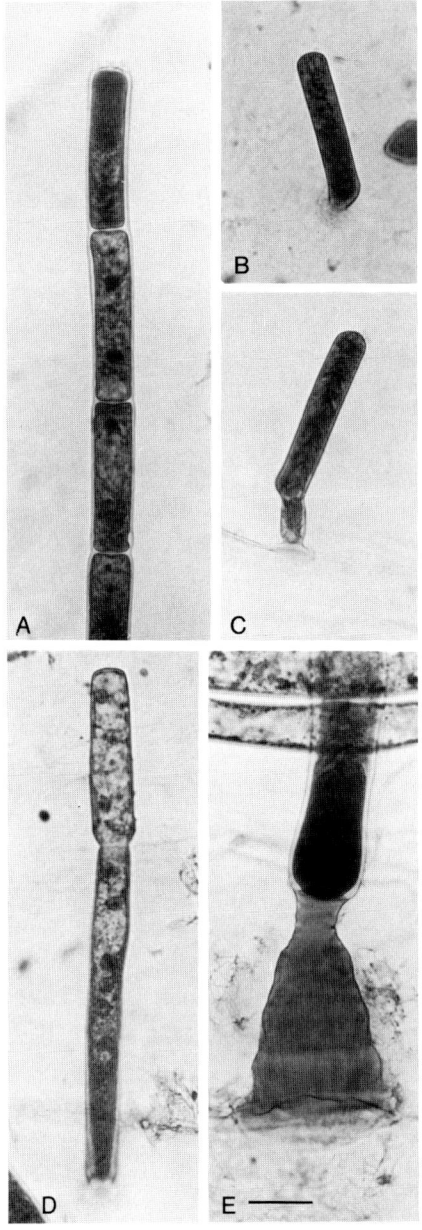

FIG. 11.34. *Enterobryus elegans*. A. Secondary infestation sporangiospores within a thallus. B. Released sporangiospore attached to millipede hindgut cuticle. C. Sporangiospore beginning to germinate. D. Further stage of thallial development, with the spore mother–cell positioned terminally. E. Large holdfast of a mature thallus (also see Fig. 7.13). Scale bar = 20 μm for all figures.

but it was restudied by Lichtwardt (1951, 1954a, 1954b, and unpublished). This species does not appear to produce primary infestation sporangiospores, which leads to the interesting question of how it is transmitted so successfully from one millipede to another, since most spirobolid millipedes are infested. One possibility is that the secondary infestation sporangiospores, which normally serve to multiply the fungus endogenously, are the propagules that transmit the fungus to other individuals, and perhaps also can reinfest molting millipedes that eat part of their molt. The large multinucleate cells produced by some thalli could conceivably serve this function, but this has not been established. The multinucleate cells may break loose, according to Leidy (1853). Surprisingly, Leidy never found sporangiospores despite doing a large number of dissections over many seasons (Leidy, 1853). The slightly swollen multinucleate cells (which appear to be what Leidy described and illustrated) do not produce sporangiospores, and can be distinguished by their numerous nuclei scattered in the cytoplasm. In contrast, the sporangia are generally narrower and cylindrical, and, as they develop, a distinct wall is laid down by the sporangiospore, and the usual 8 nuclei at maturity are lined up more or less in a row.

The attachment of *E. elegans* to parasitic nematodes living in the hindgut is not unique to this species of *Enterobryus*, and has been observed by several persons (see Table 7.2). Lichtwardt (1954b) reported that, just before molting, a modified spore type that attached to nematodes was produced in some thalli, and that this attachment could be one means by which the fungus remained in the gut during ecdysis when other thalli attached to the hindgut cuticle were being expelled.

The holdfast of *E. elegans* has been studied with the electron microscope by several investigators (Dang, 1978; Wright, 1979; Mayfield and Lichtwardt, 1980), and its interesting ultrastructural features are discussed in Chapter 7 in the section on Holdfasts.

References: Leidy, 1849a, 1853; Lichtwardt, 1951, 1954a, 1954b; Dogma, 1975; Dang, 1978; Wright, 1979; Mayfield and Lichtwardt, 1980.

□ *Enterobryus euryuri* Lichtwardt, 1954b

Thalli with tips frequently curved, up to 4 mm long by (6–)10–13(–30) μm (rarely up to 90 μm) diam. Holdfasts unevenly cylindrical, (4–)15–25(–75) × (4–)6–8(–12) μm, with or without a basal disk. Secondary infestation sporangiospores (32–)50–100 × (6–)10–13(–16) μm, with a basal bulbous swelling. Primary infestation sporangiospores 5–15 × 10–13 μm diam.

Illustrations: Figs. 7.8, 7.20C, 11.32B; also see Lichtwardt, 1954b, Figs. 17–22.

Host: In hindgut of *Euryurus erythropygus* (Brandt) (Diplopoda, Platyrhacidae).

Distribution: In rotting logs and leaf litter, Illinois, U.S.A.

Enterobryus euryuri also produces two cell types of unknown function.

One consists of series of swollen uninucleate cells often 11 μm long by 22 μm diam borne apically on wide thalli 23–30(–90) μm diam. The other cell type is filled with tightly packed nuclei, and measures ~35–60 × 11 μm diam. Although nematodes are often abundant in this millipede species, the fungus apparently is unable to attach to their cuticle, as do a few other species of *Enterobryus*.

References: Lichtwardt, 1954a, 1954b.

☐ *Enterobryus halophilus* Cronin & Johnson, 1958

Thalli sinuous and curved, 200–4500 μm long, 10–15 μm diam, usually with a wider basal part up to 25 μm diam, often with an apical cap. Holdfast 9–18 μm long, shaped like an inverted cup. Secondary infestation sporangiospores 35–58 × 8–14 μm. Primary infestation sporangiospores 8–14 × 11–17 μm diam.

Illustrations: See Cronin and Johnson, 1958, Figs. 1–36.

Hosts: Attached to hindgut cuticle (sometimes on anal hairs and consequently exposed) of *Emerita talpoida* Say and *E. analoga* (Stimpson) (Crustacea, Decapoda, Hippidae), commonly called mole crabs.

Distribution: Wave–washed sandy beaches of intertidal zone, North Carolina and California, U.S.A.

In addition to the two spore types cited above, Cronin and Johnson (1958) reported finding uninucleate endospores developing in short distal segments of some thalli. Such spore types have been occasionally reported by Manier and co–workers in a few species of *Enterobryus*, but in the writer's judgment these may be artifacts, or at least not a functional spore type. Almost certainly artifactual are the protoplasmic segments of various shapes that Cronin and Johnson observed emerging from the tips of some thalli; these are not uncommon in disintegrating, damaged thalli, or when the osmolality of the surrounding medium is not satisfactory. The writer has collected *E. halophilus* on the shores of North Carolina, but not in sufficient quantity to verify the observations of Cronin and Johnson.

An interesting point is that *E. halophilus* was found by Cronin and Johnson only in female specimens of the mole crab, *Emerita talpoida*, although males were also examined. If this proves to be constant, it would be the only recorded instance where infestation of an arthropod by a trichomycete is sex dependent. *Emerita analoga* is a mole crab found on the west coast of the U.S.A., where Whisler (1963) reported finding in its hindgut a fungus that resembled *E. halophilus*.

Although a number of species of Eccrinales are marine, *E. halophilus* is the only published marine species of *Enterobryus*. Nevertheless, it appears to be classified in the correct genus based upon current concepts of *Enterobryus*. Several mud–burrowing true crabs (Brachyura) have undescribed trichomycetes in their hindguts that also appear to be species of *Enterobryus*. These have been collected and partially studied by the writer (but not published) in several burrowing crabs, including at least three species of fiddler crabs on the east coast of the U.S.A. *(Uca pugilator, U. pugnax, U. minax)*, and another species of eccrinid in the west coast fiddler crab *(U. crenulata)*. Wagner–Merner (1979) and Mattson and TeStrake (1983) have published their investigations of the *Enterobryus* sp. from *Uca pugilator*. Wolf and Wolf (1947) reported that the mud crab, *Panopeus herbstii*, contains a gut fungus that they called *Enterobryus* sp. All of these, and several other crab fungi the writer has collected, need further study to determine if *Enterobryus* spp. are common in some kinds of marine Decapods, and to ascertain if they indeed belong to the genus *Enterobryus*.

References: Cronin and Johnson, 1958; Johnson and Sparrow, 1961; Whisler, 1961, 1963.

☐ *Enterobryus hydrophilorum* (Léger & Duboscq) Manier, 1970b (1969b)
  = *Eccrinopsis hydrophilorum* Léger & Duboscq, 1916 (basionym)
  [= *Trichella hydrophilorum* (Léger & Duboscq) Poisson, 1931a, *nom. nud.*]

Thalli coiled near the base, 3–6 mm long by 9–20 μm diam, wider near the basal region. Holdfast near cylindrical to campanulate or funnel-shaped, often longitudinally striated. Secondary infestation sporangiospores 400–600 μm long by ~9–11 μm diam. Primary infestation sporangiospores sometimes in long series (up to 51), variable in size, 25–32 × 12–16 μm, but in wider thalli spores sometimes wider than long.

Illustrations: See Léger and Duboscq, 1916, Figs. I–III.

Hosts: In anterior hindgut of *Hydrous piceus* L., *H. pistaceus* Cost., and *H. flavipes* Steph. (Coleoptera, Hydrophilidae).

Distribution: Aquatic habitats in Departments of Hérault, Pyrénées-Orientales, and Manche, France; and Germany (German Democratic Republic).

The basionym of this species is *Eccrinopsis hydrophilorum*, the type species for the genus *Eccrinopsis* Léger & Duboscq, 1916, a genus no longer valid. [*Eccrinopsis* had been proposed but not validly published earlier by Léger and Duboscq (1906)]. *Trichella hydrophilorum* is the type species for *Trichella* Poisson 1931a, likewise no longer valid.

Enterobryus helocharei also occurs in a hydrophilid beetle. Both of these species of *Enterobryus* are reported to shed their shorter and wider primary infestation spores by a mechanism whereby the elastic sporangial wall acts as an elater (Léger and Duboscq, 1916; Poisson, 1931a).

References: Manier, 1969b (1970b), Léger and Duboscq, 1916; Poisson, 1931a; Duboscq et al., 1948; Maessen, 1955.

☐ *Enterobryus isoporostrepti* Manier, Gasc & Bouix, 1975 (1974)

Thalli sinuous, curved at the base, 1–1.5(–2.1) mm long by 6–14 μm diam. Holdfast often wider than long, about as wide as the thallus. Producing two forms of secondary infestation sporangiospores: (1) long and folded within a sporangium 125–320 μm long, upon liberation being straight and tapered to a point apically, basal end rounded, 250–350 μm long, with 2–8 nuclei in a row; and (2) short, 20–30 × 15–18 μm, with 6–8 irregularly arranged nuclei.

Illustrations: See Manier et al., 1974, Figs. 2A–D.

Host: In very anterior part of hindgut of *Isoporostreptus bouixi* Demange (Diplopoda, Spirostreptidae).

Distribution: Terrestrial, Dahomey, Africa.

This species also produces terminal series of 1–5 cells packed with nuclei (or uninucleate cells?) 60–110 × 8–16 μm.

Note: Manier et al. described in 1974 a number of species of *Enterobryus* living in millipedes from Dahomey, Africa. Four of these *(Enterobryus isoporostrepti, E. adjanohouni, E. onychostrepti,* and *E. peridontopygei),* from millipede hosts representing two families, have certain characteristics that make them very similar, if one excludes certain cell types of unknown function that are seen in many other *Enterobryus* spp., and if one accepts the considerable intraspecific variation of many *Enterobryus* spp. All of these four fungi live in the very anterior part of the hindgut just posteriad to the Malpighian tubes and were found in millipedes taken in the Porto–Novo botanic garden in Dahomey. Especially notable is a unique form of secondary infestation sporangiospore produced by all four species: It is very long (often several hundred micrometers) and folded within the somewhat shorter sporangium, and upon release the sporangium usually collapses in a pleated fashion, while the straightened released spore is seen to be tapered and pointed at its distal end. The question of proper specific identity arises. Two possibilities are suggested. One is that the four mentioned species are, in fact, conspecific and transmissible among the different millipedes, but differ morphologically because of their phenotypic plasticity. The other possibility is that more than one species of eccrinid lives in the millipedes' gut, and that the thalli with the long tapered spores belong to one fungal species and is shared by the other millipede species. Manier et al. (1974) recognized and discussed the difficulty of determining the species described in that paper. Since the problem presented here cannot be resolved at this time, the writer has chosen to maintain the four species as distinct until it can be demonstrated otherwise experimentally.

Reference: Manier et al., 1974 (1975).

☐ *Enterobryus leptoiuli* Manier ex Manier, 1970b (1969b)

Thalli up to 2 mm long, curved at the base; basal region 9–15 μm diam, narrowing to 4.5–7 μm diam near the apex. Holdfast 9–19 μm long, irregularly campanulate, with only a slight basal enlargement. Secondary infestation sporangiospores 8–nucleate, 70–115 μm long, produced in series of 2–3 per thallus. Producing primary infestation sporangiospores in series of up to 65 per thallus, of three kinds: (1) 5–6 × 7 μm diam; (2) isodiametric, 7 × 7 μm; and (3) 8–11 × 4.5–6 μm diam.

Illustrations: See Manier, 1950, Figs. IV, 1–5 (but not Figs. IV, 6–10; see comments below).

Host: In hindgut of *Leptoiulus belgicus* Latzet (Diplopoda, Julidae).

Distribution: Terrestrial habitats, in Department of Hérault, France.

The original description of *E. leptoiuli* (Manier, 1950) included some peculiar thick-walled elongate to oval spores, which would proscribe placing this species in the genus *Enterobryus*. However, Manier restudied the fungus and believes those structures were artifacts (personal communication). Her 1969 description, the basis for the above diagnosis, differs from her earlier one in several other respects as well.

References: Manier, 1950 (1951), 1969b (1970b).

□ *Enterobryus moniliformis* (Leidy) Lichtwardt, 1957b
= *Eccrina moniliformis* Leidy, 1850b

Thalli up to 3 mm long by 6–12 μm diam, spiraled throughout their length, or curved or spiraled only at the base. Holdfasts 12–35 × ~10 μm, usually constricted near the attachment to the thallus, without a prominent basal disk; longer holdfasts usually striated longitudinally. Secondary infestation sporangiospores about 50–60 × 5–8 μm. Primary infestation sporangiospores about 8–15 × 6–10 μm, becoming spherical to oval upon release.

Illustrations: Fig. 7.20A; also see Lichtwardt, 1957b, Figs. 1–14.

Host: In hindgut of *Scytonotus granulatus* (Say) (Diplopoda, Polydesmidae).

Distribution: In leaf litter, Illinois and Pennsylvania, U.S.A.

Leidy's (1850b) genus *Eccrina* cannot be differentiated from his first genus *Enterobryus* merely on the basis of the number of terminal cells (sporangia) produced, as Leidy had proposed (Lichtwardt, 1954b). *Eccrina moniliformis*, collected from the same millipede species as Leidy's (he called it *Polydesmus granulatus*), was restudied by Lichtwardt (1957b) and transferred to the genus *Enterobryus*. Leidy (1850b) originally used the specific epithet *moniliforma*, but changed it to *moniliformis* in 1853. He never illustrated this species. Lichtwardt reported seeing ellipsoidal spores, 19 × 5 μm, germinating in the millipedes' gut, but the origin of the spores was not determined.

References: Leidy, 1850b, 1853; Lichtwardt, 1954a, 1957b.

□ *Enterobryus onychostrepti* Manier, Gasc, & Bouix, 1975 (1974)

Thalli up to 6 mm long by 6–28 μm diam, sinuous or looped. Holdfast basal, or formed on one side of an incurved base. Secondary infestation sporangiospores folded within sporangia 90–270 × 6–10 μm, upon release unfolding to 180–460 μm long, tapered and with a pointed distal end.

Illustrations: See Manier et al., 1974, Figs. 3A–G.

Hosts: In very anterior hindgut of *Onychostreptus aoutii* Demange and *O. assiniensis* Attems (Diplopoda, Spirostreptidae).

Distribution: Terrestrial, Dahomey, Africa.

The authors of this species also reported the production on some thalli of 1–3 terminal cells measuring 70–400 μm long; these were packed with nuclei (or round spores?).

See the note under *Enterobryus isoporostrepti* concerning the problem of identity of this and other species from Dahomey.

Reference: Manier, Gasc, and Bouix, 1974 (1975).

□ *Enterobryus oxidi* Lichtwardt, 1960a

Thalli dimorphic. Larger thalli attached to cuticle by a large holdfast with a basal disk, located in most anterior region of the hindgut, with a prominently curved base, 15–21 μm diam, tapering to 6–9 μm diam distally, producing primary infestation sporangiospores 6–13 × 6–9 μm diam. Smaller thalli more scattered, straight or curved near the base, 5–7 μm diam, small holdfasts with variable shapes, producing either primary infestation sporangiospores 6–13 × 6–9 μm, or secondary ones up to 170 μm long by ~6 μm diam.

Illustrations: See Lichtwardt, 1960a, Figs. 1–6.

Host: In anterior hindgut of *Oxidus gracilis* (Koch) (Diplopoda, Paradoxosomatidae).

Distribution: In greenhouse in Illinois, U.S.A., and forest humus in Madagascar.

Enterobryus flavus Maessen (1955), a species the writer does not recognize because of an insufficient description, was found by Maessen in the same millipede species in the German Democratic Republic (under the name *Orthomorpha gracilis* Koch). Lichtwardt did not believe her species resembled *E. oxidi,* although Tuzet and Manier (1967) thought the two might be conspecific.

Lichtwardt (1960a) reported finding some ellipsoidal spores of unknown origin measuring 18–20 × 4–5 μm and germinating in the millipede gut. He also found rare loose thalli bearing binucleate cells, which he erroneously thought were a different stage of *E. oxidi*. However, that organism is now clearly more closely related to the non–trichomycete genus *Mononema* (see last section in Chapter 12).

Tuzet and Manier (1967) found an *Enterobryus* in *Oxidus gracilis* from Madagascar that they consider to be *E. oxidi*. Their study included observations on the fine structure of vegetative thalli. They also reported finding additional cell (spore?) types: Disklike uninucleate cells and longer binucleate cells on some wide (25–30 μm diam) thalli, as well as some multinucleate cells on narrower thalli.

References: Lichtwardt, 1954a, 1960a; Tuzet and Manier, 1967; Maessen, 1955.

☐ *Enterobryus oxydesmi* Manier, Gasc, & Bouix, 1975 (1974)

Thalli up to 6.5 mm long, either (1) cylindrical and 6–10 or 16–20 μm diam, or (2) wider at the base, 21–31 μm diam, and tapering to 8–20 μm diam distally. Holdfasts large and conical. Secondary infestation sporangiospores 90–220 × 6–14 μm. Primary infestation sporangiospores more or less isodiametric. Other uninucleate cells like stacked disks formed from multinucleate cells 80–130 μm long.

Illustrations: See Manier et al., 1974, Figs. 6A–D.

Host: In hindgut of *Oxydesmus ganulosus* Palisot de Beauvois (Diplopoda, Polydesmidae).

Distribution: Terrestrial, Dahomey, Africa.

Reference: Manier et al., 1974 (1975).

☐ *Enterobryus peridontopygei* Manier, Gasc, & Bouix, 1975 (1974)

Thalli curved to spiraled at the base, 1–2(–4.5) mm long by 14–28 μm diam. Holdfast as long as, or longer than, the diameter, with a slight basal enlargement. Secondary infestation sporangiospores 130–200 × 20–24 μm; or 170–370 × 14–20 μm, tapered to an apical point having a small bulbous wall thickening.

Illustrations: See Manier et al., 1974, Figs. 4A–C.

Host: In very anterior hindgut of *Peridontopyge gasci* Demange (Diplopoda, Odontopygidae).

Distribution: Terrestrial, Dahomey, Africa.

Manier et al. also reported that this species produces some multinucleate cells 70–260 × 14–28 μm that cleave internally into uninucleate (rarely 2– to 3–nucleate) round or oval (21–28 × 18–21 μm) cells. Such "sporangia" become swollen and deformed. See comments under *Enterobryus tuzetae* relating to such structures. Also see the note under *Enterobryus isoporostrepti* concerning the problem of identity of this and other species from Dahomey.

Reference: Manier et al., 1974 (1975).

☐ *Enterobryus tuzetae* Manier, Gasc, & Bouix, 1972a

Four thallial types produced in four (anterior to posterior) zones of hindgut: (I) large thalli up to 6 mm long by 40 μm diam, with a holdfast 5–10 × 35–40 μm diam, producing (a) short multinucleate (up to 6 nuclei) secondary infestation sporangiospores with their bases curved back upon themselves, or (b) longer, 2– to 4–nucleate, secondary infestation sporangiospores 80–90 × 16–22 μm; (II) large thalli 1–5 mm long by 14–28 μm diam near the base and 11–26 μm diam distally, with a holdfast 15–40 × 10–14 μm diam with a basal disk, producing secondary infestation sporangiospores 70–110 μm long; (III) small thalli forked above the small (4–10 × 6–8 μm diam) holdfast so as to have two branches of unequal length (600–900 and 400–700 μm) and 5–8 μm diam, both branches producing 4–nucleate secondary infestation sporangiospores 40–50 μm long, or longer branch producing 1–2 uninucleate primary infestation sporangiospores; (IV) small thalli curved at their base, 1350 μm long by 5–10 μm diam, with a small holdfast, rarely producing primary or secondary infestation sporangiospores.

Illustrations: See Manier et al., 1972a, Figs. 1–4.

Host: In hindgut of the large millipede, *Pachybolus ligulatus* (Voges) (Diplopoda, Pachybolidae).

Distribution: In forests on litter or under surface of soil, Dahomey, Africa.

Thalli in zones I and II may also produce multinucleate spores, in the latter said to convert into uninucleate bodies. Upon molting of the host, some thalli of *E. tuzetae* are reported to produce throughout their length swollen and very irregularly shaped cells ("sporangia") that become filled with ovoid, thick–walled spores measuring ~2–26 × 1.5–16 μm. The bifurcated thalli in zone III are similar in their development to *Enterobryus bifurcatus*, which see.

Enterobryus tuzetae is one of the most unusual and variable species of *Enterobryus*, but it is also one of the most thoroughly studied. The host is one of the large spirobolid millipedes such as occur in some tropical regions: Mature adults may be 13 cm long by 15 mm diam, and the hindgut alone measures 7–8 cm long (Manier et al., 1972a). The location of four thallial types producing morphologically different spores within four zones of the hindgut is reminiscent of the kind of partitioning described by Hibbits (1978) for *Enteropogon sexuale* in the hindgut of the marine anomurid *Upogebia pugettensis*. In instances such as these, where intermediate forms or developmental continuity are not clearly evident among the different thallial and spore types, one can rightly ask if they may not represent more than one species. Lichtwardt (1957a, 1978a) decided that in the hindgut of passalid beetles, which also have very long hindguts divided into at least four recognizable zones, the fungal flora consists of several different species of trichomycetes (see the discussion under *Enterobryus attenuatus*). In the absence of experimental evidence to the contrary, the writer has tentatively accepted the conclusion of Manier et al. that the fungi in the millipede *Pachybolus ligulatus* represent one species, *E. tuzetae*.

Another problem presents itself, however. Are the unusual, completely segmented thalli with swollen and distorted cells containing masses of oval thick–walled spores found by Manier et al. in molts of these millipedes actually a reproductive stage of *E. tuzetae*? Terminal cells of *Enterobryus* spp. are sometimes densely packed with nuclei. These form in thalli that already contain more nuclei than other thalli normally contain, and the function of such cells is unknown (for examples, refer to *E. apheloriae, E. borariae, E. cherokiae,* and *E. euryuri*). In several other species of

Enterobryus from Dahomey, Manier et al. (1974) have described similar cells, but sometimes containing fewer nuclei than the ones just described, and these are said to produce uninucleate (rarely 2– to 3–nucleate) spore-like bodies in *E. isoporostrepti, E. onychostrepti,* and *E. peridontopygei.* In *E. peridontopygei* the "sporangia" that produce such cells become swollen and distorted, as in *E. tuzetae.* The question here is whether such hypertrophied structures are indeed a reproductive phase of the trichomycete, or whether they might result from endoparasitism. For example, some water molds (Saprolegniales, Blastocladiales) parasitized by other fungi, such as the endoholocarpic genera *Olpidiopsis* (Lagenidiales), *Rozella* (Chytridiales), and *Woronina* (Plasmodiophorales), are stimulated by the parasite to produce crosswalls in their coenocytic thalli, and these cells frequently hypertrophy. The cells or cysts within the hypertrophied and distorted cells of *E. tuzetae* do not appear to be identical to the reproductive stages of the fungal parasites cited above, but given the rich flora and fauna of millipede guts, including many kinds of protozoans, the possibility of endoparasitism should be given some consideration. On the other hand, if the *E. tuzetae* cell types under discussion are in fact a functional form of reproduction as Manier et al. claim, *E. tuzetae,* and other species with such reproduction, would probably not belong to the genus *Enterobryus,* based upon the characters of the type species, *E. elegans.*

References: Manier et al., 1972a, 1974 (1975).

Enteromyces

Lichtwardt, 1961a

Thalli dimorphic, growing mostly in tufts from a multiple holdfast system. Spore mother–cell persistent. Producing thick–walled uninucleate sporangiospores, or thin–walled uninucleate or multinucleate sporangiospores. Attached to foregut cuticle of Anomura and Branchyura (Crustacea). Monotypic.

Type species: *Enteromyces callianassae* Lichtwardt.

☐ *Enteromyces callianassae* Lichtwardt, 1961a

Thalli dimorphic, up to 1.6 mm long by 25–46 or 10–14 μm diam, with a persistent spore mother–cell that may later function as a sporangium. Usually growing in tufts, holdfast of each thallus attached to a predominant holdfast that anchors the complex to the gut cuticle. Thick–walled (up to 7 μm thick) uninucleate sporangiospores subspherical, 25–35 × 30–45 μm. Small uninucleate primary infestation sporangiospores 4–7 × 7–13 μm. Multinucleate (~16 nuclei) secondary infestation sporangiospores 30–55 × 23–45 μm. Type species.

Illustrations: Fig. 11.35; also see Lichtwardt, 1961a, Figs. 1–13, and Hibbits, 1978, Figs. 5–7.

Hosts: In foregut (stomach) of many Crustacea: *Callianassa uncinata*

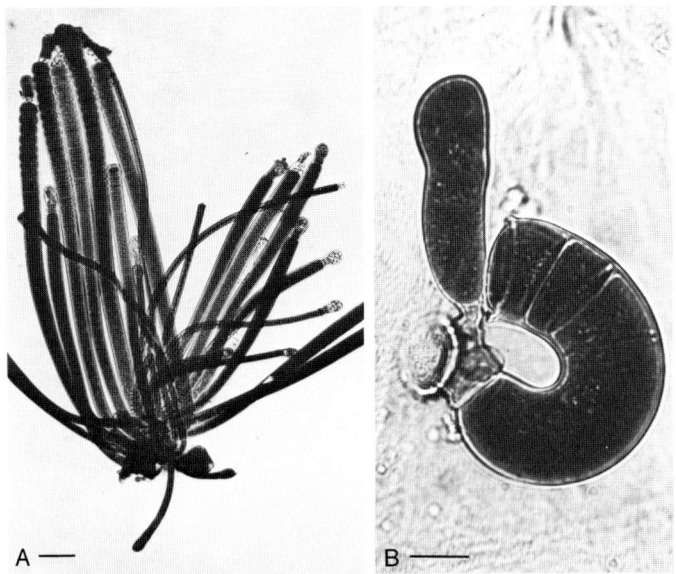

FIG. 11.35. *Enteromyces callianassae*. A. Tuft of individual thalli attached to a multiple holdfast system; note dimorphism in the thalli. B. A thallus which has become shortened due to sporulation. It has deposited a secondary infestation sporangiospore directly onto its own holdfast, and the attached spore is beginning to germinate. This process repeated many times results in tufts of thalli. Scale bars: A, 50 µm; B, 20 µm. Fig. 11.35A from Lichtwardt, 1976, by permission of Granada Publishing Ltd.

H. Milne Edwards, *C. brachyophthalma* H. Milne Edwards, *C. gigas* Dana, *C. californiensis* Dana, *Upogebia affinis* (Say), *U. pugettensis* (Dana) (Anomura, Callianassidae); *Uca pugilator* (Bosc) (Ocypodidae), *Hemigrapsus penicillatus* (de Haan) (Grapsidae) (Brachyura).

Distribution: Marine, intertidal mud or sand flats, or deeper water, in Chile, France, Japan, and U.S.A. (North Carolina, California, Washington).

Enteromyces callianassae is not only widely distributed, but is known to infest both true crabs (Brachyura) and shrimplike crabs (Anomura) commonly called ghost or mud shrimps. Most of the known hosts are intertidal, but one, *Callianassa brachyophthalma*, was taken at a depth of 35–40 m. *Arundinula galatheae* is the only other marine trichomycete found in a host approaching this depth. It is probable that the fungus is more widely distributed than present records indicate.

Enteromyces spp. appear to be restricted to the foregut of their hosts, in contrast to *Arundinula* spp., which occur also in the hindgut. The foregut in the anomurid *Upogebia pugettensis* may contain *E. callianassae* at the same time the hindgut is infested with *Enteropogon sexuale* (Hibbits, 1978).

The tufts of dimophic thalli that readily identify *E. callianassae* arise when one or more curved thalli deposit their sporangiospores directly onto the first and predominant holdfast, where the spores attach and germinate (Fig. 11.35B). Dozens of thalli (up to 160, according to Hibbits) may constitute the mature cluster, but clusters may be found in all stages of development. *Enterobryus attenuatus* in the hindgut of passalid beetles also form tufts of thalli.

References: Lichtwardt, 1961a; Tuzet and Manier, 1962; McCloskey and Caldwell, 1965; Galt, 1971; Hibbits, 1978.

Enteropogon

Hibbits, 1978

Four thallus types, each producing thin–walled uninucleate and multinucleate cells. One thallus type producing scalariform fusions. In hindgut of Anomura (Crustacea). Monotypic.

Type species: *Enteropogon sexuale* Hibbits.

The type species of *Enteropogon* is certainly distinct from all other Eccrinales. Whether the genus is distinct is less certain. In many respects it resembles *Enterobryus,* as that genus is currently conceived. The most unusual, but not unique, feature of *Enteropogon* is the fusion of thalli in a manner that could be interpreted as a sexual process; however, it has not yet been adequately demonstrated to involve karyogamy, nor has it been determined how it fits into the life cycle of the fungus. See the discussion in Chapter 7 under the section on sexual reproduction.

□ *Enteropogon sexuale* Hibbits, 1978
[= *Paraenterobryus multiformis* Galt, 1971, *nom. provis.*]

Four thallus types produced (from anterior to posterior region of hindgut): (1) Thalli 4 mm long by 11–19 μm diam, tightly coiled to slightly undulate, producing 4– or 8–nucleate secondary infestation sporangiospores 40–55 × 11–18 μm in coiled thalli or 70 × 9 μm in straighter thalli; or producing uninucleate cells 16–26 × 8–12 μm that fuse in pairs between adjacent thalli to form uninucleate spheres 32–40 μm diam. (2) Thalli 11 mm long by 7–20 μm diam with a bulbous base, producing 4–nucleate secondary infestation sporangiospores 7–15 × 8–9 μm; or uninucleate primary infestation sporangiospores 23–25 × 13–14 μm; or small uninucleate cytoplasmic segments 34–35 × 7–10 μm. (3) Thalli 8.5 mm long by 10 μm diam distally, base sigmoid and inflated to 65 μm diam, producing 4– or 8–nucleate secondary infestation sporangiospores 65–115 × 7.5–18 μm, or uninucleate primary infestation sporangiospores 35–45 × 27–29 μm. (4) Thalli 3 mm long by 5–7 μm diam, producing small cytoplasmic segments. Type species.

Illustrations: Fig. 11.36; also see Hibbits, 1978, Figs. 2–3.

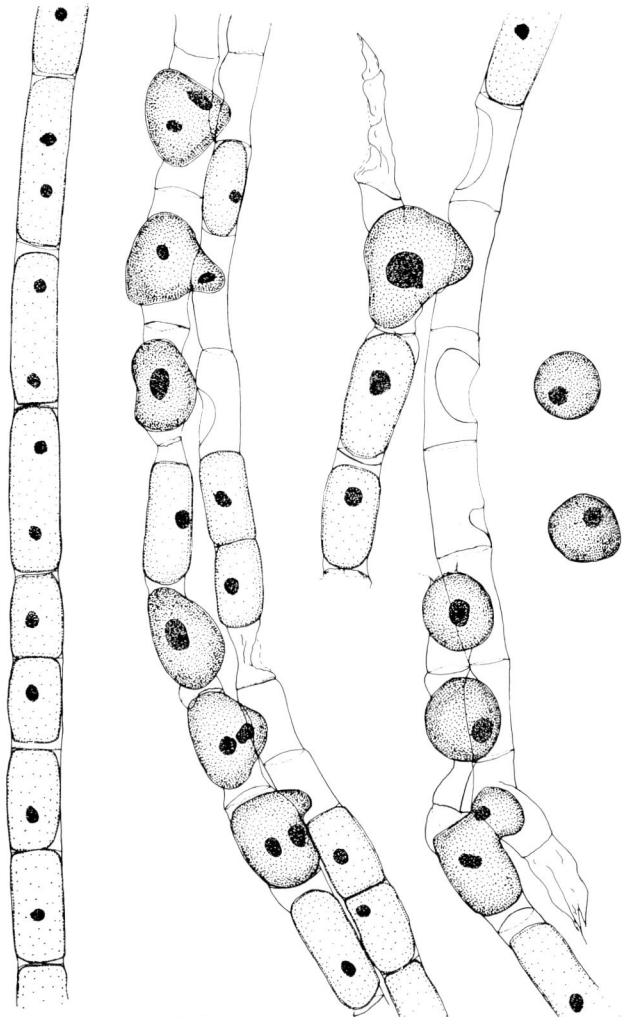

FIG. 11.36. *Enteropogon sexuale*. Scalariform fusions between thalli of the fungus. Although this resembles a sexual process, sexuality has not been confirmed (see Chapter 7). Scale bar = 10 μm. Reprinted by permission from Syesis and J. Hibbits, 1978, pp. 220.

Host: In hindgut of *Upogebia pugettensis* (Dana) (Crustacea, Callianassidae).

Distribution: Marine, mud flats, San Juan Archipelago, Washington, U.S.A.

The foregut of the mud shrimp host of *E. sexuale* may also contain thalli of *Enteromyces callianassae*.

References: Hibbits, 1978; Galt, 1971.

Paramacrinella

Manier & Grizel, 1971

Thalli dimorphic. Macrothalli unbranched, spore mother–cell persisting at the base of the mature thallus, producing primary and secondary infestation sporangiospores. Microthalli cleaving into series of small uninucleate cells. In hindgut of Amphipoda. Monotypic.

Type species: *Paramacrinella microdeutopi* Manier & Grizel.

This genus closely resembles the eccrinid genus *Ramacrinella*, whose single species also lives in an amphipod, but whose macrothalli are branched at the base.

□ *Paramacrinella microdeutopi* Manier & Grizel, 1971

Macrothalli 1.2–1.5 mm long by 12–18 μm diam distally, 18–35 μm diam at the curved base. Spore mother–cell persisting at an angle on the thallus base just above the flat holdfast disk. Apex of macrothalli may have an ampulla 50 μm diam. Producing multinucleate secondary infestation sporangiospores 50–80 μm long; or numerous thin–walled uninucleate primary infestation sporangiospores 13–18 × 4.5–8.5 μm, upon liberation becoming globose and 10–12 μm diam. Microthalli 650–750 μm long by 12–14 μm diam, producing a long series of numerous uninucleate cells. Type species.

Illustrations: Fig. 11.29; also see Manier and Grizel, 1971, Figs. 1 and 2.

Host: In hindgut of *Microdeutopus anomalus* (H. Rathke) (Amphipoda, Aoridae).

Distribution: Marine, in saltwater bay, Department of Hérault, France.

Single macrothalli of *P. microdeutopi* may produce successively both types of spores. The function of the microthallus cells is not known.

Reference: Manier and Grizel, 1971.

Ramacrinella

Manier & Ormières ex Manier, 1968
[= *Ramacrinella* Manier & Ormières, 1961a, *nom. nud.*]

Thalli dimorphic. Macrothalli branched with several fertile axes, producing primary and secondary infestation sporangiospores. Spore mother–cell persisting at the base of the mature thallus. Microthalli unbranched, cleaving into series of small uninucleate cells. In hindgut of Amphipoda. Monotypic.

Type species: *Ramacrinella raibauti* Manier & Ormières ex Manier.

The repeated branching of the coenocytic thalli of *Ramacrinella* is unusual among Eccrinales. Except for the branching, this genus resembles *Paramacrinella*, which also lives in an amphipod.

☐ *Ramacrinella raibauti* Manier & Ormières ex Manier, 1968
[= *Ramacrinella raibauti* Manier & Ormières, 1961a, *nom. nud.*]

Macrothalli with a principal fertile branch and several auxiliary fertile branches, up to 1.1 mm long by 12–13 μm diam distally and 16–18 μm diam proximally, all axes rebranching repeatedly near their bases. Spore mother–cell persistent opposite the small holdfast. Producing multinucleate secondary infestation sporangiospores 26–30 × 6.5–12 μm, or numerous uninucleate oval primary infestation sporangiospores 15–20 × 4–7 μm that become somewhat globose after release. Microthalli unbranched, 150–275 μm long by 2.5–4 μm diam distally, 5–6 μm diam proximally, producing long series of more or less globose cells 2.5–4 μm diam. Type species.

Illustrations: Fig. 11.29; also see Manier and Ormières, 1961a, Figs. 1–3 (drawings) and Figs. 1–8 (photomicrographs).

Host: In hindgut of *Microdeutopus gryllotalpa* A. Costa (Amphipoda, Aoridae).

Distribution: Littoral, saltwater bay, under masses of vegetation, Department of Hérault, France.

The macrothalli of this unusual eccrinid are coenocytic and devoid of crosswalls except for those delimiting sporangia. The subsidiary basal branches are nonfertile. Except for its branched habit, *Ramacrinella raibauti* resembles *Paramacrinella microdeutopi* found in another species of *Microdeutopus* from the same saltwater bay, the Étang de Thau near Sète. Two species of *Enterobryus* from millipede guts, *E. bifurcatus* and *E. tuzetae*, produce thalli with two divergent fertile axes, but they lack the repeated branching of *R. raibauti*.

References: Manier, 1968; Manier and Ormières, 1961a; Grizel, 1971.

Taeniella

Léger & Duboscq, 1911

Thalli producing thick–walled resistant sporangiospores with 2 appendages at both poles, and thin–walled primary and secondary infestation sporangiospores. In hindgut of Decapoda (Brachyura and Anomura). Monotypic.

Type species: *Taeniella carcini* Léger & Duboscq.

The major character that distinguishes *Taeniella* is the resistant spore with its two polar appendages. These were not mentioned in the original description by Léger and Duboscq (1911), nor have they been described from most collections of *Taeniella carcini*. The appendages are usually seen as a gelatinous mass appressed to each end of the spore within the sporangium, and such structure may be taken as putative evidence of appendages if found in Decapods. When the resistant spores of *T. carcini* are kept in seawater, this gelatinous material unfolds, upon release of the spores from the sporangia, to reveal the two long polar appendages at each pole (Moss, 1979). It is possible that some of the resistant spores described in the literature have been immature. In most instances they have not been illustrated as released spores, which is necessary for the unfurled appendages to be seen, preferably using phase–contrast microscopy. Hibbits (1978) reported finding 2 appendages at only one pole. Further studies of *Taeniella* are needed to clarify the ontogeny of these appendages, and whether the number is consistent.

☐ *Taeniella carcini* Léger & Duboscq, 1911
= *Taeniella longa* Léger & Duboscq, 1911
= *Taeniella galatheae* Manier & Ormières ex Manier, 1968
[= *T. galatheae* Manier & Ormières, 1962, *nom. nud.*]
= *Taeniella grandis* Hibbits, 1978

Thalli usually 2–4(–15) mm long by 2.5–12(–22) μm diam. Holdfast short and cylindrical. Thick–walled resistant sporangiospores oval to elongate–oval, 1– to 4–nucleate, (15–)18–21(–32) × (6–)8–12(–18) μm, bearing two long appendages at each pole upon release from the sporangium. Uninucleate thin–walled sporangia in long series, 3–5 × 3–5 μm. Secondary infestation sporangiospores 4– to 8–nucleate, 30–150 × 5–12 μm. Type species.

Illustrations: See Manier, 1961b, Fig. 3 (drawings) and Figs. 12–15 (photomicrographs); and Moss, 1979, Figs. 8.20, 8.21.

Hosts: In hindgut of Decapoda: *Carcinus maenas* Penn., *Portunus puber* L., *Cancer oregonensis* (Dana), *Pilumnus hirtellus* L., *Xantho pilipes* M.–Edw., *Hemigrapsus nudus* (Dana), *H. oregonensis* (Dana), *Gaetice depressus* (de Haan), *Sesarma hematocheir* (de Haan) (Brachyura); and *Paguristes turgidus* (Stimpson), *Pagurus kennerlyi* (Stimpson), *P. beringanus* (Benedict), *P. ganosimanus* (Stimpson), *Eupagurus excavatus* Herbst., *Galathea strigosa* L., *Callianassa gigas* Dana, *C. californiensis* Dana, *Upogebia pugettensis* (Dana), *Oedignathus inermis* (Stimpson) (Anomura).

Distribution: Marine, mostly on rocky shores and intertidal or subtidal, Mediterranean coast of France, Japan, and San Juan Archipelago, Washington, U.S.A.

The original description by Léger and Duboscq (1911) was very brief, and the species was only illustrated (but poorly) in 1948 by Duboscq, Léger, and Tuzet. Nevertheless, additional studies by Manier (1961b, 1969), Johnson (1966), and Hibbits (1978) indicate that *Taeniella carcini* is widespread in many families of Decapoda, and is also quite variable. Given the wide host range, the writer believes that all species of *Taniella* described so far are conspecific with *T. carcini,* for the morphological characters overlap considerably such that they cannot be used to differentiate the named species satisfactorily. The collections from Japan were made by the author with Drs. Yosio Kobayasi and Hiroharu Indoh (unpublished). The distribution of *T. carcini* is undoubtedly much greater than present records indicate.

The resistant spores of *T. carcini* appear to form just prior to molting of the crabs. Their appendages have been clearly demonstrated only by Moss (1979), although other investigators have observed the gelatinous pads appressed to the ends of the spores while still within their sporangia. The number of nuclei in the resistant spores is difficult to determine, but it is possible they remain uninucleate even while the thick wall develops, but become 4-nucleate at full maturity.

Several reported features of *T. carcini* are worth noting. Manier (1961b) saw thalli producing both thin-walled and resistant spores simultaneously, the former ones located distally in sections of the thalli that were considerably narrower than the proximal end with developing resistant spores. Hibbits (1978) reported one instance of two thalli conjugating in a manner resembling this process in *Enteropogon sexuale*. She also saw "segments of thalli" attached to the walls of sporangia; one wonders if these might not be a spurious form of germination *in situ,* as commonly happens in *Palavascia* spp. On some thalli she also saw multicellular filaments of what appeared to be a large blue-green alga attached to the basal region of the thalli of *T. carcini (T. grandis),* and she suggested that these might be what Johnson (1966) saw attached to the apical region of thalli from the same kinds of crabs collected in the same general locality, and which led him to describe those thalli as "*Palavascia* sp."

References: Léger and Duboscq, 1911; Duboscq, et al., 1948; Manier, 1961b, 1968, 1969 (1970b); Manier and Ormières, 1962; Johnson, 1966; Galt, 1971; Hibbits, 1978; Moss, 1979; Whisler, 1979.

Taeniellopsis

Poisson, 1927

Thalli not dimorphic, upon molting of the host producing thick-walled, oval, uninucleate sporangiospores. Also producing thin-walled uninucleate primary infestation sporangiospores, and multinucleate secondary infestation sporangiospores. In hindgut of Amphipoda (Crustacea). Three species.

Type species: *Taeniellopsis orchestiae* Poisson.

—Key to *Taeniellopsis* species—

1. Thick-walled uninucleate spores elongate-oval, 30–38 × 9–12 µm .... **T. orchestiae**
1'. Thick-walled uninucleate spores oval, usually less than 30 µm long 2
   2(1'). Thick-walled spores 25–30 × 8–10 µm .... **T. flexilis**
   2'(1'). Thick-walled spores 20–23 × 7–8 µm .... **T. susplugasi**

☐ *Taeniellopsis flexilis* Poisson, 1929

Thalli 700–1700 μm long by 3–10 μm diam. Holdfasts cylindrical, narrower than the thalli. Secondary infestation sporangiospores up to 10-nucleate and 65 μm long. Primary infestation sporangiospores variable in size, from 3 × 3 μm to 40–70 × 8–10 μm. Oval spores uninucleate 25–30 × 8–10 μm.

Illustrations: See Poisson, 1929, Figs. XV–XVII.

Host: In hindgut of *Orchestia gammerella* (Pallas) (Amphipoda, Talitridae).

Distribution: Terrestrial, near fresh, brackish or saltwater, Departments of Pyrénées–Orientales, Calvados and Finistère, France.

Although the host has a wide distribution and infestation was found in several sites by Poisson, *Taeniellopsis flexilis* was not described as fully as the type species was, and the oval, presumably resistant, spores were not illustrated. Poisson stated that some thalli had "enigmatic protruberances" on their sidewalls. He also described some thalli consisting of 2- to 4-nucleate cells up to 70 μm long that separated like arthrospores.

Reference: Poisson, 1929.

☐ *Taeniellopsis orchestiae* Poisson, 1927

Thalli up to 1.2 mm long by 9–25 μm diam. Holdfasts short and disklike. Secondary infestation sporangiospores 4- to 10-nucleate, 40–110 × 20–25 μm. Primary infestation sporangiospores uninucleate, 2–4 × 5 μm or 8–12 × 16–20 μm. Upon molting of the host producing elongate–oval, uninucleate thick–walled resistant spores 30–38 × 9–12 μm in thalli that become spiraled. Type species.

Illustrations: See Poisson, 1929, Figs. I–XIV.

Host: In posterior hindgut of *Orchestia bottae* M.–Edw. (Amphipoda, Talitridae).

Distribution: Terrestrial, border of a saline pond near Nancy, Department of Meurthe–et–Moselle, France.

Poisson illustrated and more fully described *Taeniellopsis orchestiae* in his second (1929) paper on this species. The resistant spores apparently have no appendage, as do some species of *Astreptonema* from amphipods, and they are uninucleate. Prior to their release, the sporangial crosswalls break down, and the resistant spores may emerge from an apical orifice in the spiraled thallus. Poisson (1929) claimed that the spore mother–cell may break off from the apex of a thallus, attach to the host cuticle, and develop into another thallus. He also stated that some sporangia were binucleate and became uninucleate after karyogamy, but it appears he did not actually see nuclear fusions.

Reference: Poisson, 1927, 1929.

☐ *Taeniellopsis susplugasi* Manier ex Manier, 1970b (1969b)
[= *Taeniellopsis susplugasi* Manier, 1950, *nom. nud.*]

Thalli 300–1200 μm long by 6–12 μm diam. Holdfast 3–4 μm long, narrower than holdfast. Secondary infestation sporangiospores 20–55 μm long. Primary infestation sporangiospores flat and disklike or isodiametric. Thick-walled uninucleate oval spores 20–23 × 7–8 μm.

Illustrations: See Manier, 1950, Fig. VIII 1–16; Grizel, 1971, Plate IV, Figs. 1–5.

Host: In hindgut of *Orchestia mediterranea* Costa (Amphipoda, Talitridae).

Distribution: Terrestrial, near seashore of Mediterranean coast, Hérault, France.

Taeniellopsis susplugasi has smaller oval spores than the other species of the genus. Poisson (1929) dissected hindguts of *Orchestia mediterranea*, which he found in the Department of Finistère, France, living among specimens of *O. gammarella* infested with *T. flexilis*, but found no fungi in their guts. Electronmicrographs of holdfasts and sporangia of *T. susplugasi* have been obtained by Grizel (1971), and some were published by Manier and Grizel (1972).

References: Manier, 1950 (1951), 1969b, (1970b); Grizel, 1971; Manier and Grizel, 1972.

Palavasciaceae

Duboscq, Léger & Tuzet ex Manier & Lichtwardt 1969 (1968)
[= Palavasciaceae Duboscq, Léger & Tuzet 1948, *nom. nud.*]

Only multinucleate thick–walled primary infestation sporangiospores produced. Some sporangia capable of germination *in situ* instead of producing a sporangiospore. In hindgut of Isopoda (Crustaceae). One genus.

Type genus: *Palavascia* Tuzet & Manier ex Lichtwardt.

Palavascia

Tuzet & Manier ex Lichtwardt, 1964b
[= *Palavascia* Tuzet & Manier, 1947b, *nom. nud.*]

Series of terminal sporangia producing oval to ellipsoidal, thick–walled, multinucleate primary infestation sporangiospores; or, sporangia may germinate *in situ* to produce one or more narrow filaments that segment into uninucleate cells. Spore mother–cell persistent. In hindgut of halophilic or marine Isopoda. Two species.

Type species: *Palavascia philosciae* Tuzet & Manier ex Manier & Lichtwardt.

—Key to *Palavascia* species—

1. Thalli up to 2–5 mm long; sporulating tips coiled ...... ***P. sphaeromae***
1'. Thalli shorter, up to 1 mm long; sporulating tips straight
    .... ***P. philosciae***

☐ *Palavascia philosciae* Tuzet & Manier ex Manier & Lichtwardt, 1969 (1968)
[= *Palavascia philosciae* Tuzet & Manier *in* Lichtwardt, 1964b, *nom. nud.*]
[= *Palavascia philosciae* Tuzet & Manier, 1947b, *nom. nud.*]

Thalli up to 1 mm long, 29–35 μm diam near the base, 18–26 μm diam distally. Tips straight, with a persistent narrow spore mother–cell. Primary infestation sporangiospores produced in series of up to 50 per thallus, oval or ellipsoidal, 4–nucleate, 19–25 × 7–12 μm. Sporangia may instead germinate *in situ* to produce a single filament 150–200 μm long by 4–6 μm diam that segments to form uninucleate cells containing small oval bodies 2 × 4–6 μm. Type species.

Illustrations: See Manier, 1963, Figs. 1–4.
Host: In hindgut of *Halophiloscia couchi* Kin. (Isopoda, Oniscidae).
Distribution: In saline ponds near Palavas, Hérault, France.

The illegitimate original name was provided with a Latin diagnosis by Lichtwardt (1964b), but it was not validated until a nomenclatural type was cited by Manier and Lichtwardt in 1968 (1969).

The first specimens of *Palavascia philosciae* examined by Tuzet and Manier (1947b) did not have the ellipsoidal sporangiospores. These were described by them the following year (1948b). In that second publication they (and Manier, in 1950) suggested that the multinucleate spores of both *P. philosciae* and *P. sphaeromae* form subsequent to plasmogamy and karyogamy of adjacent uninucleate protoplasts in the thallus, but this has not been substantiated by Manier, nor by the author, in subsequent studies of the more common species, *P. sphaeromae*. There is no known function for the small cells on the filaments that grow from germinated sporangia.

References: Tuzet and Manier, 1947b, 1948b; Manier and Lichtwardt, 1968 (1969); Lichtwardt, 1964b; Manier, 1950, 1963, 1969b (1970b).

☐ *Palavascia sphaeromae* Tuzet & Manier ex Manier, 1968
[= *Palavascia sphaeromae* Tuzet & Manier, 1948b, *nom. nud.*]
= *Palavascia beauforti* Lichtwardt ex Lichtwardt, 1964b
[= *Palavascia beauforti* Lichtwardt, 1961b, *nom. nud.*]

Thalli 2–5 mm long, somewhat sinuate with uneven diam along their length, 30–60 μm diam near the base, 12–35 μm diam distally. Sporulating tips coiled, with a persistent spore mother–cell. Sporangia in long series in the coiled tips, producing oval or ellipsoidal uninucleate to 4–nucleate primary infestation sporangiospores 20–40 × 9–22 μm; or sporangia ger-

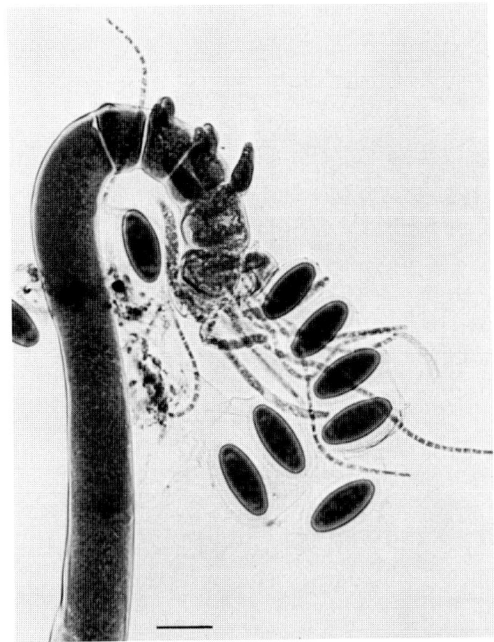

FIG. 11.37. *Palavascia sphaeromae* releasing primary infestation sporangiospores. Note some sporangia have germinated instead of producing a sporangiospore. Scale bar = 20 μm. From Lichtwardt, 1961b, by permission of J. Elisha Mitchell Sci. Soc.

minating *in situ* to produce one or more filaments ~170 μm long by 2–4 μm diam, up to 5–10 μm diam near the base, which segment to form small uninucletae cells. Sporulating tips may have sporangia with sporangiospores intermixed with germinating sporangia.

Illustrations: Figs. 7.10F, 7.22, 11.37; also see Lichtwardt, 1961b, Figs. 1–11.

Hosts: In hindgut of *Sphaeroma serratum* F., *S. quadridentatum* Say, *Exosphaeroma oregonensis* (Dana), *Tecticeps japonicus* Iwasa (Isopoda, Sphaeromidae).

Distribution: Marine, intertidal, in France (Hérault, Roscoff, Finistère), U.S.A. (North Carolina), and Japan (Hokkaido).

Palavascia beauforti was described as a new species by Lichtwardt (1961b) in *Sphaeroma quadridentatum* from the coast of North Carolina on the basis of comparing it with the descriptions provided at that time for *P. sphaeromae*. However, subsequent and more accurate descriptions, and examination of fresh material by the author from the type locality, make it apparent that the two species are synonyms. The records of *P. sphaeromae* in *Exosphaeroma oregonensis* and *Tecticeps japonicus* from northern Japan were made by the author with Drs. Yosio Kobayasi and

Hiroharu Indoh in June, 1964. It is probable that *P. sphaeromae* has a much wider distribution than present records indicate.

Tuzet and Manier (1948b) and Manier (1950) suggested that *P. sphaeromae* and *P. philosciae* produced the oval spores after fusion and karyogamy of adjacent uninucleate protoplasts in the thalli, but Manier (personal communication) does not now believe this interpretation is correct. In the writer's studies of this species, it appears that the thallus tips do not begin to coil and sporulate until they reach the anus. The coiled, sporulating tips are commonly seen protruding from the isopod's anus, and this may be one mechanism whereby younger isopods clinging to the pleopods of larger ones become infested (Lichtwardt, 1961b). The function of the small cells produced at the tips of filaments originating from germinated sporangia is unknown. Charmantier and Manier (1981) have done a study of the role of ecological parameters and population dynamics on infestation of *Sphaeroma serratum* by *P. sphaeromae*.

An interesting feature of *P. sphaeromae* is the germination of the spores after release from the sporangium. This has been described in some detail in Chapter 7 in the section on asexual reproduction in Eccrinales. Manier (1979a) has published an electron micrographic study of this species.

References: Tuzet and Manier, 1948b; Manier, 1950, 1961b, 1968, 1979a; Lichtwardt, 1961b, 1964b, 1973a; Charmantier and Manier, 1981.

Parataeniellaceae

Manier & Lichtwardt, 1969 (1968)

Thallus, or part of the thallus, developing into a sporangium containing many uninucleate primary infestation sporangiospores. Secondary infestation sporangiospores produced individually in sporangia, or absent. In hindgut of Isopoda or Insecta. Two genera.

Type genus: *Parataeniella* Poisson.

—Key to Genera of Parataeniellaceae—

1. In Isopoda; producing (usually) binucleate secondary infestation spores .... ***Parataeniella***
1'. In larval Coleoptera; no secondary infestation spores produced .... ***Lajasiella***

Lajasiella

Tuzet & Manier ex Manier, 1968
[= *Lajasiella* Tuzet & Manier, 1950b, *nom. nud.*]

Thalli short, with a persistent spore mother–cell. Producing a single large sporangium containing many uninucleate primary infestation sporangiospores, or series of terminal cells containing uninucleate bodies. In hindgut of larval Coleoptera (Insecta). Monotypic.

Type species: *Lajasiella aphodii* Tuzet & Manier ex Manier.

☐ *Lajasiella aphodii* Tuzet & Manier ex Manier, 1968
[= *Lajasiella aphodii* Tuzet & Manier, 1965, *in* Manier and Théoridès, 1965, *nom. nud.*]
[= *Lajasiella aphodii* Tuzet & Manier, 1950b, *nom. nud.*]

Thalli usually 200–300 μm long, either 4–5 or 9–10 μm diam, with a narrowed apex consisting of the spore mother–cell. Primary infestation sporangiospores oval, uninucleate, 13–19 × 3.5–4.5 μm, formed in large numbers in a sporangium consisting of the entire thallus except for the basal region. Other thalli producing apical series of cells 13–35 μm long within which form many uninucleate spherical bodies 6–7 μm diam. Type species.

Illustrations: See Tuzet and Manier, 1950b, Figs. I, 1–9.
Host: In hindgut of larval *Aphodius* sp. (Coleoptera, Scarabaeidae).
Distribution: Terrestrial, near Lattes, Hérault, France.

In 1965 Manier and Théoridès provided a Latin diagnosis for the illegitimate species, but the name was not validated until Manier (1968) cited a nomenclatural type.

The round uninucleate cells reported to form in terminal cells have no known function, and this species should be restudied to determine if they are possibly artifacts. There have been no secondary infestation sporangiospores reported in *Lajasiella aphodii*.

References: Manier, 1968; Tuzet and Manier, 1950b; Manier and Théoridès, 1965.

Parataeniella

Poisson, 1929

Thalli producing either a series of sporangia containing single, usually binucleate, secondary infestation sporangiospores, or entire thallus becoming one sporangium containing many uni– or binucleate primary infestation sporangiospores; or both types produced in one thallus. In hindgut of terrestrial Isopoda (Crustacea). Four species.

Type species: *Parataeniella mercieri* (Poisson) Poisson.

The crosswalls of the secondary infestation sporangia are characteristically somewhat oblique in most species.

—Key to *Parataeniella* species—

1. Mature thalli less than 100 μm long; primary infestation spores ellipsoidal, less than 15 μm long .... ***P. scotonisci***
1'. Mature thalli generally more than 100 μm long; primary infestation spores either not ellipsoidal, or ellipsoidal and more than 15 μm long 2
    2(1'). Primary infestation spores globose to angular in shape, ~10 μm long .... ***P. armadillidii***
    2'(1'). Primary infestation spores ellipsoidal and more than 15 μm long 3

3(2'). Primary infestation spores more than 24 μm long, sometimes produced in thalli (sporangia) with terminal series of secondary infestation sporangia .... *P. mercieri*

3'(2'). Primary infestation spores less than 24 μm long, produced in distinctly wider thalli than secondary infestation sporangia
.... *P. dilatata*

□ *Parataeniella armadillidii* Lichtwardt & Chen, 1964

Thalli up to 385 μm long by 12–15 μm diam. Secondary infestation sporangiospores binucleate, ~12 × 10 μm. Primary infestation sporangiospores uninucleate, globose to angular in shape, ~11 μm diam, produced biseriately, rarely uniseriately, in a saccate (to cylindrical) thallus, emerging from thallus (sporangium) through a longitudinal slit. Thalli may produce series of secondary infestation sporangia terminally and a primary infestation sporangium basally in the same thallus.

Illustrations: Fig. 11.38.

Hosts: In hindgut of *Armadillidum vulgare* (Latr.) and *A. nasatum* Budde–Lund (Isopoda, Armadillidae).

Distribution: Terrestrial, Kansas, U.S.A.

Reference: Lichtwardt and Chen, 1964.

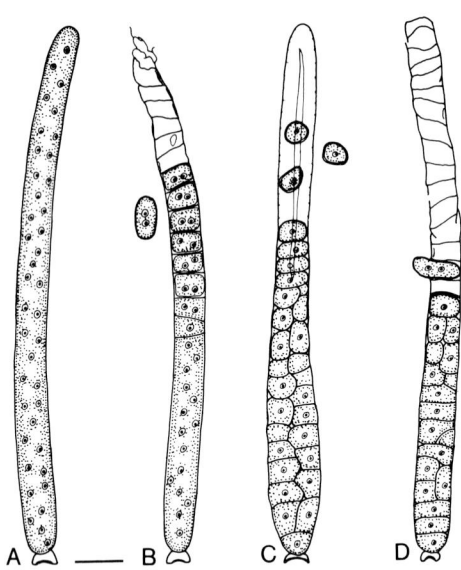

FIG. 11.38. *Parataeniella armadillidii*. A. Nonsporulating thallus. B. Binucleate secondary infestation sporangiospores each produced singly in a basipetal series of sporangia. C. Thallus converted into one sporangium bearing uninucleate primary infestation sporangiospores. D. Thallus with mixed spore types: apically it produced a series of secondary sporangiospores, and the basal portion has become a sporangium containing primary spores. Scale bar = 20 μm. Reprinted by permission from Mycologia, vol. 61: 164, Copyright 1969, R.W. Lichtwardt and The New York Botanical Garden.

☐ *Parataeniella dilatata* Poisson, 1929
= *Parataeniella binucleata* Poisson, 1929
= *Parataeniella intermedia* Poisson, 1929

Thalli of two types: (1) Thalli up to 300 μm long by (8–)12–14 μm diam, producing 2– (to 5–)nucleate secondary infestation sporangiospores (20–)45–50 μm long, or uninucleate sporangiospores 15–35 μm long; and (2) thalli 100–225 μm long by (12–)22–28 μm diam, producing uniseriate or biseriate thick–walled, oval to ellipsoidal, or uni– or binucleate primary infestation sporangiospores 16–22 × 9–11 μm, emerging from an apical orifice in the thallus (sporangium).

Illustrations: See Poisson, 1929, Figs. XXI, XXII; Manier, 1963a, Fig. 3.

Hosts: In hindgut of Isopoda: *Armadillo officinalis* Dum. (Armadillidae); *Androniscus dentiger* Verh. (= *Trichoniscus roseus* Koch) (Trichoniscidae); *Porcellio laevis* (Latr.), *P. lamellatus* (Ulj.), *Protracheoniscus occidentalis* Vandel (Porcellionidae).

Distribution: Terrestrial, in France (Pyrénées–Orientales, Hérault, Calvados), and Germany (German Democratic Republic).

Manier (1969b) considered *Parataeniella intermedia* to be a synonym of *P. dilatata,* based on her studies (1950, 1963a) of the latter species, and the author is in agreement. The author also considers *P. binucleata* to be conspecific with *P. dilatata;* although Manier did not go this far, she stated (1963a, p. 564) that the thick–walled primary infestation spores of the fungus in *Porcellio laevis* (host of *P. binucleata*) are identical with those in *Armadillo officinalis* (host of *P. dilatata*). The host of *P. intermedia* was reported by Poisson (1929) to be *Androniscus dentiger*. The other host records are those of Scheer (1976b), who collected extensively in East Germany.

References: Poisson, 1929; Manier, 1950, 1963a, 1969b (1970b); Scheer, 1976b.

☐ *Parataeniella mercieri* (Poisson) Poisson, 1929
= *Eccrinopsis mercieri* Poisson, 1928

Thalli 170–180(–300) μm long by 10–12(–15) μm diam. Secondary infestation sporangiospores 2– to 4–nucleate, 20–25(–35) μm long, or uninucleate and more or less isodiametric, 10–12 μm. Primary infestation sporangiospores uninucleate, ellipsoidal, uniseriate within the thallus (sporangium), 25–30 × 10–12 μm. Thalli may produce series of secondary infestation sporangia terminally and a primary infestation sporangium basally in the same thallus. Type species.

Illustrations: See Poisson, 1929, Figs. XVIII, XIX.

Hosts: In posterior hindgut of *Oniscus asellus* L. and *Tracheoniscus rathkii* (Brandt) (Isopoda, Oniscidae).

Distribution: Terrestrial, in France (Pyrénées–Orientales, Hérault, Calvados), and Germany (German Democratic Republic).

This fungus was first discovered, but not named, by Mercier (1914). Scheer (1976b) reported *Parataeniella mercieri* to be common in *Oniscus asellus* in many sites in the German Democratic Republic, but less common in *Tracheoniscus rathkii*.

References: Poisson, 1928, 1929; Mercier, 1914; Scheer, 1976b.

☐ *Parataeniella scotonisci* Manier ex Manier, 1970b (1969b)
[= *Parataeniella scotonisci* Manier, 1964d, *nom. nud.*]

Thalli up to 80 μm long by 3.5–9 μm diam. Secondary infestation sporangiospores more or less biconical, uni– or binucleate. Primary infestation sporangiospores becoming binucleate, thick–walled, 11–14.5 × 2.5–4.5 μm, uniseriate or biseriate in the thallus (sporangium).

Illustrations: See Manier, 1964d, Figs. 3–4.

Host: In hindgut of *Scotoniscus marcomelos* Racovitza (Isopoda, Trichoniscidae).

Distribution: In caves, Pyrenees Mountains, France.

The cave–dwelling hosts of this small eccrinid, *Parataeniella scotonisci*, were collected in several different caves. Most of the adults, but not immature specimens, were infested.

References: Manier, 1964d, 1969b (1970b).

Amoebidiales

Léger & Duboscq, 1929a

Thalli coenocytic and unbranched, entire protoplast producing amoeboid cells that encyst and form cystospores; or, in some, producing elongate spores with rigid walls. Attached by a secreted holdfast to the external cuticle (exoskeleton) or hindgut cuticle of aquatic Crustacea or Insecta.

Consisting of the single family Amoebidiaceae.

Amoebidiaceae

Lichtenstein, 1917a

Characters same as those of the order. Two genera.
Type genus: *Amoebidium* Cienkowski.

—Key to Genera of Amoebidiaceae—

1. Producing amoeboid cells, cysts, and cystospores, and sporangiospores; attached to exoskeleton of various Crustacea and Insecta, or in rectum of Cladocera (Crustacea) .... ***Amoebidium***
1'. Producing only amoeboid cells, cysts, and cystospores; in hindgut of Insecta .... ***Paramoebidium***

Amoebidium
Cienkowski, 1861

Thalli on external cuticle (exoskeleton), more rarely in rectum, of freshwater Crustacea or Insecta. Producing uninucleate sporangiospores, or uninucleate, teardrop–shaped amoeboid cells that encyst and form elongate cystospores. Two species.

Type species: *Amoebidium parasiticum* Cienkowski.

—Key to *Amoebidium* species—

1. Sporangiospores elongate, typically lunate. Thalli attached to exoskeleton of Crustacea or Insecta .... **A. parasiticum**
1'. Sporangiospores short–cylindrical to oval. In rectum of Cladocera (Crustacea) .... **A. recticola**

□ *Amoebidium parasiticum* Cienkowski, 1861
  = *Amoebidium fasciculatum* Lichtenstein, 1917a
  [= *Amoebidium poissoni* Duboscq, Léger & Tuzet, 1948, *nom. nud.*]

Thalli cylindrical to fusiform, variable in length and width, with a short holdfast. Sporangiospores typically lunate, or allantoid or fusiform, variable in number, $15–50 \times 5–10$ μm. Upon molting of the host producing amoebae that encyst. Cysts 6–32 μm diam, later forming one to many elongate cystospores 12–30 μm long. Type species.

Illustrations: Fig. 11.39.

Hosts: Widespread among Crustacea and Insecta, on external cuticle. Crustacea: Cladocera, Copepoda, Branchiopoda, Amphipoda (Gammaridae), and Isopoda (Asellidae). Insecta: larvae of Diptera (Chironomidae, Culicidae), Ephemeroptera, Trichoptera, and Odonata (in rectum).

Distribution: Probably worldwide, in freshwater, mostly lentic, habitats. Reported from Czechoslovakia, Denmark, England, France, Israel, Japan, Philippines, Poland, Puerto Rico, Singapore, Spain, Tunisia, and U.S.A.

Amoebidium parasiticum, the only ectozoic trichomycete, shows little host specificity, and has even been reported to attach to inanimate objects (see Table 7.2), but not necessarily to grow to maturity on them. Although widespread geographically and abundant in given sites, it is not always easy to find this species in nature. In the author's experience it is more likely to be found in pools or ditches that are polluted with organic matter. It was the first trichomycete to be cultured axenically (Whisler, 1960) (see Chapter 3 for methods), and as a consequence of its cultivation, the species has been used in many laboratory studies. These are discussed at length in Chapter 9. Of special interest are the factors that cause a developmental shift in thalli from sporangiospore to amoeba production (amoebagenesis) when hosts molt or are injured (also see Chapter 8).

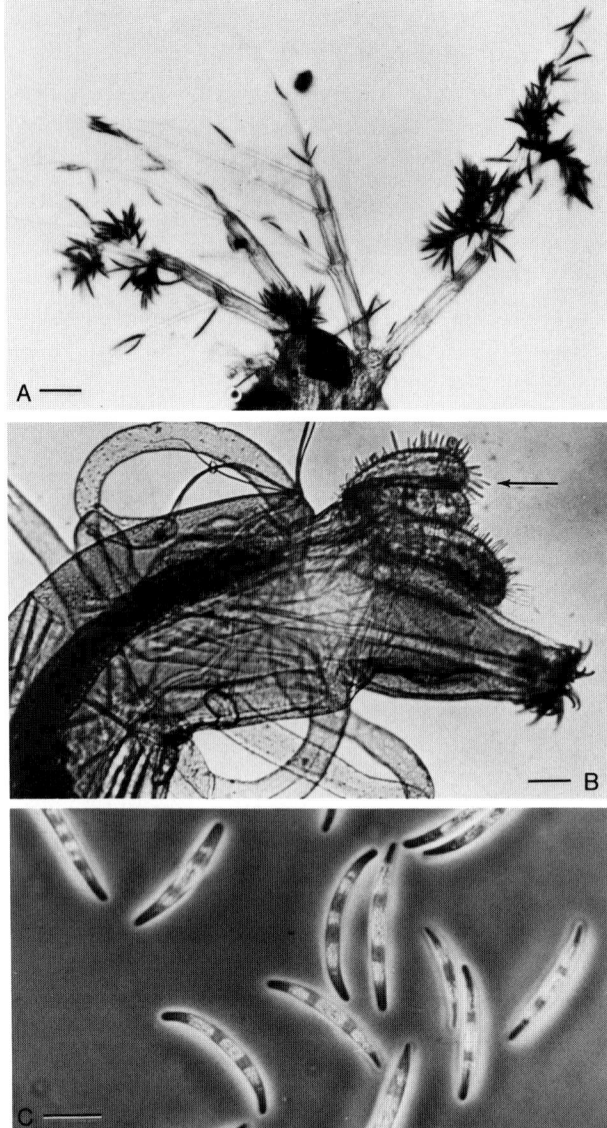

FIG. 11.39. *Amoebidium parasiticum*. A. Antennae and anterior carapace of a cladoceran bearing many attached spores and thalli of the ectocommensal; the clusters of cells are sporangia (thalli) whose spores have burst through the sporangial wall to enable them to attach to another arthropod upon contact. B. Posterior end of a bloodworm (*Chironomus* sp.) whose anal gills are covered with the ectocommensal (arrow). C. Sporangiospores released from mature thalli that were mounted in water on a slide. Scale bars: A, B, 50 μm; C, 20 μm. Figs. 11.39A and 11.39C from Lichtwardt, 1976, by permission of Granada Publishing Ltd.

Manier and Raibaut (1969) have produced a film on the life cycle of *A. parasiticum* (described in Chapter 7). Some aspects of the fine structure in this species have been studied (Whisler and Fuller, 1968; Coste–Mathiez, 1970; Dang, 1979). Poisson (1931b) claimed to have seen fusion of amoebae, but this has not been substantiated.

Cienkowski (1861) described and illustrated an interesting parasite of *A. parasiticum*, which he named *Basidiolum fimbriatum*. It somewhat resembles a *Syncephalis* van Tiegh. & Le Monnier (Zygomycetes), whose species, with one exception, are parasites of Mucorales. Despite abundant references to *A. parasiticum*, this parasite does not appear to have been collected and studied since that one report.

References: Cienkowski, 1861; Lichtenstein, 1917a, 1917b; Lieberkühn, 1856; Schenk, 1858; Moniez, 1887; Fritsch, 1895; Labbé, 1899; Chatton, 1906a, 1906b, 1908, 1920, 1925; Chatton and Roubaud, 1909; Raabe, 1911a, 1911b, 1912; Lichtenstein, 1917a, 1917b; Debaisieux, 1920; Taylor, 1928; Taylor and Colton, 1928; Poisson, 1931b; Margalef, 1946; Duboscq et al., 1948; Tuzet and Manier, 1951b; Johnson, 1952, 1963; Whisler, 1960, 1961, 1962, 1966, 1968, 1978, 1979; Whisler and Fuller, 1968; Borut, 1964; Trotter and Whisler, 1965; Manier, Rioux and Juminer, 1964; Coste–Mathiez, 1970; Sangar et al., 1972; Kuno, 1973; Lichtwardt, 1976; Manier, 1962c, 1969b (1970b); Manier and Raibaut, 1969, 1970; Coste–Mathiez, 1970; Grizel, 1971; Dogma, 1975; Cerniglia et al., 1978; Dang, 1979; Moss, 1979; Porter and Smiley, 1979.

□ *Amoebidium recticola* Chatton, 1906b

Thalli curved, the basal region curved sharply inward, averaging 100 μm long by 15–18 μm diam. Holdfast 12 × 6 μm diam. Producing short–cylindrical to oval sporangiospores 8–12 μm long, or amoeboid cells that encyst. Cysts 7–12 μm diam.

Illustrations: See Chatton, 1906b, Figs. 1–4.

Host: Attached to rectal peritrophic membrane of *Daphnia magna* Straus and *D. pulex* (de Geer) (Crustacea, Cladocera).

Distribution: Aquatic, in reptile tanks at the Paris Museum, France.

Amoebidium recticola differs from the more commonly encountered ectozoic *A. parasiticum* in the morphology of the thalli and sporangiospores, as well as its location within the host. Although species of *Paramoebidum*, an endozoic genus, are attached to the hindgut cuticle of aquatic insect larvae, *A. recticola* is clearly not a *Paramoebidium*, because it produces sporangiospores. Furthermore, its location in the rectum of the host and mode of entry into the hindgut are different from *Paramoebidium* spp.

According to Chatton (1906b, 1920), *A. recticola* is attached to the inner surface of the "rectal peritrophic membrane" of daphnids. This rectal membrane is a loose tube attached only at the very anterior region of the hindgut, and apparently is formed by the hindgut epithelium, then peeled off (except for the anterior end) by a molting process as a new hindgut cuticle develops on the epithelium. Daphnids also have the type of midgut peritrophic membrane produced by many arthropods, which develops as a loose sleeve in the midgut but originates in the foregut. In *Daphnia magna* the anterior membrane continues to grow backward into the hindgut, inside the rectal membrane, and usually projects from the anus. *Amoebidium recticola* is located between the two membranes, and is not in direct contact with either the hindgut cuticle itself or the food bolus, as are thalli of *Paramoebidium* spp. Also, *Paramoebidium* spp. infest the hindgut after ingestion of cysts (or cystospores) by the host. In *A. recticola* spores enter through the anus and attach to the rectal membrane. This is acomplished, according to Chatton, as a result of the rhythmic dilations and contractions of the rectum, which is a respiratory process. As water is drawn into the rectum, so are spores of *A. recticola*. By placing minute granules of carmine near the anus of daphnids, Chatton was able to demonstrate that the granules were aspirated into the rectal lumen between the two membranes, as with *A. recticola* spores. It is interesting that the spores were never attached to the inner (anterior) peritrophic membrane.

Several developmental stages of *A. recticola* are of interest to compare with *A. parasiticum*. Thalli of *A. recticola* occasionally attached to the outside (new) integument of daphnids, but then degenerated. Amoeboid cells were produced either from thalli in the rectum of dead daphnids or from expelled sporangiospores settled on the bottom of a container. Such expelled spores were also observed to grow and produce a limited number of sporangiospores. Chatton never saw cystospores in *A. recticola*, but they are rarely seen in *A. parasiticum* either, whose cystospores usually form after a resting period.

References: Chatton, 1906b, 1920; Chatton and Roubaud, 1909; Le Berre, 1967.

Paramoebidium

Léger & Duboscq, 1929c

Thalli attached to hindgut cuticle (or anal gills) of freshwater larvae of Insecta. Producing uninucleate, teardrop–shaped amoeboid cells that encyst and form elongate cystospores. No sporangiospores produced. Two species.

Type species: *Paramoebidium inflexum* Léger & Duboscq.

Paramoebidium species are very common in lotic populations of larval Plecoptera, Ephemeroptera, and Diptera (Simuliidae, rarely Chironomidae) throughout many parts of the world. They grow like weeds, and often can be found even when other trichomycetes (Harpellales) are not present in those kinds of insects.

Thirteen species of *Paramoebidium* have been described, but only two validly. The illegitimate species names are listed in the next section of this Chapter. In many but not all cases, they differ from each other mostly on the basis of the hosts they were discovered in. Until such time as better criteria can be found to differentiate them, it seems advisable to let them remain as *nomina nuda*. A major problem in determining the range of variation in morphological characters and host specificity is that no *Paramoebidum* sp. has been cultured axenically. The author has collected *Paramoebidium* extensively in many geographical areas, and has records of considerable morphological variation in thallus size and shape among his collections, but these are not always consistent in given species or families of hosts. The reproductive phases (amoebae, cysts, cystospores), where they have been seen, do not appear to be useful characters to differentiate species of *Paramoebidium*. Given the wide distribution and abundance of *Paramoebidium*, it is probable that at least some of the species are not very host specific.

Occasionally one finds *Paramoebidium* thalli attached to thalli of Harpellales cohabiting the hindgut, or vice versa; or thalli of *Paramoebidium* attached to thalli of its own species (see Table 7.2).

—Key to *Paramoebidium* species—

1. Mature thalli more than 300 μm long, often sharply bent; holdfast basal. In hindgut of Plecoptera larvae .... *P. inflexum*
1'. Mature thalli less than 300 μm long, curved, with a somewhat lateral holdfast. In posterior hindgut, or on anal gills, of Simuliidae larvae .... *P. curvum*

☐ *Paramoebidium curvum* Lichtwardt, 1979, *in* Dang and Lichtwardt, 1979

Thalli 140–280 × 20–60 μm, curved. Holdfast somewhat lateral on the incurved side of the thallus. Amoebae ~20 × 8 μm. Cysts spherical, enlarging to produce 1–14 fusiform to lunate cystospores 45–50 × 7–8 μm.

Illustrations: Figs. 7.24, 11.40.

Hosts: Attached to very posterior hindgut or to anal gills of larval stages of many species of Simuliidae (Diptera).

Distribution: Lotic habitats in U.S.A. (Arizona, Colorado, Kansas, Montana, Utah, Wyoming), Sweden, France, and England.

Simuliid (blackfly) larvae frequently have in their hindgut, but located more anteriad, another *Paramoebidium* sp. with larger thalli resembling *P. inflexum*. These two species may be found in the same gut, or either separately. When *P. curvum* thalli are attached to the rectractible anal gills of blackfly larvae, they may be seen without dissecting the larva, provided the gills are extended. *Paramoebidium curvum* has a sausage-shaped thallus similar to *Amoebidium recticola* in the rectum of daphnids, and it was undoubtedly *P. curvum* that Chatton and Roubaud (1909) saw on the anal gills of two *Simulium* spp., but thought it might be *A. recticola*.

While conducting an electron microscopic study of *P. curvum*, Dang

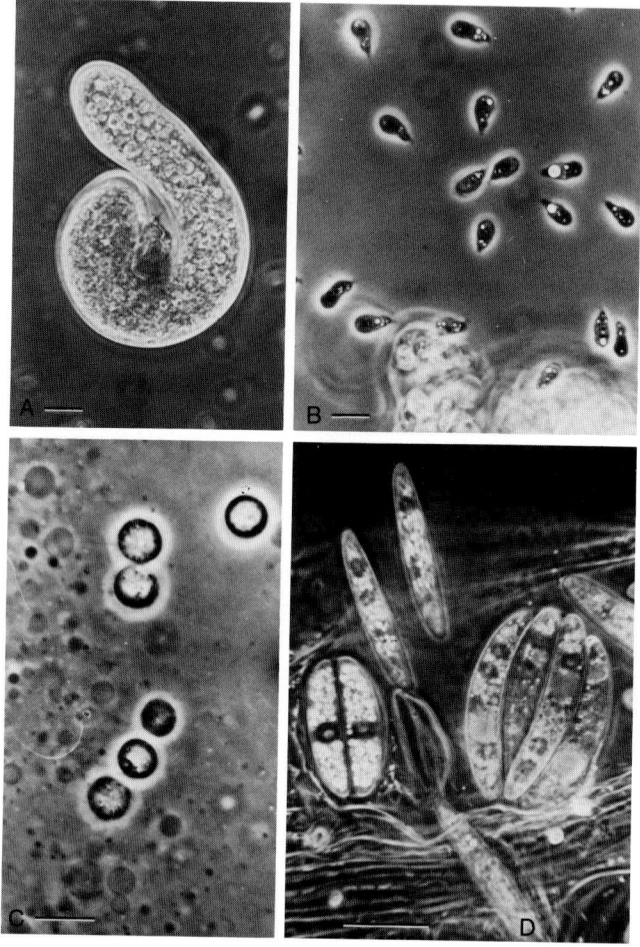

Fig. 11.40. *Paramoebidium curvum*. A. Mature thallus just prior to releasing amoeboid cells. B. Released, motile amoeboid cells. C. Encysted amoebae. D. Mature cysts with cystospores, some released. Scale bars = 20 μm.

(1978) (Dang and Lichtwardt, 1979) discovered some virus–like particles in the cytoplasm. Previously, Manier et al. (1971) had found similar but smaller, virus–like particles in *P. arcuatum (nom. nud.)*. See the last section in Chapter 7.

References: Dang and Lichtwardt, 1979; Lichtwardt, 1973a, 1976; Chatton and Roubaud, 1909; Dang, 1978.

□ *Paramoebidium inflexum* Léger & Duboscq, 1929c

Thalli 1 mm × 21 μm, 700 × 7–8 μm, or 380 × 50–55 μm, the first two types sharply bent. Holdfast basal. Amoebae ~15 μm long. Cysts spherical, (12–)25–30(–50) μm diam. Cystospores 6–36 per cyst, elongate and slightly curved, averaging 15 × 6 μm. Type species.

Illustrations: See Duboscq, Léger and Tuzet, 1948, Fig. XXXII 1–16.

Host: In hindgut of larval *Nemura variegata* Oliv. (Plecoptera, Nemuridae).

Distribution: In alpine streams, France.

Duboscq et al. (1948) provided a fuller description of the developmental stages of *Paramoebidium inflexum*.

References: Léger and Duboscq, 1929c; Duboscq et al., 1948.

Excluded and Doubtful Taxa

In this section are listed genera and species that are not acceptable, and consequently are not included in the preceding taxonomic treatment. Synonyms (both valid and illegitimate names) of valid species will not be found in the list below, but rather are placed under the accepted names.

The citation *nomen nudum (nom. nud.)* indicates that the taxon has not been validly published according to the International Code of Botanical Nomenclature, in most cases because no Latin diagnosis and nomenclatural type have been provided in accordance with Art. 36 and 37, or there is essentially no description that distinguishes it from other taxa (Art. 32). In these instances I have not validated the taxa, because they do not appear to warrant doing so without further study of the fungi. This is best done by obtaining fresh material. Unfortunately, type specimens of trichomycetes, usually consisting of prepared slides, would have limited value in resolving the taxonomic judgments in question.

The citation *nomen dubium (nom. dub.)* indicates that the taxon, although technically valid, in my opinion cannot be satisfactorily typified due to an insufficient description or questionable interpretation of the fungal material described.

These excluded taxa and their host names are also listed in Appendix B, provided they seem to be trichomycetes (a few listed below are not), as a convenience for the investigator wishing to restudy and perhaps validate or redescribe them. The last section of Chapter 12 has brief accounts of several gut organisms that have similarities to the trichomycetes but are not related.

Amoebidium cienkowskianum Moniez, 1887, *nom. nud.* No description.
    A. crassum Moniez, 1887, *nom. nud.* (non Fritsch, 1895). No description; may not be an *Amoebidium*.
    A. crassum Fritsch, 1895, *nom. dub.* (non Moniez, 1887). = *A. moniezi* Labbé, 1899.
    A. moniezi Labbé, 1899, *nom. dub.* = *A. crassum* Fritsch, 1895.
Andohaheloa Manier, 1955c, *nom. nud.*
    A. pauliani Manier, 1955c, *nom. nud.* Possibly an *Enterobryus*.

Arundinula incurvata Léger & Duboscq, 1905a, *nom. nud.*
   A. porcellanae Léger & Duboscq, 1911, *nom. dub.*
Asellaria scutigerae Manier, 1950, *nom. nud.* May not be a trichomycete.
   A. spirostrepti Tuzet, Manier & Jolivet, 1957, *nom. nud.*
Astreptonema siphonoecetis Grizel, 1971, *nom. provisorium.*
Capillus Granata, 1908, *nom. dub.*
   C. intestinalis Granata, 1908, *nom. dub.* Possibly an *Enterobryus.*
   C. iuli Maessen, 1955, *nom. dub.*
Cestodella Tuzet, Manier, & Jolivet, 1957, *nom. nud.* Type species *(C. straeleni)* and other species probably are *Enterobryus* spp.
   C. allopori Tuzet & Manier, 1957a, *nom. nud.*
   C. attenuata Tuzet & Manier, 1957a, *nom. nud.*
   C. glandulosa Tuzet, Manier, & Jolivet, 1957, *nom. nud.*
   C. operculata Tuzet, Manier, & Jolivet, 1957, *nom. nud.*
   C. pachyboli Tuzet & Manier, 1957a, *nom. nud.*
   C. parva Tuzet & Manier, 1957a, *nom. nud.*
   C. pseudonannolenis Tuzet & Manier, 1957a, *nom. nud.*
   C. rhinocrici Tuzet & Manier, 1957a, *nom. nud.*
   C. straeleni Tuzet, Manier, & Jolivet, 1957, *nom. nud.*
Daloala Tuzet, Manier & Vogeli–Zuber, 1952, *nom. nud.*
   D. mardonii Tuzet, Manier, & Vogeli–Zuber, 1952, *nom. nud.* Possibly an *Enterobryus.*
Dixidium Poisson, 1932b, *nom. dub.*
   D. dixae Poisson, 1932b, *nom. dub.* Possibly a *Smittium.*
Eccrina Leidy, 1850b. Genus indistinguishable from *Enterobryus* Leidy; refer to the valid species *Enterobryus moniliformis* (Leidy) Lichtwardt.
   E. brevis Maessen, 1955, *nom. dub.*
   E. gigantea Tuzet & Manier, 1954b, *nom. nud.*
   E. longa Leidy, 1850b, *nom. dub.* First of two species of *Eccrina* described by Leidy, but description insufficient.
   E. longipes Maessen, 1955, *nom. dub.*
   E. montana Maessen, 1955, *nom. dub.*
   E. pseudoramosa Maessen, 1955, *nom. dub.*
   E. uncigeri Maessen, 1955, *nom. dub.*
Eccrinella Léger & Duboscq, 1933. = *Astreptonema* Hauptfleisch, 1895; see valid species *A. gammari* (Léger & Duboscq) Manier, and *A. corophii* (Manier) Manier.
Eccrinoides stammeri Maessen, 1955, *nom. dub.*
Eccrinopsis Léger & Duboscq, 1916. Type species = *Enterobryus hydrophilorum* (Léger & Duboscq) Manier.
   E. leidyi Léger & Duboscq, 1916, *nom. dub.* = *Trichella leidyi* (Léger & Duboscq) Poisson, 1931a, *nom. nud.*
   E. ligidii Maessen, 1955, *nom. dub.*
   E. sciarae Tschudovskaia, 1928, *nom. dub.* = *Trichella sciarae* (Tschudovskaia) Poisson, 1931a, *nom. nud.*

Enterobryus allobrogicus Duboscq, Léger & Tuzet, 1948, *nom. nud.*
- *E. brachyspiroboli* Tuzet, Manier & Jolivet, 1957, *nom. nud.*
- *E. bresiliensis* Tuzet & Manier, 1951a, *nom. nud.* = *Pistillaria bresiliensis* (Tuzet & Manier) Jeekel, Tuzet, Manier & Jolivet, 1959, *nom. nud.*
- *E. broelemanni* Duboscq, Léger & Tuzet ex Manier, 1970b (1969b), *nom. dub.*
- *E. doscari* Manier, 1950, *nom. nud.* May not be an *Enterobryus*.
- *E. duboscqui* Tuzet & Manier ex Manier, 1968, *nom. dub.* Probably not an *Enterobryus*.
- *E. eurydesmi* Tuzet & Manier, 1957a, *nom. nud.*
- *E. fertilis* Tuzet & Manier, 1957a, *nom. nud.*
- *E. flavus* Maessen, 1955, *nom. dub.* = *Pistillaria flavus* (Maessen) Jeekel, Tuzet, Manier, & Jolivet, 1959, *nom. nud.*
- *E. gracilis* Duboscq, Léger & Tuzet ex Manier, 1970b (1969b), *nom. dub.* May not be an *Enterobryus*.
- *E. helocharei* (Poisson) Manier, 1970b (1969b), *nom. nud.* = *Trichella helocharei* Poisson, 1931a, *nom. nud.*
- *E. hyalinus* Léger & Duboscq, *nom. provisorium.*
- *E. inflatus* Duboscq, Léger & Tuzet, 1948, *nom. nud.*
- *E. iuli–terrestris* Robin, 1853, *nom. dub.*
- *E. leptodesmi* Tuzet & Manier, 1957a, *nom. nud.*
- *E. nudatus* Tuzet, Manier & Jolivet, 1957, *nom. nud.*
- *E. orthomorphae* Tuzet & Manier, 1957a, *nom. nud.*
- *E. pennatus* Tuzet, Manier & Jolivet, 1957, *nom. nud.*
- *E. pentodoni* (Manier) Manier, 1970b (1969b), *nom. nud.* Basionym illegitimate. = *Paratrichella pentodona* Manier, 1947, *nom. nud.*
- *E. philhydri* (Poisson) Manier, 1970b (1969b), *nom. nud.* = *Trichella philhydri* Poisson, 1931, *nom. nud.* Probably not an *Enterobryus*.
- *E. pseudoeurydesmi* Tuzet & Manier, 1957a, *nom. nud.*
- *E. rectus* Duboscq, Léger & Tuzet ex Manier, 1970b (1969b) *nom. dub.*
- *E. robini* Duboscq, Léger & Tuzet ex Manier, 1970b (1969b), *nom. dub.* May not be an *Enterobryus*.
- *E. sao–pauloi* Tuzet & Manier, 1951a, *nom. nud.*
- *E. schubarti* Tuzet & Manier, 1951a, *nom. nud.*
- *E. seychelloboli* Tuzet & Manier, 1957a, *nom. nud.*
- *E. spiralis* Leidy, 1849b, *nom. dub.* Description insufficient.
- *E. spirostrepti* Tuzet & Manier, 1957a, *nom. nud.*
- *E. strongylosomae* Duboscq, Léger & Tuzet, 1948, *nom. nud.*
- *E. tumidus*, Duboscq, Léger & Tuzet, 1948, *nom. nud.*
- *E. vulgaris* Tuzet, Manier, & Jolivet, 1957, *nom. nud.*

Genistella choanifera Tuzet & Manier, 1953, *nom. nud.* = *Typhella choanifera* (Tuzet & Manier) Manier & Mathiez, 1965, *nom. nud.* Probably a *Smittium*.

G. mailleti Tuzet & Manier, 1955, *nom. nud.* Probably = *Legeriomyces ramosus* Pouzar.

G. rhitrogenae Tuzet & Manier, 1955, *nom. nud.*

Lactella Maessen, 1955, *nom. dub.* Generic characters similar to *Enterobryus*.

L. aphodii Maessen, 1955, *nom. dub.*

L. cercyonis Maessen, 1955, *nom. dub.*

L. chaetarthriae Maessen, 1955, *nom. dub.*

L. coelostomatis Maessen, 1955, *nom. dub.*

L. helocharidis Maessen, 1955, *nom. dub.*

L. laccobii Maessen, 1955, *nom. dub.*

L. philhydri Maessen, 1955, *nom. dub.*

L. platystethi Maessen, 1955, *nom. dub.*

Microasellaria Tuzet, Manier, & Jolivet, 1957, *nom. nud.*

M. funicularia Tuzet, Manier, & Jolivet, 1957, *nom. nud.*

Microeccrina Maessen, 1955, *nom. dub.* Probably related to the prokaryotic genus *Arthromitus* (see last section of Chapter 12).

M. fertilis Maessen, 1955, *nom. dub.*

M. glomeri Maessen, 1955, *nom. dub.*

M. iuli Maessen, 1955, *nom. dub.*

M. leptophylli Maessen, 1955, *nom. dub.*

M. ligidii Maessen, 1955, *nom. dub.*

M. orthomorphae Maessen, 1955, *nom. dub.*

M. parva Maessen, 1955, *nom. dub.*

M. siccophila Maessen, 1955, *nom. dub.*

Microtrichella Maessen, 1955, *nom. dub.* Probably related to the prokaryotic genus *Arthromitus* (see last section of Chapter 12).

M. hydrophilorum Maessen, 1955, *nom. dub.*

Nodocrinella Scheer, 1977, *nom. nud.*

N. hylonisci Scheer, 1977, *nom. nud.* No nomenclatural type cited. Resembles a *Parataeniella*.

Opuntiella Léger & Gauthier, 1932, *nom. nud.* Description incomplete; no illustrations.

O. digitata, Léger & Gauthier, 1932, *nom. nud.* Probably a *Stachylina*.

Paramoebidium arcuatum Léger & Duboscq ex Duboscq, Léger & Tuzet, 1948, *nom. nud.* = *P. arcuatum* Léger & Duboscq, 1929c, *nom. nud.*

P. chattoni Léger & Duboscq ex Duboscq, Léger & Tuzet, 1948, *nom. nud.* = *P. chattoni* Léger & Duboscq, 1929c, *nom. nud.*

P. dispersum Léger & Duboscq ex Duboscq, Léger & Tuzet, 1948, *nom. nud.* = *P. dispersum* Léger & Duboscq, 1929c, *nom. nud.*

P. eccriniformis Léger & Duboscq ex Duboscq, Léger & Tuzet, 1948, *nom. nud.* = *P. eccriniformis* Léger & Duboscq, 1929c, *nom. nud.*

P. fuscum, Duboscq, Léger & Tuzet, 1948, *nom. nud.*

P. geniculatum Duboscq, Léger & Tuzet, 1948, *nom. nud.*

P. giganteum Duboscq, Léger & Tuzet, 1948, *nom. nud.*

P. pavillardi Manier, 1950, *nom. nud.*

P. procloeoni Manier, 1950, *nom. nud.*

P. simulii Tuzet & Manier, 1955, *nom. nud.*

P. thrauli Léger & Duboscq ex Duboscq, Léger & Tuzet, 1948, *nom. nud.* = *P. thrauli* Léger & Duboscq, 1929c, *nom. nud.*

Paratrichella Manier, 1947, *nom. nud.*

P. pentodona Manier, 1947, *nom. nud.* = *Trichella pentodoni* (Manier) Manier & Théodoridès, 1965, *nom. nud.* = *Enterobryus pentodoni* (Manier) Manier, 1970b (1969b), *nom. nud.*

Pistillaria Jeekel, Tuzet, Manier, & Jolivet, 1959, *nom. nud.* Type species, *P. plagiodesmi,* probably an *Enterobryus.*

P. bresiliensis (Tuzet & Manier) Jeekel, Tuzet, Manier & Jolivet, 1959, *nom. nud.* = *Enterobryus bresiliensis* Tuzet & Manier, 1951a, *nom. nud.*

P. flavus (Maessen) Jeekel, Tuzet, Manier, & Jolivet, 1959, *nom. nud.* = *Enterobryus flavus* Maessen, 1955, *nom. dub.*

P. plagiodesmi Jeekel, Tuzet, Manier, & Jolivet, 1959, *nom. nud.*

Recticoma Scheer, 1935, *nom. nud.*

R. cambari Scheer, 1935, *nom. nud.*

Rubetella Tuzet, Rioux, & Manier, 1961, *nom. nud.* = *Smittium* Poisson.

Trichella Poisson, 1931a, *nom. nud.* Generic name used provisionally by Léger & Duboscq, 1929a; no description. First species assigned to genus by Poisson, 1931a, but without a generic description. Type species, *T. hydrophilorum,* now an *Enterobryus.*

T. coelostomatis Maessen, 1955, *nom. nud.*

T. helocharei Poisson, 1931a, *nom. nud.* = *Enterobryus helocharei* (Poisson) Manier, 1970b (1969b), *nom. nud.*

T. hydrobii Maessen, 1955, *nom. nud.*

T. leidyi (Léger & Duboscq) Poisson, 1931a, *nom. nud.* = *Eccrinopsis leidyi* Léger & Duboscq, 1916, *nom. dub.*

T. longa Maessen, 1955, *nom. nud.*

T. pentodoni (Manier) Manier & Théodoridès, 1965, *nom. nud.* = *Paratrichella pentodona* Manier, 1947, *nom. nud.*

T. philhydri Poisson, 1931a, *nom. nud.* = *Enterobryus philhydri* (Poisson) Manier, 1970b (1969b), *nom. nud.*

T. sciarae (Tschudovskaia) Poisson, 1931a, *nom. nud.* = *Eccrinopsis sciarae* Tschudovskaia, 1928, *nom. dub.*

T. stammeri Maessen, 1955, *nom. nud.*

Trichellopsis, Maessen, 1955, *nom. dub.* Probably = *Enterobryus.*

T. polydesmi Maessen, 1955, *nom. dub.*

T. schizophylli Maessen, 1955, *nom. dub.*

Trichoceridium Poisson, 1932, *nom. dub.*

T. ramosum Poisson, 1932, *nom. dub.* May be a *Smittium.*

Typhella Léger & Gauthier, 1935a, *nom. nud.* No type species given.

T. choanifera (Tuzet & Manier) Manier & Mathiez, 1965, *nom. nud.* = *Genistella choanifera* Tuzet & Manier, 1953, *nom. nud.*

CHAPTER 12

Phylogeny

Phylogenetic deductions are always speculative to a greater or lesser extent, and any model designed at this time to interpret natural relationships among the trichomycetes and with other known fungal groups may not lie very high on a probability scale, owing to insufficient documentation (Lichtwardt, 1973a). Nevertheless, a discussion of phylogeny based on currently available evidence serves the useful purpose of providing a systematic approach to the interpretation of data on relationships and, more importantly, of developing the kinds of questions that will lead to a better understanding of the evolution of these unusual organisms.

Almost all early publications on trichomycetes included some discussion of the possible affinities of newly described species with fungi, algae, or protozoans. From time to time the taxonomic organization of known genera into families and higher taxa were suggested or revised as new information became available and taxonomic concepts developed (e.g., Léger and Duboscq, 1929a; Duboscq et al., 1948; Manier, 1950, 1955b; Manier and Lichtwardt, 1968). Less constructively, Sörgel (1953) sought to design a phylogenetic arrangement for the families of trichomycetes, but he based his scheme in part on erroneous data and ignored reproductive features of these organisms that must be considered in such an attempt. A patently unacceptable taxonomy was proposed by Locquin (1974) in which he allied the eukaryotic Amoebidiales with the prokaryotic Actinomycetes, and, curiously, the other orders of trichomycetes (and the Kickxellales) with the Endomycetes. Some of the early studies involving taxonomic judgments are of historic interest, but it would not be useful to review them here. In this chapter the author will restrict the presentation to a summary of the available data that may bear upon concepts of phylogeny. It should be noted, however, that in their 1948 monograph Duboscq, Léger, and Tuzet introduced the name Trichomycetes to include in the broad sense two classes: the Eccrinides (Amoebidiales and Eccrinales) and the Harpellides (Harpellales and Genistellales). In their view the Eccrinides were the true Trichomycetes, that is, in the strict sense. In the sections that follow, the reader will note that this original interpretation of what constitutes the "true" trichomycetes has shifted, so that some investigators now exclude the Amoebidiales and even the Eccrinales from this designation.

Relationships with Other Fungi

It is reasonable to assume that ancestral forms of trichomycetes lived independent of or in loose external association with arthropods before they evolved the specialized features demanded for successful gut habitation. It is also likely that the trichomycetes have been evolving over a considerable period of time, for reasons outlined in the introductory paragraph of Chapter 8. As a consequence of these expectations, and considering that the gut habitat has undoubtedly exerted strong selective pressures on the evolution of the trichomycetes and has thus caused them to diverge considerably from their free–living ancestral forms, there is little reason to expect to find a close and overt relationship with extant groups of fungi, which may share their ancestry but not their ecologically specialized niches.

Several lines of evidence suggest there are similarities between the Harpellales and the Zygomycetes order Kickxellales Kreisel ex Benjamin [sensu Moss and Young (1978) and Benjamin (1979) that includes only the Kickxellaceae]. The Kickxellales in this restricted sense is a relatively small order, consisting presently of eight genera with mostly saprobic species (often in soil or dung) and a few weakly mycoparasitic ones (Benjamin, 1979). The mycelium is extensively branched and septate, with some coenocytism present such as in the multinucleate sporocladia of *Linderina* spp. Unispored sporangiola are produced on the septate or nonseptate sporocladia, and spherical zygospores have been found in some of the species. The possible relationship with trichomycetes and the supporting evidence have been presented by Moss and Young (1978), together with a suggested phyletic scheme for the genera of Kickxellales, Harpellales, and Asellariales.

The similarity first observed between the Harpellales and Kickxellales—and it remains one of the strongest links between the orders—is the unusual perforate crosswall with a plug that is present in both the vegetative and reproductive parts of their thalli (Farr, 1965; Farr and Lichtwardt, 1967; Manier and Coste–Mathiez, 1968; Reichle and Lichtwardt, 1972; Lichtwardt, 1973a; Manier, 1973a,b; Moss, 1972, 1975, 1976; Moss and Lichtwardt, 1976, 1977; Young, 1969; Benny and Aldrich, 1975). The harpellid septum has been described in Chapter 7 (Fig. 7.7), and it seems to be the same as the kickxellid septum in basic structure and development. An exception is that micropores, in addition to the larger central pore, have been detected in one species of Kickxellales (*Coemansia aciculifera* Linder), although not yet in any species of Harpellales (Moss and Young, 1978).

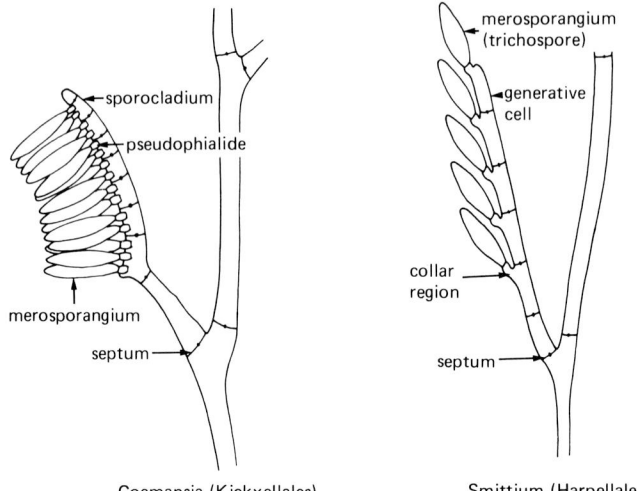

FIG. 12.1. Similarity of the asexual apparatus of the merosporangial genus *Coemansia* and the trichomycete genus *Smittium*, as interpreted by Moss and Young (1978). Reprinted by permission from Mycologia, vol. 70: 948, Copyright 1978, Stephen T. Moss and The New York Botanical Garden.

A second line of evidence for a relationship between the two orders resulted from the serological studies of Sangar et al. (1972). They found that some anti–*Smittium* antibodies precipitated when reacted with antigens of two Kickxellales, namely *Linderina pennispora* and *Dipsacomyces acuminosporus* (see Chapter 9 for data), thus indicating some immunological affinity between the two orders. Antigens of four other taxa of Zygomycetes produced no reactions, or they were negligible.

The appendaged trichospores (sporangia) of the Harpellales remain a structurally unique fungal spore type, but Moss and Young (1978) have presented a case for a degree of similarity between the harpellid asexual apparatus of a genus such as *Smittium* and that of the kickxellid genus *Coemansia* (Fig. 12.1). *Coemansia* produces branches with terminal and septate sporocladia bearing a number of unilateral pseudophialides, and from each of these a single, elongate sporangiole (unispored merosporangium) forms and later detaches. In this comparison the sporangiole–sporocladium arrangement is considered to be homologous with the trichospore–generative cell complex of the Harpellales, with the pseudophialides being comparable to the collar region of a *Smittium*. Moss and Young noted that in the harpellid *Pteromaktron protrudens* a subsidiary cell grows from the generative cell and gives rise to the trichospore, and this subsidiary cell might represent the equivalent of the pseudophialide in *Coemansia*.

Nothing comparable to the structurally and functionally distinct trichospore appendage has been found in the Kickxellales. However, Young

(1969) discovered a "labyrinthiform organelle" in *Linderina pennispora* in the pseudophialide at the base of the sporangiole, and a similar structure was found in the same species by Benny and Aldrich (1975), which they called an "abscission vacuole." It may represent a rudimentary type of appendage. At least, it seems to correspond in position and appearance with the early stages of harpellid appendage formation (Moss and Young, 1978), although it has no known function and does not develop into a well–defined structure like the typical harpellid appendage. Additional detailed studies will certainly be needed before these putative homologies are substantiated, but at least the present evidence is suggestive of some relationship.

Both the Harpellales and Kickxellales produce zygospores, signifying that they have a fundamental sexual similarity, but beyond this they differ in several significant aspects. Zygospores of the Kickxellales, like other Zygomycetes, are spherical, and develop from a homothallic fusion of either small gametangia borne on two undifferentiated sexual branches or a branch with an intercalary cell (Benjamin, 1958). Zygospores of the Harpellales typically develop from heterothallic or homothallic conjugations of mostly intercalary cells. From one of the conjugants, or from the conjugation tube between them, a cell called the zygosporophore develops. The zygosporophore produces initially a more or less spherical apical swelling, which then enlarges into a biconical zygospore. Limited fine–structural studies on *Harpella melusinae* and *Trichozygospora chironomidarum* (Moss and Lichtwardt, 1977) suggest that meiosis possibly occurs within the conjugation apparatus, resulting in zygospores having a single haploid nucleus. The biconical shape and the possibly haploid condition of the harpellid zygospore may be adaptive to rapid germination and establishment of thalli in the host gut, as discussed in Chapter 7. Unfortunately, there are no fine–structural or cytological studies of kickxellid zygospores (Benjamin, 1979); consequently, at present comparisons of these structures in the two orders are not possible beyond gross morphology.

The foregoing paragraphs have stressed the similarities that exist between the Harpellales and Kickxellales. In Chapter 9, a study by Porter and Smiley (1979) is reported in which they compared ribosomal RNA molecular weights of four species of *Smittium* and three species of Kickxellales, as well as of a variety of other Zygomycetes and *Amoebidium parasiticum*. The rRNA weights of the *Smittium* isolates were higher than those of the other test organisms. Porter and Smiley concluded that the differences were biologically significant, and that on this basis *Smittium* is not closely related to either *A. parasiticum* or any of the Zygomycetes. Additional studies, such as amino acid sequencing and DNA hybridization, would be most informative to elucidate relatedness at the macromolecular level.

As stated in the beginning of this section, close relationships between the Harpellales and other extant groups of nontrichomycetous fungi are not necessarily to be expected. Nevertheless, in aggregate, the production of zygospores, coenocytism at some stages of development, a possible sporangial homology, some serological affinity, and similar septal structures present convincing evidence that the Harpellales and Kickxellales share many common features. In addition to using new approaches to phylogeny, it would be useful to extend to other taxa of the two orders some of the studies that have been published to date, such as investigations in more depth of the cytological aspects of zygospore development, a fine–structural comparison of nuclear division in all major groups, the ontogeny and structure of trichospore appendages as compared with the "labyrinthiform organelle" of the kickxellid sporangiole apparatus, and more comprehensive serological testing.

The Asellariales and, especially, the Eccrinales exhibit fewer direct links to the Kickxellales than do the Harpellales. A consideration of their phylogeny will be taken up in the next section, in which it will be suggested that those two orders may have been derived from the Harpellales and consequently would be expected to have fewer characters in common with the kickxellid ancestral forms than the Harpellales.

The Amoebidiales present a special enigma when considering phylogeny. Their immature thalli with holdfasts can resemble those of some Eccrinales or even very young stages of Harpellales. Species of *Paramoebidium* often lie side by side in the gut with species of Harpellales, and therefore one can assume they have similar physiological requirements. Yet, their reproductive features set them apart. The actively motile amoeboid cells and the whole amoeba–cyst phase of the Amoebidiales cycle differ from any of the other trichomycete orders—and from any other fungi, for that matter. Lichtwardt (1973a) noted some resemblance between the method of production of sporangiospores in *Amoebidium* and the uninucleate primary infestation sporangiospores of the eccrinid genus *Parataeniella*, as well as a resemblance between cystopores of the Amoebidiales with resistant spore development in *Eccrinidus*, but these were not thought to be convincing evidence of a close relationship with the Eccrinales. Serologically, the Amoebidiales have shown no affinity to the Harpellales. More importantly, the walls of *Amoebidium parasiticum* appear to be devoid of any significant amounts of chitin or cellulose or other wall constituents that are characteristic of fungal taxa. Thus, the affinity of the Amoebidiales to the true fungi is in question. They have features that suggest they might be a distinct group of protozoans, and in fact have been included in many zoological treatises since the time of their discovery last century (see Duboscq et al., 1948). Emerson and Whisler (1968) called attention to some similarity between the thallus form of *Amoebidium* and the zoosporic fungus *Harpochytrium*, without implying that a close relationship exists (see last section of this chapter). For the present purpose, it can be said that the Amoebidiales are likely to have been derived from protozoan ancestors, whether or not they are now true fungi. (Theories on the origin of aquatic phycomycetes commonly invoke an ancestry from one or more protozoan groups.)

The foregoing discussion makes it evident that the trichomycetes may not have a monophyletic origin, and the class Trichomycetes is perforce an artifical one, justified only by the convenience of studying all these obligate commensals of arthropods together. Taxa of trichomycetes including the Amoebidiales do show similarities in certain aspects of their morphology, as well as in their habitats, host preferences, and integration of their development with that of their hosts. Resemblances of Amoebidiales to the other orders of trichomycetes can perhaps be attributed to convergent evolution.

Relationships Among the Trichomycetes

A suggested scheme for relationships among the families of trichomycetes is presented in Fig. 12.2, and it will be the model for the discussion that follows. Several basic assumptions have been made in considering their phylogeny. These are

1. The most primitive trichomycetes or their precursors were only superficially—and probably facultatively—associated with arthropods.

2. Their first hosts lived in freshwater habitats, because this environment was capable of providing the fungi with nutrients (as it does today for *Amoebidium parasiticum*) as well as support through bouyancy, a suitable osmolality, and prevention of desiccation.

3. The externally attached thalli may have survived better in such a nutritionally impoverished environment when they were located near the fecal–rich anal segments or near the head and therefore the food sources of the host [as suggested by Moss (1979)].

4. Detached fungal spores, especially those near the head, could have been ingested frequently (although not necessarily always by the individual

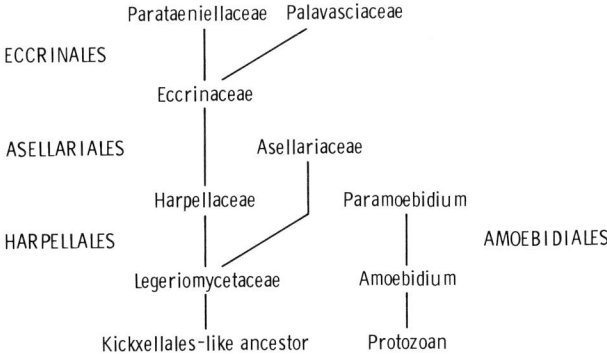

FIG. 12.2. Suggested evolutionary relationships among the Trichomycetes.

that bore the thalli), thus exposing many spores, over time, to the gut environment. Eventually, through adaptation and selection, some became attached to the gut lining and became capable of growth and sporulation in this new environment, which was favored by providing more concentrated nurients and a degree of shelter.

5. Thalli that attached either externally or internally probably had little host specificity initially, and were unspecialized in terms of their nutritional requirements and the integration of their development with that of the host.

6. The closest apparent outgroup of the trichomycetes (excluding the Amoebidiales) is the Kickxellales, and they probably share a common ancestry. For this to be true, present–day trichomycetes would not necessarily have to have many features in common with any of the extant Kickxellales. Selective pressures would have been strong once the primitive trichomycetes became established with some regularity in the gut, and, as a requirement for permanent and successful gut habitation of the respective species, they would have undergone relatively rapid phenotypic changes, both morphological and physiological.

These assumptions also apply to the Amoebidiales, except there is no outgroup with which to compare them. In this separate evolutionary line it is reasonable to suppose that species of *Amoebidium* evolved first: They are attached externally and promiscuously to a wide range of different types of aquatic hosts, and have no unusual nutritional requirements (they are culturable on simple media). They have a sporangial system for transmission of spores from host to host as well as the more resistant amoeba–cyst phase of development. The endocommensalistic thalli of *Paramoebidium* spp. probably evolved by the repeated ingestion by some arthropods of *Amoebidium* cysts containing cystospores, with the cystospores eventually establishing thalli in the hindguts of certain aquatic insects. In so evolving, *Paramoebidium* spp. lost the sporangial phase, which then had no function, and possibly became more fastidious in their nutrition (at least they are at present unculturable).

The Harpellales, as stated, show the greatest affinities to the Kickxellales. *Smittium*, the largest and most widely distributed genus of Legeriomycetaceae, has a number of features that indicate its species may have evolved earlier than other extant and currently known trichomycete genera, in view of the basic assumptions listed earlier:

a. Some species of *Smittium* have a relatively wide host range as compared with other Harpellales. For example, certain species have been found to occur naturally in two or three families of aquatic insect larvae, and in other species the host range can be extended through artifical infestation.

b. There is apparently little specialization in terms of nutrition and physiology, as evidenced by the fact that some species are culturable axenically on common mycological media, in contrast to most other trichomycete genera, which remain unculturable.

c. Thalli of *S. culisetae* are capable of attachment to the external cuticle of mosquito larvae, at least under special artificial laboratory conditions (Fig. 9.6B). Although external attachment may no longer happen naturally, it could represent a vestige of the type of attachment that occurred when trichospores commonly germinated outside the host gut.

d. The holdfast is not very elaborate in *Smittium,* as compared with some other trichomycete genera.

e. Trichospores possess a single appendage (easier to construct than multiple appendages).

f. Many species produce zygospores (a character shared with the Harpellales' outgroup, the Kickxellales), and the zygospores have an appendage (assuming complete loss of zygospore and trichospore appendages in present genera of Harpellales is a derived condition).

An offshoot of the Legeriomycetaceae, as suggested in Fig. 12.2, is the Asellariaceae. In this family the branching has been maintained, and the holdfasts have become more elaborate. The plugged septum of the Legeriomycetaceae is present in the Asellariaceae as well. The generative cells of the Asellariaceae do not produce exogenous and deciduous trichospores; rather, the cells themselves disarticulate and function as propagules. Sexual reproduction is absent or possibly rare. Whereas the Harpellales are restricted to aquatic insect larvae, the Asellariaceae have become adapted to semiterrestrial insects (Collembola) and to isopods living in freshwater, marine, and terrestrial habitats.

The Harpellaceae reproductively are almost identical to the Legeriomycetaceae, but they evolved thalli with a total lack of branching. With this reduction they also became holocarpic. Their coenocytic thalli are generally limited in size and each of the small number of cells that forms by septation now becomes a generative cell.

The unbranched Eccrinales may have evolved from Harpellaceae–like ancestors, or from Legeriomycetaceae–like species but with a loss of branching. There has been a complete loss of septation as well in the Eccrinales, except to delimit sporangia, and there has been a concomitant increase in and persistence of coenocytism in the vegetative parts of the thalli. Sporangiospores of Eccrinales emerge physically and directly from the sporangia, rather than developing within an outward growth consisting of a deciduous trichospore, and some now become thick walled to resist the environmental conditions outside of the host gut. Sexual reproduction appears to have disappeared in the order. [The conjugations that have been observed in *Enteropogon sexualis* and *Arundinula* sp. (see Chapter 7) may be a vestigial tendency to reproduce sexually.] Through adaptive radiation, species of Eccrinales became widely dispersed among different groups of arthropods. The majority of species became terrestrial, but they also invaded, perhaps secondarily, a variety of freshwater and marine hosts. The Parataeniellaceae (found in terrestrial isopods) are considered derived in the sense that septa do not form in the sporangia (consisting of the entire thallus), which produce primary infestation spores, and the

Palavasciaceae (in marine isopods) have lost the ability to produce secondary infestation spores.

Obviously, this suggested phylogeny will be modified as new discoveries are made, and even with present evidence it may not have universal appeal. Nevertheless, it is presented to stimulate thought and discussion. A different viewpoint has been published by Moss and Young (1978, their Fig. 18), who suggested a phyletic scheme for the Asellariales and Harpellales (which they believe are the only true trichomycetes) and the Kickxellales, all derived from an archetype that produced merosporangia and plugged septa. The Asellariales are shown in their 1978 paper as an evolutionary line derived from the main line leading to the Harpellales. The form of thalli (branched or unbranched) is not considered to be significant in their phyletic scheme for the genera of Harpellales, and thus a distinction between the two harpellid families is not accepted. Rather, they stress the types of zygospores produced, the presence or absence of a trichospore collar, and whether or not appendages are fully formed on trichospores. *Pteromaktron* is considered the most primitive genus on the basis that the subsidiary cell of the trichospore in that genus is thought to be homologous with the pseudophialide of the Kickxellales.

Most investigators today seem to accept the concept that the Harpellales and Asellariales are closely related but that the Amoebidiales have a questionable affinity to the "true" trichomycetes. There is less agreement about the placement of the Eccrinales with respect to the other orders. In fact, recent investigators such as Moss (1977, 1979) and Moss and Young (1978) have argued that the Eccrinales have no natural affinity to the Harpellales and Asellariales, but the present writer sees certain similarities between the two orders that are suggestive of a basic relationship. Some major problems present themselves, however.

First and foremost is that there are almost no fine-structural studies on the Eccrinales of a comparative nature that address the problem of development of reproductive bodies, septal structure, etc. Second, the adaptation of the Eccrinales to the gut habitat parallels that of the other orders of endocommensals, and consequently some caution must be exercised in interpreting structures and functions that are similar but may have arisen through convergent evolution. An obvious example of the latter is the presence of a holdfast, which is not only an essential structure in all trichomycetes, but also in many other eukaryotic thallophytes and prokaryotes where displacement from the substratum would be disadvantageous. Another example, which may or may not involve convergent evolution, is the presence of appendages on the sporangiospores of certain freshwater and marine species of Eccrinales (see Chapter 7). Careful electron microscopic studies of their ontogeny and structure should reveal whether they are homologous with trichospore appendages; if they are not, appendages may have evolved separately and *de novo* in the Eccrinales to aid in spore transmission as the present species shifted and adapted to aquatic hosts from terrestrial ones.

The differences between the Harpellales and Eccrinales in the chemical composition of their cell walls is often cited when relationships are discussed. Such differences could be significant if they are indeed present, yet no sufficient analysis of wall composition has been done on any species of Eccrinales (see Table 7.1), primarily because no species has been cultured to provide enough wall material for chemical analysis. Thus, the data base is often limited when making comparisons between the Eccrinales and other taxa, and extrapolations should be cautious.

Two lines of evidence may point to a morphological relationship between the Harpellales and Eccrinales, in the view of this writer. First, in both groups series of sporangia develop basipetally at the terminal end of the thalli or their branches. The major difference is that the harpellid sporangium (the trichospore) grows out of a generative cell so that the uninucleate trichospore becomes detachable (Lichtwardt, 1973a). The harpellid generative cell does not merely give rise to the trichospore, it also is the cell within which the appendages form, and thus is an integral part of the sporangial complex. The development of these sporangia (or sporangial complexes) in both orders requires the formation of new septa specifically to separate uninucleate segments of the thallus for reproduction. (In the Eccrinales the uninucleate condition does not persist in some types of spores.) Such basipetally formed serial chains of uninucleate sporangia or generative cells are not apparently produced in the normal sequence of development in other sporangial fungi (see *Oedogoniomyces* in the next section), and therefore this type of development could mean there is a fundamental natural relationship between the two orders. Convergent evolution need not be invoked in this instance.

Second, septa of both the Harpellales and Eccrinales have a central pore that is initially open, and there is some flaring of the septum around the pore. Our knowledge of the development of septa in the Eccrinales is based on a study by Moss (1972, 1975) of a single species, *Astreptonema gammari*. In the Harpellales this perforate septum becomes plugged with an electron–opaque lenticular structure, although plasmalemma continuity is maintained between the adjacent cells. In the Eccrinales (at least in *A. gammari*), the pore becomes completely occluded with wall material (see Chapter 7 for a summarization of the process), but only after its initial stage of formation, which closely resembles the harpellid septal structure. Moss interprets this development in the Eccrinales to result in a septum that is substantially different from the plugged harpellid septum. The present author interprets the two septal types as being, at least initially, remarkably similar, and believes the comparison strengthens, rather than weakens, the concept of an affinity between the two orders.

There is little doubt that a large number of trichomycete taxa remain to be discovered. Perhaps some of them will provide "missing links" of great interest, but in the meantime additional studies on known species could provide valuable clues to the kinds of phyletic problems discussed in this section. Almost certainly, some species yet to be discovered will prove to contradict a few of the generalizations made in this treatise, for such is to be expected in any large group of diverse organisms living under conditions in which special adaptive features are crucial for survival and where considerable variability within some taxa is the norm.

Morphologically Similar but Unrelated Organisms

Attached to the guts or external cuticle of many arthropods are various kinds of microorganisms, some of which have a superficial resemblance to trichomycetes. These, as well as certain free-living forms, have from time to time been compared with trichomycetes or sometimes have been confused with them taxonomically. There follows a list of a few such organisms and a discussion of their characteristics.

Arthromitus

The hindguts of many arthropods, especially the millipedes and certain beetles, may have numerous narrow unbranched filaments attached in clusters by means of a small common holdfast. They appear to be prokaryotes. The filaments divide into cells, and the more terminal series of cells at one stage may each contain an oval endospore. The genus *Arthromitus* was established for such microorganisms by Leidy (1849a) in the same paper in which he described the first trichomycete genus, *Enterobryus*. The type species is *Arthromitus cristatus*. Leidy later (1850b) described a second species, *A. nitidus,* but in his illustrated major paper of 1853 on gut organisms he made the second species a synonym of the first.

There is considerable morphological variation in these filaments. They are most often attached directly to the cuticle of the hindgut, but may as well attach to nematodes and even to eccrinid thalli. Their taxonomy is uncertain and confused. The 8th edition (1974) of Bergey's Manual of Determinative Bacteriology places *Arthromitus,* along with some other septate filamentous genera of endospore-producing bacteria, in the category of "uncertain taxonomic position." However, previous editions of Bergey's Manual placed several species of *Arthromitus* in a separate family of the order Caryophanales Peshkoff. Manier (1961a) studied *Arthromitus* in the guts of species belonging to several millipede families, and described phases of its development and cytology. The present writer has frequently encountered one or more species of the genus, and has been unable to culture them axenically. The filaments one observes among hosts may be so variable that it is possible there are different taxa involved. The most distinctive feature of the nonsporulating phase is the minute but obvious holdfast that connects the cluster of unbranched filaments at their base. In some cases the filaments are wound together in tight helical strands very much like a piece of rope, but they tend to uncoil when slides are prepared. The individual thalli are 0.6–2.3 μm wide and up to 1 mm or more long (Manier, 1961a), although usually considerably shorter. They are more robust than most of the other, often numerous, filamentous, and sometimes branched bacterial thalli that proliferate in the guts of certain arthropods.

Maessen (1955) apparently confused *Arthromitus* with eccrinid thalli. She established the genus *Microeccrina* and eight new species taken from millipede guts and one isopod. The writer believes, like Manier (1961a), that these are almost certainly filaments of *Arthromitus* species. The writer also believes that Maessen's new genus and species, *Microtrichella hydrophilorum*, found in a species of hydrophilid beetle, may be an *Arthromitus*.

Harpochytrium

Harpochytrium Lagerheim is a genus of zoosporic fungi normally attached epiphytically to living freshwater algae but occasionally to other substrates as well. The thalli have some resemblance to *Amoebidium*, and could possibly be confused with that genus. Reproduction in *Harpochytrium* does not resemble that of *Amoebidium*, however, and it is doubtful the two genera are related. *Harpochytrium* species had been classified by some authors as achlorophyllous algae, but Jane (1946) showed that they were indeed fungi, and in his revision of the genus Jane recognized six species, some with several different forms. The type species is *H. hyalothecae* Lagerheim. Emerson and Whisler (1968) cultured *H. hedinii* Wille axenically and established a new order of posteriorly uniflagellate lower fungi, the Harpochytriales, to contain this genus and another trichomycete look-alike, *Oedogoniomyces*.

Oedogoniomyces

Shortly after Kobayashi and Ôkubo (1954) discovered and described the new genus and species *Oedogoniomyces lymnaeae*, which they found attached to living freshwater snails in Japan, Emerson and Whisler (1959) reported the discovery and axenic isolation of several strains of *Oedogoniomyces* in Costa Rica and later (1968) in California as well. Their isolates came from freshwater ponds and soil, and were found growing naturally on substrates in the water (banana, twigs) or were obtained by baiting with apples and hemp seeds. Emerson and Whisler included this genus, along with *Harpochytrium*, in the new order of posteriorly uniflagellate fungi, the Harpochytriales, established in 1968.

The unbranched (or rarely branched) thalli of *Oedogoniomyces* with their basal holdfasts and chains of zoosporangia formed in basipetal succession have a remarkable gross similarity to an eccrinid thallus such as *Enterobryus* sp. Emerson and Whisler called attention to this similarity (1959, 1968), but on the basis of the production of zoospores and the chitinous nature of the thallus walls, they did not conclude there was a natural affinity between the two groups. In addition to the matter of motility of spores in *Oedogoniomyces*, it should be noted that while both types of coenocytic thalli form sporangia in basipetal succession, those of the eccrinids normally delimit uninucleate sporangia within which develop single uni- or multinucleate aplanospores, while the *Oedogoniomyces* sporangia presumably are multinucleate from the start (the karyological status at this stage has not been described), and each sporangium produces many uninucleate zoospores.

Mononema

This genus was established by Balbiani (1889) for a curious fungus–like growth attached to the stomodeum (foregut) of centipedes (Chilopoda). Mature specimens of *Mononema moniliforme* Balbiani are unbranched filaments of up to 1 cm or more in length consisting of swollen pairs of cells, each pair connected to the others by a narrow zone or separated from them by an isthmus. The basal cell is slightly tapered and is attached to the cuticle by the narrower end. The pairs of cells appear to arise by intercalary division, and according to Manier and Ormières (1980), the cells are uninucleate and there is no coenocytic stage of development. The pairs of cells apparently disarticulate in the manner of arthrospores and function as propagules. This species has been found in *Cryptops hortensis* L. by Balbiani and by Manier and Ormières, and in *C. anomolans lusitanicus* Verg by Léger and Duboscq (1903).

A number of millipedes (Diplopoda) harbor similar organisms in their oesophagus (foregut), which Manier and collaborators have placed in the genus *Mononema*, but further studies are needed to determine whether they are not, in fact, sufficiently different to warrant a separate generic designation. Two species from millipedes have been named: *M. perapolydesmi* Manier 1964 ex Manier, Gasc & Bouix 1972b from *Perapolydesmus progressus* (Bröl.) (a small cave–dwelling millipede) and from *Polydesmus complanatus* L.; and *M. demangei* Manier, Gasc & Bouix 1972b from *Orthomorpha coarctata* (Saus.) and *Cordyloporus ornatus* (Pet.). Mature filaments are unbranched and tapered toward the base, and consist of swollen binucleate cells that appear to develop at least partially by intercalary divisions. Filaments of *M. demangei* may attain a length of 7 mm (Manier et al., 1972b). The tapered basal cell of both species has a distinctive holdfast and is multinucleate. They propagate, like *M. moniliforme*, by separation and dispersal of cells. Chains of attached cells may also break loose.

Mononema species from millipedes have some resemblance to eccrinid thalli, although their development by intercalary divisions and their arthrospore–like disarticulating cells would seem to make them unrelated. Lichtwardt (1954, 1960) found detached chains of cells and even some entire detached filaments of *Mononema* sp. among thalli of *Enterobryus apheloriae* in the hindgut of *Apheloria iowa* and *E. oxidi* in *Oxidus gracilis*, and he mistakenly thought they were uncommon forms of development of those respective species. Subsequently, Lichtwardt (unpublished) has found *Monomena* sp. in the foregut of a number of neotropical millipedes. These same hosts are often infested at the same time with *Enterobryus* spp.

Several other fungus–like organisms have been found in the foregut of centipedes, but they have less of a resemblance to the trichomycetes. These include *Omphalocystis plateaui* Balbiani 1889, *Rhabdomyces lobjoyi* Balbiani 1889, and *Oesophagomyces lithobii* Manier and Ormières 1980.

Part IV Appendices

Appendix A
List of Fungi and Their Arthropod Hosts

Note: This list contains only accepted species of Trichomycetes. Refer to Chapter 11 for authors of species names and more detailed taxonomy of the arthropod hosts.

Trichomycetes

Harpellales

Harpellaceae

Carouxella
  C. scalaris   *Dasyhelea* spp. larvae (Diptera, Ceratopogonidae)

Harpella
  H. leptosa   *Simulium* spp. larvae (Diptera, Simuliidae)
  H. melusinae   Simuliidae larvae (Diptera)

Harpellomyces
  H. eccentricus   Thaumaleidae larvae (Diptera)

Stachylina
  S. chironomidarum   Chironomidae larvae (Diptera)
  S. euthena   Chironomidae larvae (Diptera)
  S. grandispora   Chironomidae larvae (Diptera)
  S. longa   *Tanytarsus* sp. larvae (Diptera, Chironomidae)
  S. lotica   *Maruina* sp. larvae (Diptera, Psychodidae)
  S. macrospora   Diamesinae larvae (Diptera, Chironomidae)
  S. manicata   Chironomini larvae (Diptera, Chironomidae)
  S. minuta   Tanytarsini larvae (Diptera, Chironomidae)
  S. nana   Chironomidae larvae (Diptera)
  S. pedifer   *Boreoheptagyia* sp. larvae (Diptera, Chironomidae)

S. penetralis   *Diamesa* spp. larvae (Diptera, Chironomidae)

Legeriomycetaceae

Capniomyces
   C. stellatus   *Allocapnia* sp. larvae (Plecoptera, Capniidae)

Gauthieromyces
   G. microsporus   *Baetis* sp. larvae (Ephemeroptera, Baetidae)

Genistelloides
   G. hibernus   *Allocapnia* spp. and *Mesocapnia* sp. larvae (Plecoptera, Capniidae)

Genistellospora
   G. homothallica   *Simulium* spp. and *Prosimulium* spp. larvae (Diptera, Simuliidae)

Glotzia
   G. centroptili   *Centroptilum* sp. larvae (Ephemeroptera, Baetidae)
   G. ephemeridarum   Baetidae larvae (Ephemeroptera)

Graminella
   G. bulbosa   *Baetis* sp. larvae (Ephemeroptera, Baetidae)
   G. microspora   Baetidae larvae (Ephemeroptera)

Legeriomyces
   L. aenigmaticus   *Drunella* sp. larvae (Ephemeroptera, Ephemerellidae)
   L. ramosus   Baetidae and Ephemerellidae larve (Ephemeroptera)

Orphella
   O. coronata   Nemouridae larvae (Plecoptera)

Pennella
   P. angustispora   *Simulium* spp. larvae (Diptera, Simuliidae)
   P. arctica   *Prosimulium* spp. and *Simulium* sp. larvae (Diptera, Simuliidae)
   P. grassei   *Simulium* sp. larvae (Diptera, Simuliidae)
   P. hovassi   *Simulium* spp. and *Prosimulium* sp. larvae (Diptera, Simuliidae)
   P. simulii   *Simulium* spp. larvae (Diptera, Simuliidae)

Pteromaktron
   P. protrudens   *Callibaetis* sp. and *Cloeon* sp. *larvae (Ephemeroptera, Baetidae)*

Simuliomyces
   S. microsporus   Simuliidae larvae (Diptera)
   S. spica   *Allocapnia* sp. larvae (Plecoptera, Capniidae)

Smittium
   S. alpinum   *Diamesa* spp. larvae (Diptera, Chironomidae)
   S. arcticum   Chironomidae larvae (Diptera)
   S. arvernense   *Smittia* sp. larvae (Diptera, Chironomidae)
   S. bisporum   *Psectrotanypus* sp. larvae (Diptera, Chironomidae)
   S. cellaspora   *Sympotthastia* sp. larvae (Diptera, Chironomidae)
   S. chironomi   Orthocladiinae larvae (Diptera, Chironomidae)
   S. culicis   Culicidae, Simuliidae, and Chironomidae larvae (Diptera)
   S. culisetae   Culicidae, Simuliidae, Chironomidae, and Ceratopogonidae larvae (Diptera)
   S. dimorphum   *Boreoheptagyia* sp. larvae (Diptera, Chironomidae)
   S. elongatum   Chironomidae larvae (Diptera)
   S. gigasporus   *Pagastia* sp. larvae (Diptera, Chironomidae)
   S. incrassatum   Chironomidae larvae (Diptera)
   S. longisporum   Chironomidae larvae (Diptera)
   S. macrosporum   Chironomidae larvae (Diptera)
   S. megazygosporum   *Syncricotopus* sp. larvae (Diptera, Chironomidae)
   S. morbosum   *Anopheles* spp. larvae (Diptera, Culicidae)
   S. mucronatum   *Psectrocladius* sp. larvae (Diptera, Chironomidae)
   S. orthocladii   Chironomidae larvae (Diptera)
   S. ouseli   *Eukiefferiella* sp. larvae (Diptera, Chironomidae)
   S. pennelli   *Prosimulium* spp. and *Simulium* spp. larvae (Diptera, Simuliidae)
   S. pusillum   *Procladius* sp. larvae (Diptera, Chironomidae)
   S. simulii   Simuliidae, Chironomidae, Culicidae, and Tipulidae larvae (Diptera)
   S. typhellum   *Chironomus* sp. larvae (Diptera, Chironomidae)

Spartiella
   S. barbata   Baetidae larvae (Ephemeroptera)

Stipella
   S. vigilans   Simuliidae larvae (Diptera)

Trichozygospora
   T. chironomidarum   Chironomidae larvae (Diptera)

Zygopolaris
   Z. borealis   *Epeorus* sp. larvae (Ephemeroptera, Heptaginiidae)

Z. ephemeridarum	Baetidae and Ephemerellidae larvae (Ephemeroptera)

Asellariales

Asellariaceae

Asellaria	
A. armadillidii	Armadillidiidae and Porcellionidae (Isopoda)
A. aselli	*Asellus* sp. (Isopoda, Asellidae)
A. caulleryi	*Asellus* spp. (Isopoda, Asellidae)
A. gramenei	*Asellus* sp. (Isopoda, Asellidae)
A. ligiae	*Ligia* spp. (Isopoda, Ligiidae)
A. unguiformis	*Lirceus* sp. and *Caecidotea* sp. (Isopoda, Asellidae)
Orchesellaria	
O. lattesi	*Orchesella* sp. (Collembola, Entomobryidae)
O. mauguioi	Isotomidae (Collembola)
O. pelta	*Hydroisotoma* sp. (Collembola, Isotomidae)
O. podurae	*Podura* sp. (Collembola, Poduridae)

Eccrinales

Eccrinaceae

Alacrinella	
A. limnoriae	*Limnoria* sp. (Isopoda, Limnoidae)
A. sanjuanensis	*Limnoria* sp. (Isopoda, Limnoidae)
Arundinula	
A. capitata	Paguridae (Decapoda)
A. galatheae	*Galathea* sp. (Decapoda, Galatheidae)
A. haplogaster	*Haplogaster* sp. (Decapoda, Lithodidae)
A. orconectis	*Orconectis* sp. (Decapoda, Astacidae)
A. washingtoniensis	*Paguristes* sp. (Decapoda, Paguridae)
Astreptonema	
A. corophii	*Corophium* sp. (Amphipoda, Corophiidae)
A. gammari	Gammaridae (Amphipoda)
A. longispora	*Gammarus* sp. (Amphipoda, Gammaridae)
A. pacificum	*Orchestia* sp. (Amphipoda, Talitridae)
A. typica	*Gammarus* sp. (Amphipoda, Gammaridae)
Eccrinidus	
E. flexilis	*Glomeris* spp. (Diplopoda, Glomeridae)

Eccrinoides
  E. broelemanni   *Protoglomeris* sp. (Diplopoda, Glomeridae)
  E. helleriae   *Helleria* sp. (Isopoda, Tylidae)
  E. henneguyi   *Loboglomeris* spp. (Diplopoda, Glomeridae)
  E. monticolae   *Porcellio* sp. (Isopoda, Porcellionidae)

Enterobryus
  E. adjanohouni   *Plethocrossus* sp. (Diplopoda, Odontopygidae)
  E. ahlesi   *Apheloria* sp. (Diplopoda, Xystodesmidae)
  E. apheloriae   *Apheloria* sp. (Diplopoda, Xystodesmidae)
  E. attenuatus   Passalidae (Coleoptera)
  E. bifurcatus   *Californiobolus* sp. (Diplopoda, Spirobolidae)
  E. borariae   *Boraria* sp. (Diplopoda, Xystodesmidae)
  E. cherokiae   *Cherokia* sp. (Diplopoda, Xystodesmidae)
  E. cingaloboli   *Cingalobolus* sp. (Diplopoda, Pachybolidae)
  E. compressus   Passalidae (Coleoptera)
  E. cylindroiuli   *Cylindroiulus* sp. (Diplopoda, Julidae)
  E. dixidesmi   *Dixidesmus* sp. (Diplopoda, Polydesmidae)
  E. elegans   *Narceus* spp. (Diplopoda, Spirobolidae)
  E. euryuri   *Euryurus* sp. (Diplopoda, Platyrhacidae)
  E. halophilus   *Emerita* spp. (Decapoda, Hippidae)
  E. hydrophilorum   *Hydrous* spp. (Coleoptera, Hydrophilidae)
  E. isoporostrepti   *Isoporostreptus* sp. (Diplopoda,
                                       Spirostreptidae)
  E. leptoiuli   *Leptoiulus* sp. (Diplopoda, Julidae)
  E. moniliformis   *Scytonotus* sp. (Diplopoda, Polydesmidae)
  E. onychostrepti   *Onychostreptus* spp. (Diplopoda,
                                       Spirostreptidae)
  E. oxidi   *Oxidus* sp. (Diplopoda, Paradoxosomatidae)
  E. oxydesmi   *Oxydesmus* sp. (Diplopoda, Polydesmidae)
  E. peridontopygei   *Peridontopyge* sp. (Diplopoda, Odontopygidae)
  E. tuzetae   *Pachybolus* sp. (Diplopoda, Pachybolidae)

Enteromyces
  E. callianassae   Anomura and Brachyura (Crustacea)

Enteropogon
  E. sexuale   *Upogebia* sp. (Crustacea, Anomura)

Paramacrinella
  P. microdeutopi   *Microdeutopus* sp. (Amphipoda, Aoridae)

Ramacrinella
  R. raubauti   *Microdeutopus* sp. (Amphipoda, Aoridae)

Taeniella
  T. carcini   Brachyura and Anomura (Crustacea)

Taeniellopsis
  T. flexilis   *Orchestia* sp. (Amphipoda, Talitridae)

T. orchestiae   *Orchestia* sp. (Amphipoda, Talitridae)
T. susplugasi   *Orchestia* sp. (Amphipoda, Talitridae)

Palavasciaceae

Palavascia
  P. philosciae   *Halophiloscia* sp. (Isopoda, Oniscidae)
  P. sphaeromae   Sphaeromidae (Isopoda)

Parataeniellaceae

Lajasiella
  L. aphodii   *Aphodius* sp. larvae (Coleoptera, Scarabaeidae)

Parataeniella
  P. armadillidii   *Armadillidium* spp. (Isopoda, Armadillidae)
  P. dilatata   Armadillidae, Trichoniscidae and Porcellionidae (Isopoda)
  P. mercieri   Oniscidae (Isopoda)
  P. scotonisci   *Scotoniscus* sp. (Isopoda, Trichoniscidae)

Amoebidiales

Amoebidiaceae

Amoebidium
  A. parasiticum   On external cuticle of many Crustacea and Insecta
  A. recticola   *Daphnia* spp. (Cladocera)

Paramoebidium
  P. curvum   Simuliidae larvae (Diptera)
  P. inflexum   *Nemura* sp. larvae (Plecoptera, Nemuridae)

APPENDIX B
List of Arthropods and Their Fungal Associates

Note: The names of trichomycetes in parentheses are rejected taxa, but are listed here in order to record their hosts; these fungi will be found listed alphabetically with author references at the end of Chapter 11. Refer also to the taxonomic treatment of accepted species in Chapter 11 for the authors of fungal and arthropod names. The habitats given below pertain particularly to the trichomycete hosts and not necessarily to all arthropods in those respective taxa.

Myriapoda

Diplopoda (millipedes)—all terrestrial

Blaniulidae

Blaniulus hirsutus	*(Enterobryus broelemanni)*

Chelodesmidae

Eurydesmus acutus	*(Enterobryus eurydesmi)*
Leptodesmus gasparae	*(Enterobryus sao–pauloi)*
Leptodesmus juncudus	*(Enterobryus bresiliensis)*
Leptodesmus paulistus	*(Enterobryus fertilis)*
Leptodesmus ramosus	*(Enterobryus leptodesmi)*
Leptodesmus vagans	*(Enterobryus schubarti)*
Pseudoeurydesmus baguassuensis	*(Enterobryus pseudoeurydesmi)*

Glomeridae

Glomeris connexa	*(Eccrina pseudoramosa)*
Glomeris hexasticha	*(Eccrinoides stammeri)*

Glomeris spp. (many)	*Eccrinidus flexilis*
Loboglomeris spp.	*Eccrinoides henneguyi*
Protoglomeris vasconica	*Eccrinoides broelemanni*

Julidae

Brachyiulus pusillus	(*Enterobryus spiralis*)
Cylindroiulus friseus	(*Eccrina brevis*)
Cylindroiulus londinensis	*Enterobryus cylindroiuli*
Cylindroiulus nitidus	(*Eccrina longipes*)
Julus albipes	(*Enterobryus rectus*)
Julus scandinavius	(*Capillus iuli*)
Julus terrestris	(*Enterobryus iuli–terrestris*)
Leptoiulus belgicus	*Enterobryus leptoiuli*
Pachyiulus communis	(*Capillus intestinalis*)
Palloiulus distinctus	(*Enterobryus inflatus*)
Schizophyllum sabulosum	(*Enterobryus duboscqui*)
	(*Trichellopsis schizophylli*)
Schizophyllum spp.	(*Enterobryus robini*)
Unciger foetidus	(*Eccrina uncigeri*)
Unidentified	(*Enterobryus allobrogicus*)
	(*Enterobryus doscari*)

Odontopygidae

Peridontopyge gasci	*Enterobryus peridontopygei*
Plethocrossus acutiformis	*Enterobryus adjanohouni*

Oxydesmidae

Oxydesmus granulosus	*Enterobryus oxydesmi*
Plagiodesmus aotypus	(*Enterobryus pennatus*)
Plagiodesmus occidentalis	(*Pistillaria plagiodesmi*)
Plagiodesmus sp.	(*Enterobryus nudatus*)

Pachybolidae

Brachyspirobolus spp.	(*Enterobryus brachyspiroboli*)
Cingalobolus carli	*Enterobryus cingaloboli*
Pachybolus ligulatus	*Enterobryus tuzetae*
Pachybolus sp.	(*Cestodella pachyboli*)

Paradoxosomatidae

Orthomorpha coarctata	(*Enterobryus orthomorphae*)
Oxidus gracilis	*Enterobryus oxidi*
	(*Enterobryus flavus*)
Strongylosoma guerini	(*Enterobryus strongylosomae*)

Platyrhacidae

Euryurus erythropygus *Enterobryus euryuri*

Polydesmidae

Dixidesmus tallulanus *Enterobryus dixidesmi*
Polydesmus complanatus *(Enterobryus gracilis)*
Polydesmus denticulatus *(Trichellopsis polydesmi)*
Polydesmus rubromarginatus *(Enterobryus hyalinus)*
 (Enterobryus tumidus)
Polydesmus virginiensis *(Eccrina longa)*
Scytonotus granulatus *Enterobryus moniliformis*

Pseudonannolenidae

Pseudonannolene tricolor *(Cestodella pseudonannolenis)*

Rhinocricidae

Rhinocricus divaricatus *(Cestodella attenuata)*
Rhinocricus spp. *(Cestodella rhinocrici)*

Sphaerotheriidae

Globotherium actaeon *(Eccrina gigantea)*

Spirobolellidae

Sechellobolus dictyonotus *(Enterobryus seychelloboli)*

Spirobolidae

Californiobolus uncigerus *Enterobryus bifurcatus*
Narceus spp. *Enterobryus elegans*

Spirostreptidae

Alloporus spp. *(Cestodella allopori)*
Cladostreptus castaneus *(Cestodella parva)*
Ischiotrichus fossulatus *(Cestodella operculata)*
Isoporostreptus bouixi *Enterobryus isoporostrepti*
Mardonius piceus *(Daloala mardonii)*
Onychostreptus aoutii *Enterobryus onychostrepti*
Scaphiostreptus sjoestedti *(Cestodella glandulosa)*
Spirostreptus fossulatus *(Microasellaria funicularia)*
Spirostreptus ibanda *(Enterobryus vulgaris)*
 (Asellaria spirostrepti)
Spirostreptus virgator *(Cestodella straeleni)*

Spirostreptus spp.   (*Cestodella operculata*)
Spirostreptus sp.   (*Enterobryus spirostrepti*)
Unidentified   (*Andohaheloa pauliani*)

Xystodesmidae

Apheloria iowa   *Enterobryus apheloriae*
Apheloria montana   *Enterobryus ahlesi*
Boraria carolina   *Enterobryus borariae*
Cherokia georgiana   *Enterobryus cherokiae*

Crustacea

Branchiopoda (fairy shrimps)—aquatic, lentic

Branchipus sp.   *Amoebidium parasiticum*

Cladocera (water fleas)—aquatic, lentic

Daphnia magna   *Amoebidium recticola*
Daphnia pulex   *Amoebidium recticola*
Daphnia spp. and other
   cladocerans   *Amoebidium parasiticum*
Several genera and species of   (*Amoebidium cienkowskianum*)
   cladocerans   (*Amoebidium crassum*)

Copepoda (water fleas)—aquatic, lentic

Diaptomus gracilis   *Amoebidium parasiticum*
Eucyclops agilis   *Amoebidium parasiticum*

Amphipoda

Aoridae—marine

Microdeutopus anomalus   *Paramacrinella microdeutopi*
Microdeutopus gryllotalpa   *Ramacrinella raibauti*

Corophiidae—marine, muddy sand

Corophium volutator   *Astreptonema corophii*

Gammaridae (scuds)—aquatic, lotic, and lentic

Echinogammarus berilloni	*Astreptonema gammari*
Gammarus duebeni	*Astreptonema typica*
Gammarus pulex	*Astreptonema gammari*
	Astreptonema longispora
	Amoebidium parasiticum
Gammarus roeselii	*Astreptonema gammari*

Talitridae (sand fleas, beach hoppers)—terrestrial or marine, near shoreline or saline ponds

Orchestia bottae	*Taeniellopsis orchestiae*
Orchestia gammarella	*Taeniellopsis flexilis*
Orchestia mediterranea	*Taeniellopsis susplugasi*
Orchestia traskiana	*Astreptonema pacificum*

Isopoda

Armadillidae—terrestrial

Armadillo officinalis	*Parataeniella dilatata*

Armadillidiidae (pill bugs)—terrestrial

Armadillidium nasatum	*Asellaria armadillidii*
	Parataeniella armadillidii
Armadillidium vulgare	*Asellaria armadillidii*
	Parataeniella armadillidii
Armadillidium simoni	*Asellaria armadillidii*

Asellidae—aquatic, lotic, and lentic

Asellus aquaticus	*Amoebidium parasiticum*
	Asellaria aselli
	Asellaria caulleryi
Asellus meridianus	*Asellaria caulleryi*
	Asellaria gramenei
Caecidotea laticauda	*Asellaria unguiformis*
Lirceus hoppinae	*Asellaria unguiformis*

Ligiidae (rock louse)—marine shoreline, or terrestrial

Ligia spp.	*Asellaria ligiae*
Ligidium hypnorum	(*Eccrinopsis ligidii*)

Limnoideae (gribbles)—marine

Limnoria lignorum	*Alacrinella sanjuanensis*
Limnoria tripunctata	*Alacrinella limnoriae*

Oniscidae—terrestrial, or saline ponds

Chaetophiloscia elongata	*Asellaria armadillidii*
Halophiloscia couchii	*Palavascia philosciae*
Hyloniscus riparius	(*Nodocrinella hylonisci*)
Oniscus asellus	*Parataeniella mercieri*
Tracheoniscus rathkii	*Parataeniella mercieri*

Porcellionidae (sow bugs)—terrestrial

Acaeroplastes melanurus	*Asellaria armadillidii*
Porcellio laevis	*Parataeniella dilatata*
Porcellio monticola	*Eccrinoides monticolae*
Porcellio spp.	*Asellaria armadillidii*
Protracheoniscus occidentalis	*Asellaria armadillidii*
	Parataeniella dilatata

Sphaeromidae—marine, intertidal

Exosphaeroma oregonensis	*Palavascia sphaeromae*
Sphaeroma spp.	*Palavascia sphaeromae*
Tecticeps japonicus	*Palavascia sphaeromae*

Trichoniscidae—terrestrial

Androniscus dentiger	*Parataeniella dilatata*
Scotoniscus marcomelos	*Parataeniella scotonisci*

Tylidae—terrestrial

Helleria brevicornis	*Eccrinoides helleriae*

Decapoda

(Macrura)

Astacidae (crayfish)—aquatic, lotic, and lentic

Cambarus affinis	(*Recticoma cambari*)
Orconectis nais	*Arundinula orconectis*

(Anomura)

Callianassidae (ghost and mud shrimps)—marine, mostly intertidal

Callianassa spp.	*Enteromyces callianassae*
	Taeniella carcini
Upogebia spp.	*Enteromyces callianassae*
	Taeniella carcini

Galatheidae—marine, subtidal

Galathea strigosa	*Arundinula galatheae*
	Taeniella carcini

Porcellanidae (porcelain crabs)—marine

Porcellana platycheles	*Arundinula porcellanae*

Hippidae (mole crabs)—marine, sandy shores

Emerita spp.	*Enterobryus halophilus*

Paguridae (hermit crabs)—marine, intertidal or subtidal

Eupagurus cuanensis	*Arundinula capitata*
Eupagurus excavatus	*Taeniella carcini*
Eupagurus prideuxi	(*Arundinula incurvata*)
Paguristes turgidus	*Arundinula washingtoniensis*
	Taeniella carcini
Pagurus maculatus	*Arundinula capitata*
Pagurus spp.	*Taeniella carcini*

Lithodidae—marine, intertidal

Haplogaster mertensii	*Arundinula haplogaster*
Oedignathus inermis	*Taeniella carcini*

(Brachyura)

Canceridae—marine, under rocks

Cancer oregonensis	*Taeniella carcini*

Portunidae (swimming crabs)—marine

Carcinus maenas	*Taeniella carcini*
Portunus puber	*Taeniella carcini*

Xanthidae (mud crabs)—marine, intertidal

Pilumnus hirtellus	*Taeniella carcini*
Xantho pilipes	*Taeniella carcini*

Ocypodidae (fiddler crabs)—marine, intertidal

Uca pugilator	*Enteromyces callianassae*

Grapsidae—marine, intertidal

Hemigrapsus penicillatus	*Enteromyces callianassae*
Hemigrapsus spp.	*Taeniella carcini*
Sesarma hematocheir	*Taeniella carcini*

Insecta

Collembola (springtails)—aquatic, lentic, or lotic

Entomobryidae

Orchesella villosa	*Orchesellaria lattesi*

Isotomidae

Agrenia bidenticulata	*Orchesellaria mauguioi*
Hydroisotoma schaefferi	*Orchesellaria pelta*
Isotoma sp.	*Orchesellaria mauguioi*
Isotomurus palustris	*Orchesellaria mauguioi*

Poduridae

Podura aquatica	*Orchesellaria podurae*

Ephemeroptera larvae (mayflies)—aquatic, mostly lotic

Baetidae

Baetis atrebatinus	(*Paramoebidium pavillardi*)
Baetis bioculatus	(*Genistella mailleti*)
Baetis pumilus	*Gauthieromyces microsporus*
Baetis tricaudatus	*Glotzia ephemeridarum*
Baetis sp.	*Graminella bulbosa*
Baetis spp.	*Legeriomyces ramosus*
Baetis spp.	(*Paramoebidium arcuatum*)

Baetis tricaudatus and other Baetidae	
Baetis spp. and other Baetidae	*Graminella microspora*
Baetis spp. and other Baetidae	*Spartiella barbata*
Centroptilum luteolum	*Zygopolaris ephemeridarum*
Callibaetis pacificus	*Glotzia centroptili*
Cloeon dipterum	*Pteromaktron protrudens*
	Pteromaktron protrudens
	Amoebidium parasiticum
Procloeon bifidum	*(Paramoebidium procloeoni)*

Ecdyonuridae

Ecdyonurus sp.	*(Paramoebidium geniculatum)*

Ephemerellidae

Drunella spinifera	*Legeriomyces aenigmaticus*
Ephemerella infrequens	*Legeriomyces ramosus*
Ephemerella spp.	*Zygopolaris ephemeridarum*

Heptageniidae

Epeorus longimanus	*Zygopolaris borealis*
Rhitrogena alpestris	*(Genistella rhitrogeniae)*

Leptophlebiidae

Habrophlebia fusca	*(Paramoebidium dispersum)*
Paraleptophlebia sp.	*(Paramoebidium eccriniformis)*
Thraulus bellus	*(Paramoebidium thrauli)*

Odonata larvae (dragonflies)—aquatic, lentic

Aeschna sp.	*Amoebidium parasiticum*
Anax imperator	*Amoebidium parasiticum*

Plecoptera larvae (stoneflies)—aquatic, lotic

Capniidae (winter–emerging stoneflies)

Allocapnia spp.	*Capniomyces stellatus*
Allocapnia spp.	*Genistelloides hibernus*
Allocapnia sp.	*Simuliomyces spica*
Mesocapnia sp.	*Genistelloides hibernus*

Chloroperlidae

Chloroperla sp.   (*Paramoebidium giganteum*)

Nemuridae

Nemura cinerea	*Orphella coronata*
Nemura variegata	*Paramoebidium inflexum*
Nephelopteryx nebulosa	(*Paramoebidium fuscum*)
Protonemura humeralis	*Orphella coronata*

Trichoptera larvae (caddis worms)—aquatic, lentic

Phryganeidae

Unidentified   *Amoebidium parasiticum*

Coleoptera (beetles)

Hydrophilidae—aquatic, mostly lentic

Cercyon obsoletus	(*Lactella cercyonis*)
Chaetarthria siminulum	(*Lactella chaetarthriae*)
Coelostoma orbiculare	(*Lactella coelostomatis*)
	(*Trichella coelostomatis*)
Dicyrtocercyon ustulatus	(*Lactella cercyonis*)
Helochares lividus	(*Enterobryus helocharei*)
Helochares spp.	(*Lactella helocharidis*)
	(*Trichella longa*)
Hydrobius fuscipes	(*Trichella hydrobii*)
Hydrochus sp.	(*Eccrinopsis leidyi*)
Hydrous spp.	*Enterobryus hydrophilorum*
Laccobius nigriceps	(*Lactella laccobii*)
Philhydrus testaceus	(*Trichella stammeri*)
Philhydrus spp.	(*Lactella philhydri*)
	(*Trichella philhydri*)

Passalidae (bess beetles)—terrestrial

Popilius disjunctus and other Passalidae	*Enterobryus attenuatus*
Various species of Passalidae	*Enterobryus compressus*

Scarabaeidae—terrestrial

Aphodius spp.	(*Lactella aphodii*)
Aphodius sp. (larvae)	*Lajasiella aphodii*

Pentodon punctatus (larvae)   (*Enterobryus pentodoni*)
Platystethus cornutus   (*Lactella platystethi*)

Diptera (flies)

Ceratopogonidae larvae (biting midges)—aquatic, lentic

Dasyhelea spp.	*Carouxella scalaris*
	Smittium culisetae

Chironomidae larvae (nonbiting midges)—aquatic, lotic or lentic

Boreoheptagyia lurida	*Stachylina pedifer*
	Smittium dimorphum
Chironomus plumosus	(*Genistella choanifera*)
	Stachylina euthena
Chironomus spp.	*Amoebidium parasiticum*
	Smittium culicis
	Smittium typhellum
Cricotopus spp.	*Smittium elongatum*
	Smittium longisporum
	Trichozygospora chironomidarum
Diamesa valkanovi	*Trichozygospora chironomidarum*
Diamesa spp.	*Smittium alpinum*
	Smittium elongatum
	Smittium orthocladii
	Stachylina macrospora
	Stachylina penetralis
Eukiefferiella sp.	*Smittium ouseli*
Orthocladius spp.	*Smittium orthocladii*
	Trichozygospora chironomidarum
Orthocladiinae	*Smittium chironomi*
Pagastia sp.	*Smittium gigasporus*
Polypedilum sp.	*Stachylina manicata*
Procladius sp.	*Smittium pusillum*
Psectrocladius sordidellus	*Smittium mucronatum*
Psectrotanypus varius	*Smittium bisporum*
	Stachylina euthena
Pseudochironomus sp.	*Stachylina manicata*
Smittia sp.	*Smittium arvernense*
Sympotthastia sp.	*Smittium cellaspora*
Syncricotopus rufiventris	*Smittium megazygosporum*
Syndiamesa macronyx	*Stachylina macrospora*
Tanytarsus sp.	*Stachylina longa*
Tanytarsini	*Stachylina minuta*
Trissocladius sp.	(*Opuntiella digitata*)

Bloodworms (unidentified)　　*Stachylina chironomidarum*
Unidentified　　　　　　　　　*Smittium arcticum*
　　　　　　　　　　　　　　　Smittium culisetae
　　　　　　　　　　　　　　　Smittium incrassatum
　　　　　　　　　　　　　　　Smittium macrosporum
　　　　　　　　　　　　　　　Stachylina nana
Many genera and species　　　*Smittium simulii*
　　　　　　　　　　　　　　　Stachylina grandispora

Culicidae larvae (mosquitoes)—aquatic, lentic

Anopheles annulipes　　　　*Smittium morbosum*
Anopheles hilli　　　　　　　*Smittium morbosum*
Many genera and species　　　*Amoebidium parasiticum*
　　　　　　　　　　　　　　　Smittium culicis
　　　　　　　　　　　　　　　Smittium culisetae
　　　　　　　　　　　　　　　Smittium simulii

Dixidae larvae (meniscus midges)—aquatic, lentic

Dixa sp.　　　　　　　　　　(*Dixidium dixae*)

Psychodidae larvae (moth flies)—aquatic, lotic

Maruina sp.　　　　　　　　*Stachylina lotica*

Sciaridae larvae (root gnats)—terrestrial

Sciara sp.　　　　　　　　　(*Eccrinopsis sciarae*)

Simuliidae larvae (blackflies)—aquatic, lotic

Prosimulium exigens　　　　　*Pennella arctica*
　　　　　　　　　　　　　　　Smittium pennelli
Prosimulium ferrugineus　　　*Pennella arctica*
Prosimulium onychodactylum　 *Smittium pennelli*
Prosimulium spp.　　　　　　 *Genistellospora homothallica*
　　　　　　　　　　　　　　　Pennella hovassi

Simulium arcticum　　　　　　*Pennella arctica*
Simulium canonicolum　　　　*Harpella leptosa*
Simulium equinum　　　　　　(*Paramoebidium simulii*)
　　　　　　　　　　　　　　　Pennella grassei
Simulium venustum　　　　　 *Harpella leptosa*
Simulium vittatum　　　　　　*Smittium culisetae*
Simulium spp.　　　　　　　　*Genistellospora homothallica*
　　　　　　　　　　　　　　　(*Paramoebidium chattoni*)
　　　　　　　　　　　　　　　Pennella hovassi
　　　　　　　　　　　　　　　Smittium pennelli

Many genera and species
Harpella melusinae
Paramoebidium curvum
Pennella angustispora
Pennella simulii
Simuliomyces microsporus
Smittium culicis
Smittium simulii
Stipella vigilans

Thaumaleidae larvae (solitary midges)—aquatic, lotic

Thaumalea sp.
Harpellomyces eccentricus

Tipulidae larvae (crane flies)—aquatic, lotic; or terrestrial

Elliptra astigmatica
Trichocera sp.
Smittium simulii
(Trichoceridium ramosum)

APPENDIX C
Axenic Isolates of Trichomycetes and Their Sources

The living trichomycete cultures listed in this appendix are maintained by the author at the Department of Botany, University of Kansas, Lawrence, Kansas 66045–2106, U.S.A. Those with ATCC numbers are available from the American Type Culture Collection, 12301 Parklawn Drive, Rockville, Maryland 20852–1776, U.S.A. Listed under each trichomycete species are the isolate designation, the host from which the culture was made, its geographic locality and year of isolation, and the isolator (if not indicated, the isolations were made by the author). Abbreviations of host families are as follows: Cap. = Capniidae; Cer. = Ceratopogonidae; Chi. = Chironomidae; Cul. = Culicidae; Sim. = Simuliidae. All insect hosts were larval stages. For culturing and maintenance techniques refer to Chapter 3.

Harpellales

Capniomyces stellatus
ARK–12–133	*Allocapnia* sp.—Cap.	Arkansas, 1983	S.W. Peterson
MIS–10–106 (ATCC 46884)	*Allocapnia* sp.	Missouri, 1981	S.W. Peterson
MIS–10–108	*Allocapnia* sp.	Missouri, 1981	S.W. Peterson
MIS–21–127	*Allocapnia* sp.	Missouri, 1982	S.W. Peterson

Genistelloides hibernus
ARK–12–136	*Allocapnia* sp.—Cap.	Arkansas, 1983	S.W. Peterson
RC–7	*Allocapnia vivipara*	Kansas, 1982	S.W. Peterson
SWP–1	*Allocapnia vivipara*	Kansas, 1982	S.W. Peterson

Smittium culicis
CAL–X–1	*Aedes melanimon*—Cul.	California	Clark, Kellen & Lindegren, 1963
CAN–X–1	*Prosimulium* sp.—Sim.	Canada, 1962	H.C. Whisler
FRA–6–16 (ATCC 32712)	Bloodworm—Chi.	France, 1968	
FRA–8–7	Culicidae	France, 1968	
FRA–9–1	Culicidae	France, 1968	
FRA–15–2	Chironomidae	France, 1968	

FRA-15-5	Bloodworm—Chi.	France, 1968	
FRA-15-7	Bloodworm	France, 1968	
LEA-7-10	*Simulium* sp.—Sim.	Kansas, 1981	
WYO-51-11 (ATCC 26070)	*Aedes sticticus*—Cul.	Wyoming, 1965	
WYO-51-14	*Aedes sticticus*	Wyoming, 1965	
ZEA-6-3	*Culex pervigilans*—Cul.	New Zealand, 1983	

Smittium culisetae

AUS-2-8	*Chironomus alternans*—Chi.	Australia, 1983	
AUS-2-9	*Chironomus alternans*	Australia, 1983	
AUS-2-10	*Chironomus alternans*	Australia, 1983	
CAL-X-2	*Culiseta incidens*—Cul.	California	Clark, Kellen & Lindegren, 1963
COL-18-3 (type) (ATCC 16244)	*Culiseta impatiens*—Cul.	Colorado, 1963	
FRA-7-1 (ATCC 32710)	*Dasyhelia* sp.—Cer.	France, 1968	
HAW-2-E	Culicidae	Hawaii, 1982	M.C. Williams
HAW-5-5	*Aedes albopictus*—Cul.	Hawaii, 1964	
HAW-5-7	*Aedes albopictus*	Hawaii, 1964	
HAW-5-8	*Aedes albopictus*	Hawaii, 1964	
HAW-5-9 (ATCC 32711)	*Aedes albopictus*	Hawaii, 1964	
HAW-13-3	*Aedes vexans*	Hawaii, 1964	
HAW-14-5	*Aedes albopictus*	Hawaii, 1964	
HAW-14-7	*Aedes albopictus*	Hawaii, 1964	
HAW-14-8	*Aedes albopictus*	Hawaii, 1964	
JAP-30-3	*Culex* sp.—Cul.	Japan, 1964	
JAP-30-4	*Culex* sp.	Japan, 1964	
JAP-30-8	*Culex* sp.	Japan, 1964	
JAP-57-1	Culicidae	Japan, 1964	
JAP-77-3	Culicidae	Japan, 1967	
JAP-77-7	Culicidae	Japan, 1967	
JAP-77-8	Culicidae	Japan, 1967	
JAP-77-9	Culicidae	Japan, 1967	
KAU-W-4	Culicidae	Hawaii, 1982	M.C. Williams
LEA-7-2	*Simulium* sp.—Sim.	Kansas, 1981	
LEA-7-4	*Simulium* sp.	Kansas, 1981	
LEA-7-106	*Simulium* sp.	Kansas, 1981	S.W. Peterson
NE-27-1 (ATCC 46885)	*Aedes vexans*—Cul.	Nebraska	Williams, 1983b
WYO-42-26	Chironomidae	Wyoming, 1965	
WYO-44-1	Culicidae	Wyoming, 1965	
WYO-44-5	Culicidae	Wyoming, 1965	
WYO-44-8	Culicidae	Wyoming, 1965	
WYO-44-9	Culicidae	Wyoming, 1965	
WYO-44-11	Culicidae	Wyoming, 1965	
WYO-44-15	Culicidae	Wyoming, 1965	

Smittium morbosum

Sweeney	*Anopheles hilli*—Cul.	Australia	Sweeney, 1981a

Smittium mucronatum

FRA–12–3 (ATCC 26071)	*Psectrocladius sordidellus*—Chi.	France, 1968	

Smittium simulii

CAL–8–1 (type) (ATCC 16245)	*Simulium argus*—Sim.	California, 1963	
CLW–2–44a	*Simulium vittatum*	Minnesota	Chapman, 1966
CLW–9–15	*Simulium* sp.	Minnesota, 1965	M.E. Chapman
CLW–10–5	*Simulium* sp.	Minnesota, 1965	M.E. Chapman
CLW–10–6	*Simulium* sp.	Minnesota, 1965	M.E. Chapman
FRA–18–4	Chironomidae	France, 1968	
JAP–31–7	*Simulium* sp.	Japan, 1964	
JAP–32–4	Chironomidae	Japan, 1964	
JAP–33–2 (ATCC 32713)	Chironomidae	Japan, 1964	
JAP–51–5	*Simulium* sp.	Japan, 1964	
JAP–51–6	*Simulium* sp.	Japan, 1964	
JAP–51–11 (ATCC 32714)	*Simulium* sp.	Japan, 1964	
JAP–51–12	*Simulium* sp.	Japan, 1964	
JAP–51–18	*Simulium* sp.	Japan, 1964	
JAP–57–5	Culicidae	Japan, 1964	
SWE–8–4 (ATCC 32716)	*Diamesa* sp.—Chi.	Sweden, 1971	

Smittium spp.

DGK–2–3	Chironomidae	Kansas, 1980	S.W. Peterson
SWE–3–5 (ATCC 32715)	*Orthocladius* sp.—Chi.	Sweden, 1971	
SWI–4–8	Orthocladiinae—Chi.	Switzerland, 1971	
SWI–4–13 (ATCC 32717)	Orthocladiinae—Chi.	Switzerland, 1971	
SWI–7–6 (ATCC 32718)	*Pseudokiefferiella* sp.—Chi.	Switzerland, 1971	

Unnamed Legeriomycetaceae

ARK–8–107	*Allocapnia rickeri*—Cap.	Kansas, 1982	S.W. Peterson
MIS–20–127	*Allocapnia* sp.	Missouri, 1982	S.W. Peterson

Amoebidiales

Amoebidium parasiticum

A1a	Cladocera	California	Whisler, 1960
FRA–1–14 (ATCC 32708)	Cladocera	France, 1968	
JAP–7–2 (ATCC 32709)	*Chironomus* sp.	Japan, 1964	

References

In order to give as complete a bibliography of the trichomycete literature as possible, some references are included here that have not been directly cited in the text, including references to papers describing illegitimate or doubtful species that have not been validated or accepted in this monograph. Excluded are some references to trichomycetes in zoological treatises, mycological textbooks, and research papers that merely mention these fungi or speculate on their taxonomic placement without any other original contribution. However, textbooks that give some coverage to the subject are included.

Akov, S. 1962. A qualitative and quantitative study of the nutritional requirements of *Aedes aegypti* L. larvae. J. Ins. Physiol. 8: 319–335.

Alexopoulos, C.J., and C.W. Mims. 1979. Introductory mycology, 3rd ed. John Wiley & Sons, New York.

Arvy, L., and W.L. Peters. 1973. Phorésies, biocoenoses et thanatocoenoses chez les Éphéméroptères. Pp. 254–312 *in* W.L. Peters and J. Peters (eds.), Proc. First Int. Conf. Ephemeroptera, 1970, Tallahassee. E.–J. Brill, Leyde.

Arvy, L., and W.L. Peters. 1976. Liste des Éphéméroptères–hôtes de parasites, de commensaux et autres associés. Ann. Parasitol. (Paris) 51: 121–141.

Balbiani, E.–G. 1889. Sur trois entophytes nouveaux du tube digestif des Myriapodes. J. Anat. Physiol., Paris 25: 5–45.

Bartnicki–Garcia, S. 1968. Cell wall chemistry, morphogenesis, and taxonomy of fungi. Ann. Rev. Microbiol. 22: 87–108.

Bartnicki–Garcia, S. 1970. Cell wall composition and other biochemical markers in fungal phylogeny. Pp. 81–103 *in* J.B. Harborne (ed.), Phytochemical phylogeny. Academic Press, London.

Benjamin, R.K. 1958. Sexuality in the Kickxellaceae. Aliso 4: 149–169.

Benjamin, R.K. 1979. Zygomycetes and their spores. Pp. 573–621 *in* B. Kendrick (ed.), The whole fungus, Vol. 2. Nat. Mus. Nat. Sci., Ottawa.

Benny, G.L., and H.C. Aldrich. 1975. Ultrastructural observations on septal and merosporangial ontogeny in *Linderina pennispora* (Kickxellales; Zygomycetes). Can. J. Bot. 53: 2325–2335.

Borut, S. 1961. *Amoebidium parasiticum* Cienkowski—a Trichomycete growing on *Daphnia* sp. Israel J. Bot. 10D: 142–147.

Brain, A.P.R., P. Jeffries, and T.W.K. Young. 1982. Ultrastructure of septa in *Tieghemiomyces californicus*. Mycologia 74: 173–181.

Brassard, G.R., S. Frost, M. Laird, O.A. Olsen, and D.H. Steele. 1971. Studies of the spray zone of Churchill Falls, Labrador. Biol. Conserv. 4: 13–18.

Cerniglia, C.E., R.L. Hebert, P.J. Szaniszlo, and D.T. Gibson. 1978. Fungal transformation of naphthalene. Arch. Microbiol. 117: 135–143.

Chadefaud, M., and L. Emberger. 1960. Les Trichomycètes: Amoebidiales, Eccrinales et Harpellales, Vol. 1, Sec. 5. Pp. 895–902 in Traité de Botanique systématique. Masson et Cie, Paris.

Chapman, M.E. 1966. Isolation and experimental studies on some Trichomycetes. Master of Arts Thesis, University of Kansas. 42 pp.

Charmantier, G., and J.-F. Manier. 1981. Relations écologiques entre *Sphaeroma serratum* (Fabricius, 1787) et son commensal intestinal *Palavascia sphaeromae*. Vie Milieu 31: 101–111.

Chatton, E. 1906a. Sur la biologie, la spécification et la position systématique des *Amoebidium*. Arch. Zool. Exp. Gen., Ser. 4, 5: 17–31.

Chatton, E. 1906b. Sur la morphologie et l'évolution de l'*Amoebidium recticola*, nouvelle espèce commensale des Daphnies. Arch. Zool. Exp. Gen., Ser. 4, 5: 33–38.

Chatton, E. 1908. Revue des parasites et des commensaux des Cladocères. Observations sur des formes nouvelles ou peu connues. C.R. Assoc. Fr. Avance. Sci., Reims Congr., 1907, pp. 797–811.

Chatton, E. 1920. Les membranes péritrophiques des Drosophiles (Diptères) et des Daphnies (Cladocères); leur genèse et leur role à l'égard des parasites intestinaux. Bull. Soc. Zool. Fr. 45: 265–280.

Chatton, E. 1925. *Pansporella perplexa* amoebien à spores protégées parasite des Daphnies. Ann. Sci. Nat. Zool., Ser. 10, 8: 5–84.

Chatton, E., and E. Roubaud. 1909. Sur un *Amoebidium* du rectum des larves de Simulies (*Simulium argyreatum* Meig. et *S. fasciatum* Meig). C.R. Soc. Biol. 66: 701–703.

Cienkowski, L. 1861. Ueber parasitische Schläuche auf Crustaceen und einigen Insektenlarven (*Amoebidium parasiticum* m.). Bot. Zeit. 19: 169–174.

Clark, T.B., W.R. Kellen, and J.E. Lindegren. 1963. Axenic culture of two Trichomycetes from Californian mosquitoes. Nature 197: 208–209.

Clements, A.N. 1963. The physiology of mosquitoes. Intern. Ser. Monogr. Pure Appl. Biol., Vol. 17. Pergamon Press Ltd., New York.

Coluzzi, M. 1966. Experimental infections with *Rubetella* fungi in *Anopheles gambiae* and other mosquitoes. Proc. First Int. Congr. Parasitol., 1964 (Rome), Vol. 1, pp. 592–593.

Cooke, R. 1977. The biology of symbiotic fungi. John Wiley & Sons, London.

Coste-Mathiez, F. 1970. Parasites de larves de Chironomides (Diptères, Nematocères) des environs de Montpellier. Docteur de Spécialité Thesis, University of Montpellier.

Cronin, E.T., and T.W. Johnson, Jr. 1958. A halophilic *Enterobryus* in the mole crab *Emerita talpoida* Say. J. Elisha Mitchell Sci. Soc. 74: 167–172.

Crosby, T.K. 1974. Trichomycetes (Harpellales) of New Zealand *Austrosimulium* larvae (Diptera: Simuliidae). J. Nat. Hist. 8: 187–192.

Dang, S. 1979. Electron–microscope studies on the holdfast structure of some Trichomycetes. Master of Arts Thesis, University of Kansas. 78 pp.

Dang, S., and R.W. Lichtwardt. 1979. Fine structure of *Paramoebidium* (Trichomycetes) and a new species with viruslike particles. Amer. J. Bot. 66: 1093–1104.

Davies, B.R. 1976. The dispersal of Chironomidae larvae: a review. J. Entomol. Soc. S. Afr. 39: 39–62.

Debaisieux, P. 1920. *Coelomycidium simulii* nov. gen., nov. spec., et remarques sur l'*Amoebidium* des larves de *Simulium*. Cellule 30: 249–276.

Dogma, I.J., Jr. 1975. Of Philippine mycology and lower fungi. Kalikasan, Philipp. J. Biol. 4: 69–105.

Dollfus, R.–P. 1952. Quelques Oxyuroidea de Myriapodes. Ann. Parasitol. (Paris) 27: 143–236.

Drechsler, C. 1946. A nematode–destroying Phycomycete forming immotile spores in aerial evacuation tubes. Bull. Torrey Bot. Club: 73: 1–17.

Dubitskii, A.M. 1978. Biological control of blood sucking Diptera in the USSR. Inst. Zool., Kazakhstan Acad. Sci., Alma Ata. (pp. 92–93.).

Duboscq, O., L. Léger, and O. Tuzet. 1948. Contribution à la connaissance des Eccrinides: les Trichomycètes. Arch. Zool. Exp. Gen. 86: 29–144.

El–Buni, A.M. 1972. Spore germination in axenic cultures of *Smittium* spp. (Trichomycetes). Master of Arts Thesis, University of Kansas. 47 pp.

El–Buni, A.M. 1975. Factors affecting sporulation, growth and spore germination in species of *Smittium* (Trichomycetes). Ph.D. Dissertation, University of Kansas. 136 pp.

El–Buni, A.M., and R.W. Lichtwardt. 1976a. Asexual sporulation and mycelial growth in axenic cultures of *Smittium* spp. (Trichomycetes). Mycologia 68: 559–572.

El–Buni, A.M., and R.W. Lichtwardt. 1976b. Spore germination in axenic cultures of *Smittium* spp. (Trichomycetes). Mycologia 68: 573–582.

Emerson, R., and H.C. Whisler. 1959. The nature and relationships of *Oedogoniomyces*. Proc. IX Int. Bot. Congr. 2: 103–104 (Abstr.).

Emerson, R., and H.C. Whisler. 1968. Cultural studies of *Oedogoniomyces* and *Harpochytrium*, and a proposal to place them in a new order of aquatic Phycomycetes. Arch. Mikrobiol. 61: 195–211.

Farr, D.F. 1965. Some nutritional and electron microscopic observations on *Smittium culisetae* (Trichomycetes). Master of Arts Thesis, University of Kansas. 40 pp.

Farr, D.F., and R.W. Lichtwardt. 1967. Some cultural and ultrastructural aspects of *Smittium culisetae* (Trichomycetes) from mosquito larvae. Mycologia 59: 172–182.

Fritsch, A. 1895. Über Parasiten bei Crustaceen und Raederthieren der süssen Gewässer. Bull. Acad. Sci. Prague 2: 79–85.

Frost, S., and J.–F. Manier. 1971. Notes on Trichomycetes (Harpellales: Harpellaceae and Genistellaceae) in larval blackflies (Diptera: Simuliidae) from Newfoundland. Can. J. Zool. 49: 776–778.

Gauthier, M. 1936. Sur un nouvel Entophyte du groupe des Harpellacées Lég. et Dub., parasite des larves d'Éphémérides. C.R. Acad. Sci. Paris 202: 1096–1098.

Gauthier, M. 1960. Un nouveau Trichomycète rameux parasite des larves de *Baëtis pumilus* (Burm). Trav. Lab. Hydrobiol. Piscic. Univ. Grenoble 51: 225–227.

Gauthier, M. 1961. Une nouvelle espèce de *Stachylina: St. minuta* n. sp., parasite des larves de Chironomides Tanytarsiens. Trav. Lab. Hydrobiol. Piscic. Univ. Genoble 52, 53: 1–4 (1960–1961).

Goldberg, L., and B. De Meillon. 1948. The nutrition of the larva of *Aedes aegypti* Linnaeus. 3. Lipid requirements. Biochem. J. 43: 372–379.

Granata, L. 1908. Di un nuovo parassita dei millepiedi (*Capillus* n. g. *intestinalis* n. sp.). Biologica (Torino) 2: 3–16.

Grizel, H. 1971. Le parasitisme chez les Amphipodes de la region de Montpellier. Docteur de Spécialité Thesis, Université des Sciences et Techniques du Languedoc. 125 pp.

Hardy, D.E. 1960. Insects of Hawaii. Vol. 10. Diptera: Nematocera–Brachycera. University of Hawaii Press, Honolulu.

Hauptfleisch, P. *Astreptonema longispora* n. g. n. sp., eine neue Saprolegniacee. Ber. Deut. Bot. Ges. 13: 83–88.

Heymons, R., and H. Heymons. 1934. *Passalus* und seine intestinale Flora. Biol. Zentralbl. 54: 40–51.

Hibbits, J. 1978. Marine Eccrinales (Trichomycetes) found in crustaceans of the San Juan Archipelago, Washington. Syesis 11: 213–261.

Hitchcock, S.W. 1974. Guide to the insects of Connecticut. Part VII. The Plecoptera or stoneflies of Connecticut. Bull. Conn. Geol. Nat. Hist. Survey 107: 1–262 (p. 8).

Hollingsworth, L.A. 1978. Immunotaxonomy of selected species of the fungal classes Trichomycetes and Zygomycetes. Master of Science Thesis, Pittsburg State University (Kansas). 31 pp.

Horn, B.W. 1980. Studies on the nutritional relationship of larval *Aedes aegypti* (Diptera: Culicidae) with *Smittium culisetae* (Trichomycetes). Master of Arts Thesis, University of Kansas. 79 pp.

Horn, B.W., and R.W. Lichtwardt. 1981. Studies on the nutritional relationship of larval *Aedes aegypti* (Diptera: Culicidae) with *Smittium culisetae* (Trichomycetes). Mycologia 73: 724–740.

Hynes, H.B.N. 1976. Biology of Plecoptera. Ann. Rev. Entomol. 21: 135–153.

Ingold, C.T. 1967. Why not look for Harpellales. Bull. Brit. Mycol. Soc. 1: 43–44.

Jane, F.W. 1946. A revision of the genus *Harpochytrium*. J. Linn. Soc. London 53: 28–40.

Jeekel, C.A.W., O. Tuzet, J.–F. Manier, and P. Jolivet. 1959. Myriapodes et leurs parasites. Nat. Albert Park, Ser. 2, 9: 3–32.

Johnson, D.S. 1952. *Amoebidium parasiticum*, an epibiote of fresh–water Crustacea, not previously recorded in Britain. J. Quekett Micr. Cl. Ser. 4: 387–391.

Johnson, D.S. 1963. The occurrence of *Amoebidium parasiticum* Cienkowski in Singapore. Bull. Nat. Mus. State Singapore 32: 158–159.

Johnson, T.W., Jr. 1966. Trichomycetes in species of *Hemigrapsus*. J. Elisha Mitchell Sci. Soc. 82: 1–6.

Johnson, T.W., Jr., and F.K. Sparrow. 1961. Fungi in oceans and estuaries. J. Cramer, Weinheim.

Kazama, F.Y. 1979. Ultrastructural evidence for viruses in lower fungi. Pp. 405–439 *in* P.A. Lemke (ed.), Viruses and plasmids in fungi. Marcel Dekker, New York.

Kermarrec, A., and J.–F. Manier. 1971. Sur un thallophyte parasite intestinal de *Neodiplogaster rühmi* Laumond, 1970 et de *Neodiplogaster* n. sp. (Nematoda–Rhabditida). Ann. Parasitol. (Paris) 46: 749–756.

Kobayashi, Y., and M. Ôkubo. 1954. On a new genus *Oedogoniomyces* of the Blastocladiaceae. Bull. Nat. Sci. Mus. (Tokyo) 1: 59–66.

Kobayasi, Y., N. Hiratsuka, R.P. Korf, K. Tubaki, K. Aoshima, M. Soneda, and J. Sugiyama. 1967. Mycological studies of the Alaskan Arctic. Ann. Rep. Inst. Fermentation, Osaka 3: 1–138.

Kobayasi, Y., N. Hiratsuka, Y. Otani, K. Tubaki, S. Udagawa, and M. Soneda. 1969. Bull. Nat. Sci. Mus. (Tokyo) 12: 311–430.
Kobayasi, Y., N. Hiratsuka, Y. Otani, K. Tubaki, S. Udagawa, J. Sugiyama, and K. Konno. 1971. Bull. Nat. Sci. Mus. (Tokyo) 14: 1–96.
Kreisel, H. 1969. Grundzüge eines natürlichen Systems der Pilze. Gustav Fisher Verlag, Jena.
Kuno, G. 1973. Biological notes of *Amoebidium parasiticum* found in Puerto Rico. J. Invert. Pathol. 21: 1–8.
Labbe, A. 1899. Das Tierreich, Vol. 5. R. Friedländer und Sohn, Berlin.
Lang, C.A., K.J. Basch, and R.S. Storey. 1972. Growth, composition and longevity of the axenic mosquito. J. Nutrition 102: 1057–1066.
Lea, A.O., J.B. Dimond, and D.M. DeLong. 1956. A chemically defined medium for rearing *Aedes aegypti* larvae. J. Econ. Entomol. 49: 313–315.
Le Berre, R. 1967. Les membranes péritrophiques chez les Arthropodes leur rôle dans la digestion et leur intervention dans l'évolution d'organismes parasitaires. Cah. O.R.S.T.O.M., Ser. Ent. Med. 5: 146–204.
Léger, L., and O. Duboscq. 1903. Recherche sur les Myriapodes de Corse et leur parasites. Arch. Zool. Exp. Gen. 1: 307–311.
Léger, L., and O. Duboscq. 1905a. Les Eccrinides, nouveau groupe de Protophytes parasites. C.R. Acad. Sci. Paris 141: 425–427.
Léger, L., and O. Duboscq. 1905b. Les Eccrinides, nouveau groupe de végétaux inférieurs, parasites des Arthropodes. Bull. Assoc. Fr. Avance. Sci., pp. 331–332.
Léger, L., and O. Duboscq. 1906. L'évolution des *Eccrina* des *Glomeris*. C.R. Acad. Sci. Paris 142: 590–592.
Léger, L., and O. Duboscq. 1911. Sur les Eccrinides des Crustacés Decapodes. Ann. Univ. Grenoble 23: 139–141.
Léger, L., and O. Duboscq. 1916. Sur les Eccrinides des Hydrophilides. Arch. Zool. Exp. Gen. 56: 21–31.
Léger, L., and O. Duboscq. 1929a. *Eccrinoïdes henneguyi* n. g. n. sp. et la systématique des Eccrinides. Arch. Anat. Microscop. 25: 309–324.
Léger, L., and O. Duboscq. 1929b. *Harpella melusinae* n. g. n. sp. Entophyte eccriniforme parasite des larves de Simulie. C.R. Acad. Sci. Paris 188: 951–954.
Léger, L., and O. Duboscq. 1929c. L'évolution des *Paramoebidium*, nouveau genre d'Eccrinides, parasite des larves aquatiques d'Insectes. C.R. Acad. Sci. Paris 189: 75–77.
Léger, L., and O. Duboscq. 1933. *Eccrinella (Astreptonema?) gammari* Lég. et Dub. Eccrinide des Gammares d'eau douce. Arch. Zool. Exp. Gen. 75: 283–292.
Léger, L., and M. Gauthier. 1931. *Orphella coronata* n. g. n. sp. Entophyte parasite des larves de Némurides. Trav. Lab. Hydrobiol. Piscic. Univ. Grenoble 23: 67–72.
Léger, L., and M. Gauthier. 1932. Endomycètes nouveaux des larves aquatiques d'Insectes. C.R. Acad. Sci. Paris 194: 2262–2265.
Léger, L., and M. Gauthier. 1935a. La spore des Harpellacées (Léger et Duboscq), Champignons parasites des Insectes. C.R. Acad. Sci. Paris 200: 1458–1460.
Léger, L., and M. Gauthier. 1935b. La spore des Harpellacees (Léger et Duboscq) Champignons parasites des Insectes. Trav. Lab. Hydrobiol. Piscic. Univ. Grenoble 27: 3–6.

Léger, L., and M. Gauthier. 1937. *Graminella bulbosa* nouveau genre d'Entophyte parasite des larves d'Éphémérides du genre *Baetis*. C.R. Acad. Sci. Paris 202: 27–29.
Leidy, J. 1849a. *Enterobrus*, a new genus of Confervaceae. Proc. Acad. Nat. Sci. Philadelphia 4: 225–233.
Leidy, J. 1849b. Descriptions (accompanied by drawings,) of new genera and species of Entophyta. Proc. Acad. Nat. Sci. Philadelphia 4: 249–250.
Leidy, J. 1850a. (Observations upon an entophytic forest.) Proc. Acad. Nat. Sci. Philadelphia 5: 8–9.
Leidy, J. 1850b. Descriptions of new Entophyta growing within animals. Proc. Acad. Nat. Sci. Philadelphia 5: 35–36.
Leidy, J. 1853. A flora and fauna within living animals. Smithsonian Contr. Knowledge 5: 1–67.
Lichtenstein, J.L. 1917a. Sur un *Amoebidium* à commensalisme interne du rectum des larves d'*Anax imperator* Leach: *Amoebidium fasciculatum* n. sp. Arch. Zool. Exp. Gen. 56: 49–62.
Lichtenstein, J.L. 1917b. Sur un mode nouveau de multiplication chez les Amoebidiacées. Arch. Zool. Exp. Gen. 56: 95–99.
Lichtwardt, R.W. 1951. Studies on some species of Eccrinales inhabiting the intestinal tract of millipedes. Master of Science Thesis, University of Illinois. 50 pp.
Lichtwardt, R.W. 1954a. Morphological, cytological, and taxonomic observations on species of *Enterobryus* from the hindgut of certain millipedes and beetles. Ph.D. Dissertation, University of Illinois. 241 pp.
Lichtwardt, R.W. 1954b. Three species of Eccrinales inhabiting the hindguts of millipedes, with comments on the Eccrinids as a group. Mycologia 46: 564–585.
Lichtwardt, R.W. 1957a. *Enterobryus attenuatus* from the Passalid beetle. Mycologia 49: 463–474.
Lichtwardt, R.W. 1957b. An *Enterobryus* occurring in the milliped *Scytonotus granulatus* (Say). Mycologia 49: 734–739.
Lichtwardt, R.W. 1958. An *Enterobryus* from the milliped *Boraria carolina* (Chamberlin). Mycologia 50: 550–561.
Lichtwardt, R.W. 1960a. An *Enterobryus* (Eccrinales) in a common greenhouse milliped. Mycologia 52: 248–254.
Lichtwardt, R.W. 1960b. Taxonomic position of the Eccrinales and related fungi. Mycologia 52: 410–428.
Lichtwardt, R.W. 1960c. New species of *Enterobryus* from southeastern United States. Mycologia 52: 743–752.
Lichtwardt, R.W. 1961a. A stomach fungus in *Callianassa* spp. (Decapoda) from Chile. Reports of the Lund University Chile Expedition 1948–49. 41. Lunds Univ. Arsskr. 57: 3–10.
Lichtwardt, R.W. 1961b. A *Palavascia* (Eccrinales) from the marine isopod *Sphaeroma quadridentatum* Say. J. Elisha Mitchell Sci. Soc. 77: 242–249.
Lichtwardt, R.W. 1962. An *Arundinula* (Trichomycetes, Eccrinales) in a crayfish. Mycologia 54: 440–447.
Lichtwardt, R.W. 1964a. Axenic culture of two new species of branched Trichomycetes. Amer. J. Bot. 51: 836–842.
Lichtwardt, R.W. 1964b. Validation of the genus *Palavascia* (Trichomycetes). Mycologia 56: 318–319.
Lichtwardt, R.W. 1967. Zygospores and spore appendages of *Harpella* (Trichomycetes) from larvae of Simuliidae. Mycologia 59: 482–491.

Lichtwardt, R.W. 1968. Why stop with the animals? Turtox News 46: 194–196.
Lichtwardt, R.W. 1972. Undescribed genera and species of Harpellales (Trichomycetes) from the guts of aquatic insects. Mycologia 64: 167–197.
Lichtwardt, R.W. 1973a. The Trichomycetes: what are their relationships? Mycologia 65: 1–20.
Lichtwardt, R.W. 1973b. Trichomycetes. Pp. 237–243 in G.C. Ainsworth, F.K. Sparrow, and A.S. Sussman (eds.), The fungi, an advanced treatise, Vol. IVB. Academic Press, New York.
Lichtwardt, R.W. 1974. Trichomycetes. Pp. 106–119 in R.B. Stevens (ed.), Mycology guidebook. University of Washington Press, Seattle.
Lichtwardt, R.W. 1976. Trichomycetes. Pp. 651–671 in E.B.G. Jones (ed.), Recent advances in aquatic mycology. Elek Science, London.
Lichtwardt, R.W. 1978a. Taxonomic problems in Trichomycetes related to their growth in arthropod guts. Pp. 109–114 in C.V. Subramanian (ed.), Proc. Int. Symp. Tax. Fungi Univ. Madras, 1973. University of Madras, India.
Lichtwardt, R.W. 1978b. *Smittium culisetae*. Pp. 167–168 in M.S. Fuller (ed.), Lower fungi in the laboratory. University of Georgia, Athens.
Lichtwardt, R.W. 1982. Trichomycetes. Pp. 195–197 in S.P. Parker (ed.), Synopsis and classification of living organisms. McGraw–Hill Book Co., New York.
Lichtwardt, R.W. 1983. *Gauthieromyces*, a new genus of Harpellales based on *Genistella microspora*. Mycotaxon 17: 213–215.
Lichtwardt, R.W. 1984a. Species of Harpellales living within the guts of aquatic Diptera larvae. Mycotaxon 19: 529–550.
Lichtwardt, R.W. 1984b. Validation of *Eccrinoides helleriae* (Eccrinales). Mycotaxon 20: 519–520.
Lichtwardt, R.W., and A.W. Chen. 1964. A *Parataeniella* (Trichomycetes, Eccrinales) in an isopod. Mycologia 56: 163–169.
Lichtwardt, R.W., and J.-F. Manier. Validation of the Harpellales and Asellariales. Mycotaxon 7: 441–442.
Lichtwardt, R.W., and S.T. Moss. 1981. Vegetative propagation in a new species of Harpellales, *Graminella microspora*. Trans. Brit. Mycol. Soc. 76: 311–316.
Lichtwardt, R.W., and S.T. Moss. 1984a. New Asellariales (Trichomycetes) from the hindguts of aquatic isopods and springtails. Mycotaxon 20: 259–274.
Lichtwardt, R.W., and S.T. Moss. 1984b. *Harpellomyces eccentricus*, an unusual Harpellales from Sweden and Wales. Mycotaxon 20: 511–517.
Lichtwardt, R.W., and M.C. Williams. 1983a. Two unusual Trichomycetes in an aquatic midge larva. Mycologia 75: 728–734.
Lichtwardt, R.W., and M.C. Williams. 1983b. A new *Legeriomyces* (Harpellales) with variable trichospore size. Mycologia 75: 757–761.
Lichtwardt, R.W., and M.C. Williams. 1984. *Zygopolaris borealis*, a new gut fungus (Trichomycetes) living in aquatic mayfly larvae. Can. J. Bot. 62: 1283–1286.
Lieberkühn, N. 1856. Ueber parasitische Schläuche auf einigen Insectenlarven. Arch. Anat. Physiol. 25: 494–495.
Locquin, M.V. 1974. De taxia fungorum, Vol. 1, Syllabus. Published by the author (U.A.E. Mondedition, Paris).
Madelin, M.F. 1966. Fungal parasites of insects. Ann. Rev. Entomol. 11: 423–448.
Maessen, K. 1955. Die zooparasitären Eccrinidales. Parasitol. Schr. 2: 1–129.
Manier, J.-F. 1947. *Paratrichella pentodona* n. g., n. sp. Entophyte parasite des larves de *Pentodon punctatus* de Vill. Ann. Sci. Nat., Zool., Ser. 11, 9: 275–279.

Manier, J.–F. 1950 (1951). Recherches sur les Trichomycètes. Ann. Sci. Nat., Bot., Ser. 11, 11: 53–162.

Manier, J.–F. 1954. Essais de culture des *Eccrina flexilis* Léger et Duboscq Trichomycètes endocommensaux des *Glomeris marginata* Villers. Ann. Parasitol. (Paris) 29: 265–270.

Manier, J.–F. 1955a. Nouvelles observations sur *Stipella vigilans* Léger et Gauthier et sur *Paramoebidium chattoni* Duboscq, Léger et Tuzet. Leurs cultures. Ann. Sci. Nat., Zool., Ser. 11, 17: 63–66.

Manier, J.–F. 1955b. Classification et nomenclature des Trichomycètes. Ann. Sci. Nat., Zool., Ser. 11, 17: 395–397.

Manier, J.–F. 1955c. *Andohaheloa pauliani* n. g. n. sp. Trichomycète commensal de Myriapodes–Diplopodes de Madagascar. Son évolution en cultures. Natur. Malgache 7: 83–90.

Manier, J.–F. 1958. *Orchesellaria lattesi* n. g. n. sp. Trichomycète rameux Asellariidae commensal d'un Aptérigote Collembole *Orchesella villosa* L. Ann. Sci. Nat., Zool., Ser. 11, 20: 131–139.

Manier, J.–F. 1961a. Arthromitaceae Schizophytes symbiotes de l'intestin postérieur des Myriapodes Diplopodes. Ann. Parasitol. (Paris) 36: 1–16.

Manier, J.–F. 1961b. Eccrinides de Crustacés récoltés sur les côtes du Finistère (*Eccrinella corophii* n. sp., *Palavascia spheromae* Tuz. et Man., *Toeniella carcini* Lég. et Dub., *Arundinula* sp.). Cahiers Biol. Marine 2: 313–326.

Manier, J.–F. 1962a. Présence de Trichomycètes dans le rectum des larves d'Éphémères des torrents du Massif du Néouvieille (Hautes–Pyrénées). Bull. Soc. Hist. Nat. Toulouse 97: 241–254.

Manier, J.–F. 1962b. Révision du genre *Spartiella* Tuzet et Manier 1950 (sa place dans la classe des Trichomycètes). Ann. Sci. Nat., Zool., Ser. 12, 4: 517–525.

Manier, J.–F. 1962c. État actuel de la connaissance des Trichomycètes (1962). Pp. 201–210, Groupement des Protistologues de langue Française.

Manier, J.–F. 1963a. Trichomycètes parasites d'Isopodes Oniscoidea. Ann. Sci. Nat., Bot., Ser. 12, 4: 557–577.

Manier, J.–F. 1963b. Trichomycètes de larves de Simulies (Harpellales du proctodeum). Ann. Sci. Nat., Bot., Ser. 12, 4: 737–750.

Manier, J.–F. 1964a. Position systématique des Trichomycètes. Arch. Zool. Exp. Gen. 104: 95–98.

Manier, J.–F. 1964b. Nouvelle contribution à l'étude des Trichomycètes (Eccrinales parasites d'Amphipodes). Ann. Sci. Nat., Bot., Ser. 12, 5: 767–772.

Manier, J.–F. 1964c. *Orchesellaria mauguioi* n. sp., Trichomycète Asellariale parasite du rectum de *Isotomurus palustris* (Müller) 1776, (Insecte Aptérygote Collembole). Rev. Ecol. Biol. Sol 1: 443–449.

Manier, J.–F. 1964d. Endophytes parasites d'Arthropodes cavernicoles récoltés dans des grottes de l'Ariège et de la Haute–Garonne. Ann. Spéléologie 19: 803–812.

Manier, J.–F. 1964e. Un groupe the Thallophytes parasite d'Arthropodes Mandibulates: les Trichomycètes. Proc. 1st. Int. Congr. Parasitol., pp. 593–594.

Manier, J.–F. 1965. Les Amoebidiales Protistes de position systématique incertaine. Exc. Med. Int. Congr. Ser. No. 91, paper no. 246, 2nd Int. Conf. Protozool. (Abstr.).

Manier, J.–F. 1968. Validation de Trichomycètes par leur diagnose latine. Ann. Sci. Nat., Bot., Ser. 12, 9: 93–108.

Manier, J.–F. 1969a (1970a). Changement de nom pour *Eccrina flexilis* Léger et Duboscq, 1906. Ann. Sci. Nat., Bot., Ser. 12, 10: 469–471.

Manier, J.–F. 1969b (1970b). Trichomycètes de France. Ann. Sci. Nat., Bot., Ser. 12, 10: 565–672.

Manier, J.–F. 1970. Sur la fréquence de Trichomycètes Eccrinales dans le proctodeum des Myriapodes Diplopodes. Bull. Mus. Nat. Hist. Natur. 41: 91–95.

Manier, J.–F. 1973a. L'ultrastructure de la trichospore de *Genistella ramosa* Léger et Gauthier, Trichomycète Harpellale parasite du rectum des larves de *Baetis rhodani* Pict. C. R. Acad. Sci. Paris 276: 2159–2162.

Manier, J.–F. 1973b. Quelques aspects ultrastructuraux du Trichomycète Asellariale, *Asellaria ligiae* Tuzet et Manier, 1950 ex Manier, 1968. C.R. Acad. Sci. Paris 276: 3429–3431.

Manier, J.–F. 1978. Mycoses de Crustacés. Arch. Inst. Pasteur Tunis 55: 401–417.

Manier, J.–F. 1979a. Étude ultrastructurale de *Palavascia sphaeromae* (Trichomycète Eccrinale) parasite du proctodeum de *Sphaeroma serratum* (Crustacé Isopode). Ann. Parasitol. (Paris) 54: 537–554.

Manier, J.–F. 1979b. *Orchesellaria podurae* n. sp. (Trichomycète, Asellariale) parasite de *Podura aquatica* L. (Insecte, Apterygote, Collembole). Rev. Mycol. 43: 341–350.

Manier, J.–F., and F. Coste. 1971. Trichomycètes Harpellales de larves de Diptères Chironomidae; création de cinq nouvelles espèces. Bull. Soc. Mycol. France 87: 91–99.

Manier, J.–F., and F. Coste-Mathiez. 1968. L'ultrastructure du filament de la spore de *Smittium mucronatum* Manier, Mathiez 1965 (Trichomycète, Harpellale). C.R. Acad. Sci. Paris 266: 341–342.

Manier, J.–F., and H. Grizel. 1971. *Paramacrinella microdeutopi* n. g., n. sp., Trichomycète parasite de *Microdeutopus anomalus* H. Rathke (Amphipode). Ann. Sci. Nat., Bot., Ser. 12, 12: 1–8.

Manier, J.–F., and H. Grizel. 1972. L'ultrastructure de l'enveloppe et du "pavillon" des Trichomycètes Eccrinales. C.R. Acad. Sci. Paris 274: 1159–1160.

Manier, J.–F., and R.W. Lichtwardt. 1968 (1969). Révision de la systématique des Trichomycètes. Ann. Sci. Nat., Bot., Ser. 12, 9: 519–532.

Manier, J.–F., and F. Mathiez. 1965. Deux Trichomycètes Harpellales Genistellacées, parasites de larves de Chironomides. Ann. Sci. Nat., Bot., Ser. 12, 6: 183–196.

Manier, J.–F., and R. Ormières. 1961a. *Ramacrinella raibauti* n. g. n. sp. Eccrinide ramifié commensal de l'intestin postérieur de *Microdeutopus gryllotalpa* A. Costa (Amphipodes—Aoridae). Ann. Sci. Nat., Bot., Ser. 12, 2: 625–634.

Manier, J.–F., and R. Ormières. 1961b. *Alacrinella limnoriae* n. g., n. sp. Trichomycète Eccrinidae parasite du rectum de *Limnoria tripunctata* Menziès (Isopode). Vie Milieu 12: 285–295.

Manier, J.–F., and R. Ormières. 1962. *Arundinula galatheae* n. sp. et *Toeniella galatheae* n. sp. Trichomycètes Eccrinacées parasites de *Galathea strigosa* L. (Crustacés Decapodes). Vie Milieu 13: 453–466.

Manier, J.–F., and R. Ormières. 1980. Champignons du stomodeum des Myriapodes. Ann. Sci. Nat., Zool., Ser. 13, 2: 151–165.

Manier, J.–F., and A. Raibaut. 1969. Cycle biologique du Trichomycète *Amoebidium parasiticum* (Cienkowski). 16–mm film. Service du Film de Recherche Scientifique, Paris.

Manier, J.-F., and A. Raibaut. 1970. Évolution des kystes de *Amoebidium parasiticum* Cienkowski, 1861 (Trichomycète, Amoebidiale). Bull. Soc. Zool. France 95: 31–33.

Manier, J.-F., and J. Théodoridès. 1957. Eccrinida d'un *Gargilius* sp. (Coléoptère Ténébrionide). Nat. Albert Park, Ser. 2, 5: 3–6.

Manier, J.-F., and J. Théodoridès. 1965. À propos d'une Eccrinale parasite de Coléoptère Passalide du Laos. Ann. Parasitol. (Paris) 40: 497–504.

Manier, J.-F., M. Akbarieh, and G. Bouix. 1976. *Coelosporidium chydoricola* Mesnil et Marchoux, 1897: observations ultrastructurales, données nouvelles sur le cycle et la position systématique. Protistologica 12: 599–612.

Manier, J.-F., C. Gasc, and G. Bouix. 1972a. *Enterobryus tuzetae* n. sp. (Trichomycètes—Eccrinales) de l'intestin postérieur de *Pachybolus ligulatus* (Voges) (Diplopodes—Spirobolidae) récoltés au Dahomey (Afrique). Biol. Gabonica 3–4: 305–322.

Manier, J.-F., C. Gasc, and G. Bouix. 1972b. *Mononema demangei* Thallophyte de l'oesophage de *Orthomorpha coarctata* (Saussure) et de *Cordyloporus ornatus* (Peters) Myriapodes Polydesmides du Dahomey. Biol. Gabonica 3–4: 323–331.

Manier, J.-F., C. Gasc, and G. Bouix. 1974 (1975). Sur quelques *Enterobryus* (Trichomycètes Eccrinales) parasites de Myriapodes Diplopodes du Sud–Dahomey. Bull. Inst. Fondamen. Afr. Noire 36: 614–641.

Manier, J.-F., J.-A. Rioux, and B. Juminer. 1964. Présence en Tunisie de deux Trichomycètes parasites de larves de Culicides. Arch. Inst. Pasteur Tunis 41: 147–152.

Manier, J.-F., J.-A. Rioux, and H.C. Whisler. 1961. *Rubetella inopinata* n. sp. et *Carouxella scalaris* n. g., n. sp., Trichomycètes parasites de *Dasyhelea lithotelmatica* Strenzke, 1951 (Diptera Ceratopogonidae). Nat. Monspeliensia, Ser. Bot., 13: 25–38.

Manier, J.-F., J.-A. Rioux, and H.C. Whisler. 1965. Validation du genre *Carouxella* et de l'espèce–type *Carouxella scalaris* Manier, Rioux et Whisler, 1961. Nat. Monspeliensia 16: 87.

Manier, J.-F., C. Vago, G. Devauchelle, and J.-L. Duthoit. 1971. Infection virale chez les Trichomycètes. C.R. Acad. Sci. Paris 273: 1241–1243.

Mattson, R.A., and D. (Wagner–Merner) TeStrake. 1983. Trichomycetes in Brachyurans (true crabs) from Tampa Bay, Florida. 3rd Int. Mycol. Congr., Tokyo. Abstracts, p. 546.

Mayfield, S. Dang, and R.W. Lichtwardt. 1980. Comparative study of the holdfast structure in four Trichomyctes. Can. J. Bot. 58: 1074–1087.

McCloskey, L.R., and S.P. Caldwell. 1965. *Enteromyces callianassae* Lichtwardt (Trichomycetes, Eccrinales) in the mud shrimp *Upogebia affinus* (Say). J. Elisha Mitchell Sci. Soc. 81: 114–117.

Mercier, L. 1914. Sur un Protophyte du rectum d'*Oniscus asellus* L. C.R. Soc. Biol. 76: 600–602.

Mesnil, F., and E. Marchoux. 1897. Sur un Sporozoaire nouveau (*Coelosporidium chydoricola* n. g. n. sp.) intermédiaire entre les Sarcosporidies et les *Amoebidium* Cienkowsky. C.R. Soc. Biol. 4: 839–841.

Moniez, R. 1887. Sur des parasite nouveaux des Daphnies. C.R. Acad. Sci. Paris 104: 183–185.

Moore–Landecker, Elizabeth. 1982. Fundamentals of the fungi, 2nd ed. Prentice–Hall, Inc., New Jersey.

Moss, S.T. 1970. Trichomycetes inhabiting the digestive tract of *Simulium equinum* larvae. Trans. Brit. Mycol. Soc. 54: 1–13.
Moss, S.T. 1972. Occurrence, cell structure and taxonomy of the Trichomycetes, with special reference to electron microscope studies of *Stachylina*. Ph.D. Dissertation, University of Reading. 304 pp.
Moss, S.T. 1974. A note on the nuclear cytology of *Stachylina grandispora* (Trichomycetes, Harpellales). Mycologia 66: 173–178.
Moss, S.T. 1975. Septal structure in the Trichomycetes with special reference to *Astreptonema gammari* (Eccrinales). Trans. Brit. Mycol. Soc. 65: 115–127.
Moss, S.T. 1976. Formation of the trichospore appendage in *Stachylina grandispora* (Trichomycetes). Pp. 279–294 *in* R. Fuller and D.W. Lovelock (eds.), Microbial ultrastructure. The use of the electron microscope. Academic Press, New York.
Moss, S.T. 1979. Commensalism of the Trichomycetes. Pp. 175–227 *in* L.R. Batra (ed.), Insect–fungus symbiosis. Nutrition, mutualism, and commensalism. Allanheld, Osmun & Co., Montclair.
Moss, S.T., and R.W. Lichtwardt. 1976. Development of trichospores and their appendages in *Genistellospora homothallica* and other Harpellales and fine-structural evidence for the sporangial nature of trichospores. Can. J. Bot. 54: 2346–2364.
Moss, S.T., and R.W. Lichtwardt. 1977. Zygospores of the Harpellales: an ultrastructural study. Can. J. Bot. 55: 3099–3110.
Moss, S.T., and R.W. Lichtwardt. 1980. *Harpella leptosa,* a new species of Trichomycetes substantiated by electron microscopy. Can. J. Bot. 58: 1035–1044.
Moss, S.T., and T.W.K. Young. 1978. Phyletic considerations of the Harpellales and Asellariales (Trichomycetes, Zygomycotina) and the Kickxellales (Zygomycetes, Zygomycotina). Mycologia 70: 944–963.
Moss, S.T., R.W. Lichtwardt, and J.–F. Manier. 1975. *Zygopolaris,* a new genus of Trichomycetes producing zygospores with polar attachment. Mycologia 67: 120–127.
Müller, E., and W. Loeffler. 1976. Mycology: an outline for science and medical students. Translated by B. Kendrick and F. Bärlocher. Georg Thiem Publishers, Stuttgart.
Müller–Kögler, E. 1971. Ein Beitrag zur axenishchen Kultur von zwei insektenbewohnenden Trichomyceten. Entomophaga 16: 5–9.
Patrick, M.A., V.K. Sangar, and P.R. Dugan. 1973. Lipids of *Smittium culisetae*. Mycologia 65: 122–127.
Peterson, S.W. 1984. Systematic studies of the Harpellales (Trichomycetes) from winter–emerging stoneflies (Plecoptera). Ph.D. Dissertation, University of Kansas. 122 pp.
Peterson, S.W., and R.W. Lichtwardt. 1983. *Capniomyces stellatus* and *Simuliomyces spica:* new taxa of Harpellales (Trichomycetes) from winter–emerging stoneflies. Mycologia 75: 242–250.
Peterson, S.W., R.W. Lichtwardt, and B.W. Horn. 1981. *Genistelloides hibernus:* a new Trichomycete from a winter–emerging stonefly. Mycologia 73: 477–485.
Poisson, R. 1927. Sur une Eccinide nouvelle: *Taeniellopsis orchestiae* nov. gen., nov. sp., Protophyte parasite du rectum de l'*Orchestia bottae* M. Edw. (Crust. Amphipode). Son cycle évolutif. C.R. Acad. Sci. Paris 185: 1328–1329.
Poisson, R. 1928. *Eccrinopsis mercieri* n. sp., Eccrinide parasite du rectum de l'*Oniscus asellus* L. Son cycle évolutif. C.R. Acad. Sci. Paris 186: 1765–1767.

Poisson, R. 1929. Recherches sur quelques Eccrinides parasites de Crustacés Amphipodes et Isopodes. Arch. Zool. Exp. Gen. 69: 179–216.
Poisson, R. 1931a. Recherches sur les Eccrinides. Deuxième contribution. Arch. Zool. Exp. Gen. 74: 63–68.
Poisson, R. 1931b. À propos du cycle évolutif des *Amoebidium* (Eccrinideae Amoebidina). C.R. Soc. Biol. 106: 354–358.
Poisson, R. 1932a. *Asellaria caulleryi* n. g., n. sp., type nouveau d'Entophyte parasite intestinal des Aselles (Crustacés Isopodes). Bull. Biol. France Belgique 66: 232–254.
Poisson, R. 1932b. Sur deux Entophytes parasites intestinaux de larves de Diptères. Ann. Parasitol. (Paris) 10: 435–443.
Poisson, R. 1936. Sur un Endomycète nouveau: *Smittium arvernense* n. g., n. sp., parasite intestinal de larves de *Smittia* sp. (Diptères Chironomides) et description d'une nouvelle espèce du genre *Stachylina* Lég. et Gauth. 1932. Pp. 75–86 *in* Mélanges dédiés au Professeur Lucien Daniel. Université de Rennes.
Porter, D., and R. Smiley. 1979. Ribosomal RNA molecular weights of Trichomycetes and Zygomycetes. Exp. Mycol. 3: 188–193.
Pouzar, Z. 1972. *Genistella* Léger et Gauthier vs. *Genistella* Ortega; a nomenclatural note. Folia Geobot. Phytotax., Praha 7: 319–320.
Preisner, T.R. 1973. Studies on the morphology and ultrastructure of vegetative hyphae, trichospores, and trichospore development of *Smittium* spp. and *Genistellospora homothallica*. Ph.D. Dissertation, University of Kansas. 212 pp.
Raabe, H. 1911a. *Amoebidium parasiticum* Cienk. Cześć I. Jadro, budowa jego i podzial. Sprawozdania Towarz. Nauk. Warsz. 4: 229–252.
Raabe, H. 1911b. *Amoebidium parasiticum* Cienk. Cześć II. Cialka metachromatyczne. Sprawozdania Towarz. Nauk. Warsz. 4: 252–263.
Raabe, H. 1912. Les divisions du noyau chez *Amoebidium parasiticum* Cienk. Arch. Zool. Exp. Gen., Ser. 5, 10: 371–398.
Rajagopalan, C. 1967. An *Enterobryus* (Trichomycetes, Eccrinales) in a milliped. Curr. Sci. 36: 20–22.
Reichle, R.E. 1978. *Enterobryus* sp., Eccrinales, Trichomycetes. Pp. 169–172 *in* M.S. Fuller (ed.), Lower fungi in the laboratory. University of Georgia, Athens.
Reichle, R.W., and R.W. Lichtwardt. 1972. Fine structure of the Trichomycete, *Harpella melusinae*, from black–fly guts. Arch. Mikrobiol. 81: 103–125.
Reynolds, D.R. 1967. New records of Philippine fungi. Philip. Agr. 50: 784–790.
Roberts, D.W., and M.A. Strand (eds.). 1977. Pathogens of medically important arthropods. Suppl. No. 1–Vol. 55, Bull. World Health Organ. 419 pp.
Robin, C. 1853. Histoire naturelle des végétaux parasites qui croissent sur l'homme et sur les animaux vivants. Vol. 1, pp. 395–404, Vol. 2, pp. 9–10. J. B. Baillière, Paris.
Sangar, V.K. 1969. Immunological and electrophoretic studies on the fungal genus *Smittium* (Trichomycetes). Ph.D. Dissertation, University of Kansas. 59 pp.
Sangar, V.K., and P.R. Dugan. 1973. Chemical composition of the cell wall of *Smittium culisetae* (Trichomycetes). Mycologia 65: 421–431.
Sangar, V.K., R.W. Lichtwardt, J.A.W. Kirsch, and R.N. Lester. 1972. Immunological studies on the fungal genus *Smittium* (Trichomycetes). Mycologia 64: 342–358.
Scheer, D. 1935. Vorläufige Mitteilung über einen Pilz aus dem Darm von *Cambarus affinis* Say. Zool. Anzeig. 109: 268–269.
Scheer, D. 1944. Ein neuer parasitärer Pilz aus dem Darm der Wasserassel (*Asellus aquaticus* L.). Z. Parasitenk. 13: 275–282.

Scheer, D. 1972a. Eingliederung des Pilzes *Recticharella aselli* Scheer 1944 in die Asellariaceae (Eccrinales, Endomycetes). Arch. Protistenk. 114: 343–348.

Scheer, D. 1972b. Über Pilze (Asellarien) aus dem Darm von Wasserasseln des Süsswassers. Z. Binnenfischerei 12: 369–373.

Scheer, D. 1976a. Der wahre Wirt von *Astreptonema longispora* Hauptfleisch (Trichomycetes, Eccrinales) und die Konsquenzen aus seiner Ermittelung. Arch. Protistenk. 118: 11–17.

Scheer, D. 1976b. *Parataeniella mercieri* (Poisson) (Trichomycetes, Eccrinales) und ihre Wirte in der Deutschen Demokratischen Republik. Arch. Protistenk. 118: 202–208.

Scheer, D. 1977. *Nodocrinella hylonisci* n. g., n. sp., eine neue Eccrinacee (Trichomycetes, Eccrinales) aus dem Darm von *Hyloniscus riparius* (C.L. Koch) (Crustacea, Isopoda). Arch. Protistenk. 119: 163–177.

Schenk, A. 1858. Ueber parasitische Schläuche auf Crustaceen. Pp. 252–259 *in* Physikalisch–Medizinische Gesellschaft, Vol. 8, Würzburg.

Singh, K.R.P., and A.W.A. Brown. 1957. Nutritional requirements of *Aedes aegypti* L. J. Ins. Physiol. 1: 199–220.

Sörgel, G. 1953. Zur Systematik der Trichomyceten. Arch Mikrobiol. 18: 391–396.

Starr, A.M. 1976. Sterol analysis of cultured Trichomycetes. Master of Arts Thesis, University of Kansas. 92 pp.

Starr, A.M., R.W. Lichtwardt, J.D. McChesney, and T.A. Baer. 1979. Sterols synthesized by cultured Trichomycetes. Arch. Microbiol. 120: 185–189.

Steelman, C.D. 1976. Effects of external and internal arthropod parasites on domestic livestock production. Ann. Rev. Entomol. 21: 155–178.

Sweeney, A.W. 1981a. An undescribed species of *Smittium* (Trichomycetes) pathogenic to mosquito larvae in Australia. Trans. Brit. Mycol. Soc. 77: 55–60.

Sweeney, A.W. 1981b. Fungal pathogens of mosquito larvae. Pp. 403–424 *in* E.W. Davidson (ed.), Pathogenesis of invertebrate microbial diseases. Allanheld, Osmun & Co., Totowa, New Jersey.

Taylor, W.R. 1928. Observations on *Amoebidium parasiticum* Cienkowski. J. Elisha Mitchell Sci. Soc. 44: 126–132.

Taylor, W.R., and H.S. Colton. 1928. The phytoplankton of some Arizona pools and lakes. Amer. J. Bot. 15: 596–614.

Thaxter, R. 1920. Second note on certain peculiar fungus–parasites of living insects. Bot. Gaz. 69: 1–27.

Théodoridès, J. 1955. Contribution à l'étude des parasites et phoretiques de Coleoptères terrestre. Vie Milieu, Suppl. No. 4, pp. 9–312.

Thomas, L.J. 1930. *Rhigonema nigella* spec. nov., a nematode and its plant commensal, *Enterobrus* sp? from the milliped. J. Parasitol. 17: 30–34.

Torrey, G.S. 1954. The classification of the Eccrinales, fungi *incertae sedis*, and their descriptive terminology. Pp. 80–82 *in* Communications parvenus avant le Congrès aux Sections 18, 19, et 20, 8me Congrès International de Botanique, Paris.

Trager, W. 1935. The culture of mosquito larvae free from living microorganisms. Amer. J. Hyg. 22: 18–25.

Trotter, M.J., and H.C. Whisler. 1965. Chemical composition of the cell wall of *Amoebidium parasiticum*. Can. J. Bot. 43: 869–876.

Tschudovskaia, I. 1928. Über einige Parasiten aus dem Darmkanal der *Sciara*–Larven. Arch. Protistenk. 60: 287–304.

Tuzet, O., and J.-F. Manier. 1947a. *Orphella culici* n. sp., Entophyte parasite du rectum des larves de *Culex hortensis* Fclb. C.R. Acad. Sci. Paris 225: 264–266.

Tuzet, O., and J.-F. Manier. 1947b. *Palavascia philoscii* n. g. n. sp., Entophyte eccriniforme parasite de *Philoscia Couchii* Kin. C.R. Acad. Sci. Paris 224: 1854–1856.

Tuzet, O., and J.-F. Manier. 1948a. La sexualité et les spores durables des Eccrinides du genre *Enterobryus*. C.R. Acad. Sci. Paris 226: 1312–1314.

Tuzet, O., and J.-F. Manier. 1948b. La reproduction sexuée chez *Palavascia philoscii* Tuzet et Manier et chez *Palavascia sphaeromae,* nouvelle espèce de Palavasciées parasite de *Sphaeroma serratum* F. C.R. Acad. Sci. Paris 226: 2177–2178.

Tuzet, O., and J.-F. Manier. 1949. Les Eccrinides du genre *Enterobryus*. P. 212 in XIIIe Congrès International de Zoologie, Paris, 1948.

Tuzet, O., and J.-F. Manier. 1950a. Les Trichomycètes. Revision de leur diagnose. Raisons qui nous font y joindre les Asellariées. Ann. Sci. Nat., Zool., Ser. 11, 12: 15–23.

Tuzet, O., and J.-F. Manier. 1950b. *Lajassiella aphodii* n. g., n. sp. Palavascide parasite d'une larve d'*Aphodius* (Coléoptère Scarabaeidae). Ann. Sci. Nat., Zool., Ser. 11, 12: 465–470.

Tuzet, O., and J.-F. Manier. 1951a. Sur quelques Eccrinides du Brésil. Ann. Sci. Nat., Zool., Ser. 11, 13: 145–147.

Tuzet, O., and J.-F. Manier. 1951b. Le cycle de l'*Amoebidium parasiticum* Cienk. Revision du genre *Amoebidium*. Ann. Sci. Nat., Zool., Ser. 11, 13: 351–364.

Tuzet, O., and J.-F. Manier. 1952. Trichophytes commensaux de l'intestin postérieur de Diplopodes du Brésil. Quelques considérations sur les Trichophytes déja décrits infestant les Diplopodes. Ann. Sci. Nat., Zool., Ser. 11, 14: 249–262.

Tuzet, O., and J.-F. Manier. 1953. Recherches sur quelques Trichomycètes rameux. *Asellaria armadillidii* n. sp. *Genistella choanifera* n. sp. *Genistella chironomi* n. sp. *Spartiella barbata* Tuzet et Manier. Ann. Sci. Nat., Zool., Ser. 11, 15: 373–391.

Tuzet, O., and J.-F. Manier. 1954a. Importance des cultures de Trichomycètes pour l'étude du cycle et de la classification de ces organismes. C.R. Acad. Sci. Paris 238: 1904–1905.

Tuzet, O., and J.-F. Manier. 1954b. Trichomycètes commensaux de l'intestin postérieur de Myriapodes Diplopodes récoltés dans la forêt de la Mandraka (Madagascar). Mem. Inst. Sci. Madagascar, Ser. A, 9: 1–13.

Tuzet, O., and J.-F. Manier. 1955a. Étude des Trichomycètes de l'intestin des larves de *Simulium equinum* Linné récoltés aux Eyzies (Dordogne). Ann. Sci. Nat., Zool., Ser. 11, 17: 55–62.

Tuzet, O., and J.-F. Manier. 1955b. Sur deux nouvelles espèces de Génistellales: *Genistella rhitrogenae,* n. sp., et *Genistella mailleti* n. sp., observées dans les larves de *Rhitrogena alpestris* Eat. et *Boetis bioculatus* L. récoltés aux Eyzies (Dordogne). Ann. Sci. Nat., Zool., Ser. 11, 17: 67–71.

Tuzet, O., and J.-F. Manier. 1957a. Troisième contribution à la connaissance des Eccrinida commensaux de l'intestin postérieur de Myriapodes Diplopodes du Brésil. Révision des Eccrinida déja identifiés chez les Diplopodes. Arch. Zool. Exp. Gen. 94: 121–147.

Tuzet, O., and J.-F. Manier. 1957b. Écologie parasitaire chez *Glomeris marginata* Villers. Vie Milieu 8: 58–71.

Tuzet, O., and J.-F. Manier. 1962. *Enteromyces callianassae* Lichtwardt Trichomycète Eccrinale commensal de l'estomac de *Uca pugilator* Latreille. Ann. Sci. Nat., Bot., Ser. 12, 3: 615–617.

Tuzet, O., and J.-F. Manier. 1967. *Enterobryus oxidi* Lichtwardt, Trichomycète Eccrinale parasite du Myriapode Diplopode *Oxidus gracilis* (Koch) (cycle, ultrastructure). Protistologica 3: 413–421.

Tuzet, O., J.-F. Manier, and P. Jolivet. 1957. Trichomycètes monoaxes et rameux de l'intestin postérieur de Polydesmida, Spirostreptida et Spirobolida. Nat. Alberta Park, Ser. 2, 5: 21–38.

Tuzet, O., J.-F. Manier, and M. Vogeli–Zuber. 1952. Sur quelques parasites intestinaux de *Mardonius piceus* Attems 1952, Myriapode–Diplopode de Daloa (Côte d'Ivoire). Bull. Inst. Fr. Afrique Noire 14: 1143–1151.

Tuzet, O., J.-F. Manier, and M. Vogeli–Zuber. 1953. Trichophytes et Ciliés parasites intestinaux de *Pachybolus* sp., *Scaphiostreptus obesus* Attems et *Termatodiscus nimbanus* Attems (Myriapodes Diplopodes) récoltés par l'expédition Française au Mont Nimba (Guinée) en Août 1951. Bull. Inst. Fr. Afrique Noire 15: 133–142.

Tuzet, O., J.-A. Rioux, and J.-F. Manier. 1961. *Rubetella culicis* (Tuzet et Manier 1947), Trichomycète rameux parasite de l'ampoule rectale des larves de Culicides (morphologie et spécificité). Vie Milieu 12: 167–187.

Udekem, J.d'. 1859. Notice sur quelques parasites du *Julus terrestris*. Bull. Acad. Sci. Lettr. Belgique 7: 552–567.

Voss, E.G. 1973. General Committee Report, 1970–1971. Taxon 22: 153–163.

Whisler, H.C. 1960. Pure culture of the Trichomycete, *Amoebidium parasiticum*. Nature 186: 732–733.

Whisler, H.C. 1961. Cultural studies of the Trichomycetes. Ph.D. Dissertation, University of California, Berkeley. 141 pp.

Whisler, H.C. 1962. Culture and nutrition of *Amoebidium parasiticum*. Amer. J. Bot. 49: 193–199.

Whisler, H.C. 1963. Observations on some new and unusual enterophilous Phycomycetes. Can. J. Bot. 41: 887–900.

Whisler, H.C. 1966. Host–integrated development in the Amoebidiales. J. Protozool. 13: 183–188.

Whisler, H.C. 1968. Developmental control of *Amoebidium parasiticum*. Develop. Biol. 17: 562–570.

Whisler, H.C. 1978. *Amoebidium parasiticum*. Pp. 165–166 *in* M.S. Fuller (ed.), Lower fungi in the laboratory. University of Georgia, Athens.

Whisler, H.C. 1979. The fungi versus the arthropods. Pp. 1–32 *in* L.R. Batra (ed.), Insect–fungus symbiosis. Nutrition, mutualism, and commensalism. Allanheld, Osmun & Co., Montclair.

Whisler, H.C., and M.S. Fuller. 1968. Preliminary observations on the holdfast of *Amoebidium parasiticum*. Mycologia 60: 1068–1079.

Williams, M.C. 1971. Studies on Trichomycetes and their relationships to arthropod hosts. Ph.D. Dissertation, University of Kansas. 102 pp.

Williams, M.C. 1982. *Smittium cellaspora*, a new Harpellales (Trichomycetes) from a chironomid hindgut. Mycotaxon 16: 183–186.

Williams, M.C. 1983a. Spore longevity of *Smittium culisetae* (Harpellales, Legeriomycetaceae). Mycologia 75: 171–174.

Williams, M.C. 1983b. Zygospores in *Smittium culisetae* (Trichomycetes) and observations on trichospore germination. Mycologia 75: 251–256.

Williams, M.C., and R.W. Lichtwardt. 1971. A new *Pennella* (Trichomycetes) from *Simulium* larvae. Mycologia 63: 910–914.

Williams, M.C., and R.W. Lichtwardt. 1972a. Infection of *Aedes aegypti* larvae by axenic cultures of the fungal genus *Smittium* (Trichomycetes). Amer. J. Bot. 59: 189–193.

Williams, M.C., and R.W. Lichtwardt. 1972b. Physiological studies on the cultured Trichomycete, *Smittium culisetae*. Mycologia 64: 806–815.

Williams, M.C., and R.W. Lichtwardt. 1984. Two *Stachylina* and two *Smittium* species (Trichomycetes) from Montana. Mycologia 76: 204–210.

Williams, M.C., R.W. Lichtwardt, and S.W. Peterson. 1982. *Smittium longisporum*, a new Harpellales (Trichomycetes) from chironomid guts. Mycotaxon 16: 167–171.

Williams, M.C., and H.G. Nagel. 1980. Occurrence of Trichomycete fungi in mosquito larvae near Kearney, Nebraska. Mosquito News 40: 445–447.

Wolf, F.A., and F.T. Wolf. 1947. The fungi. Vols. I and II. John Wiley & Sons, New York.

Wright, K.A. 1979. Trichomycetes and oxyuroid nematodes in the millipede, *Narceus annularis*. Proc. Helminthol. Soc. Wash. 46: 213–223.

Young, T.W.K. 1969. Ultrastucture of aerial hyphae in *Linderina pennispora*. Ann. Bot. 33: 211–216.

Glossary

The words below are defined in terms of their usage in this book. See the Index for references or illustrations that appear in the text.

amoebagenesis	Development of amoeboid cells in a species of Amoebidiales.
appendages	Nonmotile, elongated, and often very fine structures attached to the end(s) of spores. In the Harpellales, one or more appendages are formed within the generative cell, and these remain attached to the base of the trichospore after its release. In some species of Eccrinales, released sporangiospores have either one or two appendages at both poles.
arthrospore	A specialized cell functioning as a spore and derived from the disarticulation of cells of a formerly vegetative branch.
axenic	Living without the presence of other organisms. May refer to a pure culture of one organism, or to an organism that has no microorganisms either internally or externally.
basipetal	Development from the apical part of a thallus or branch toward the base.
ectocommensal	A commensal living on the outer surface of its host.
endocommensal	A commensal living within its host.
coenocytic	Containing many nuclei within one cell or nonseptate thallus.
collar	An outgrowth from a generative cell that remains attached to the base of certain trichospores after their detachment.

commensal	An organism that benefits from its association with another organism but neither benefits nor harms the host.
conjugant	Either cell of a pair that conjugates (copulates), a process which eventually produces a zygospore.
cuticle	In the gut of arthropods, a lining secreted by epithelial cells of the foregut and hindgut that consist of chitinous and nonchitinous materials. Also refers to the outer layers of nematode bodies and to the exoskeleton of arthropods.
cyst	A stage developing from amoebae in the Amoebidiales consisting of a thick-walled cell within which cystospores are produced.
dimorphic	Having two distinct growth forms.
eccrinid	Any species of Eccrinales.
ecdysis	The developmental process of shedding the exoskeleton, including cuticular linings of the gut and respiratory system, in arthropods.
eucarpic	Only part of the thallus converts to reproductive structures.
exuvia	In arthropods, the cast-off exoskeleton (see ecdysis).
generative cell	A specialized cell that produces a trichospore externally and within which one or more trichospore appendages may be formed.
harpellid	Any species of Harpellales.
heterothallic	A condition in which sexual reproduction will not occur except with another compatible thallus.
holdfast	A specialized structure that attaches the thallus to the substrate. It may consist of a secreted substance with a characteristic shape or the entire basal cell of the thallus (the *holdfast cell*) modified for attachment.
holocarpic	Entire thallus converts to reproductive structures.
homothallic	A condition in which sexual reproduction can proceed within one thallus without requiring a sexual partner.
hypha	One of the filaments of a fungus.
lentic	Pertaining to standing or still waters.
lotic	Pertaining to flowing waters, such as a stream.

Malpighian tubes [or tubules]	A set of excretory diverticula located at the anterior end of the hindgut of insects and some other arthropods.
molt	In arthropods, to shed the exoskeleton and any associated cuticular linings during ecdysis. Also, as a noun, refers to the shed skin (exoskeleton) or exuvia.
multivoltine	Reproducing several times each year.
nomen dubium [*nom. dub.*]	A name which, although technically valid, cannot be satisfactorily typified due to an insufficient description or questionable interpretation of the fungal material described.
nomen nudum [*nom. nud.*]	A name that has not been validly published according to the International Code of Botanical Nomenclature.
peritrophic membrane	A thin, tubular membrane of chitin and protein continuously secreted by a ring of cells surrounding the anterior end of the midgut of an insect, serving to protect delicate epithelial cells of the midgut from abrasion.
primary infestation sporangiospore	A nonsexual spore serving to disseminate species of Eccrinales from one host to another. The spore is uni–, bi– or quadrinucleate, and thick walled in most species.
resistant spore	A thick–walled spore believed to remain viable under adverse conditions, such as desiccation.
secondary infestation sporangiospore	A nonsexual spore of the Eccrinales that germinates within the same gut where it was produced, thus serving to increase the population of thalli in that gut. The spore is usually thin walled and contains more than one nucleus at maturity.
septum	A crosswall in a thallus.
shaken cultures	Cultures grown in liquid medium contained in a flask and kept on an orbital or reciprocal shaker throughout the period of growth.
spore mother–cell	A germinated sporangiospore of an Eccrinales which has given rise to a new thallus. It is usually seen at the apex of the developing thallus, and may persist to maturity of the thallus or it may be ephemeral and disappear during thallial growth.

sporangiospore	A nonsexual spore produced within a sporangium.
substrate	Material on or in which an organism is growing or to which it is attached.
thallus	Entire vegetative (assimilative) body of a fungus or other thallophyte.
trichospore	A deciduous sporangium of the Harpellales that usually has one or more appendages attached to its base, serving to disseminate the fungus from host to host. It contains a single uninucleate sporangiospore, and develops externally as an outgrowth from a generative cell.
univoltine	Reproducing only once a year.
zygospore	A thick-walled sexual spore resulting from the fusion of two nuclei (karyogamy) usually preceeded by the fusion of two cells (plasmogamy).
zygosporophore	A specialized branch that bears a zygospore.

Index

Abscission vacuole, 277
Acaeroplastes melanurus, 201
Aedes, 181
   aegypti, 193
   albopictus, 311
   detritus, 33
   melanimon, 310
   sticticus, 311
   vexans, 182, 311
Agrenia bidenticulata, 208
Alacrinella, 211
   key to species, 211
   limnoriae, 212–213
   sanjuanensis, 213
Allocapnia, 156, 173, 310, 312
   granulata, 156, 158
   rickeri, 158, 312
   vivipara, 156, 158, 310
Alloporus, 299
Amoebagenesis, 100–101, 126–127
Amoebidiaceae, 262
   key to genera, 262
Amoebidiales, 262
   asexual reproduction, 86–87
   key to genera, 262
   life cycle, 46–47
   list of species, 296
Amoebidium, 263
   cienkowskianum, 269, 300
   crassum, 269, 300
   fasciculatum, 263
   key to species, 263
   moniezi, 269
   parasiticum, 64, 263–265, 312

   poissoni, 263
   recticola, 265–266
Andohaheloa, 269
   pauliani, 269, 300
Androniscus dentiger, 261
Apheloria
   iowa, 229
   montana, 229
Aphodius, 259, 306
Apical spore bodies, 75–76, 121
Appendage
   formation, 60–74
   function, 76
   phylogeny, 276–277
   types, 71
Anopheles, 181
   annulipes, 186
   hilli, 186, 188, 312
Armadillidium
   nasatum, 260
   simoni, 201
   vulgare, 30, 201, 260
Armadillo officinalis, 261
Arthromitus, 284–285
Arthropods, *see also* Hosts
   collection of, 11–14
   dissection techniques, 16–19
   habitats, 29–34
   list of species, 297–309
   maintenance and preservation, 14–16
Arundinella capitata, 214–215
Arundinula, 213–214
   capitata, 214–215

Arundinula (cont.)
  galatheae, 215
  haplogaster, 216
  incurvata, 270, 303
  key to species, 214
  orconectis, 216–217
  porcellanae, 270
  washingtoniensis, 217–218
Asellaria, 200
  armadillidii, 201–202
  aselli, 56, 202–203
  caulleryi, 202, 203
  gramenei, 202, 203–204
  key to species, 200–201
  ligiae, 59, 202, 204–206
  scutigerae, 270
  spirostrepti, 270, 299
  unguiformis, 202, 206
Asellariaceae, 200
  key to genera, 200
Asellariales, 199
  asexual reproduction, 77–78
  key to genera, 200
  life cycle, 43–44
  list of species, 294
Asellus
  aquaticus, 202, 203
  meridianus, 203, 204
Asexual reproduction, 68–87
  in Amoebidiales, 86–87
  in Asellariales, 77–78
  in Eccrinales, 78–86
  in Harpellales, 69–77
Astreptonema, 218
  corophii, 219
  gammari, 219–221
  key to species, 218–219
  longispora, 221–222
  pacificum, 222
  siphonoecetis, 270
  typica, 222
Austrosimulium, 143
  australense, 193

Baetis, 163, 304
  atrebatinus, 304
  bicaudatus, 198
  bioculatus, 165, 304
  gemellus, 194
  parvus, 198
  pumilus, 157
  rhodani, 165, 194
  tricaudatus, 162, 163, 198
Basidiolum fimbriatum, 265
Beach hoppers, list of species, 301
Beetles, 31
  list of species, 306–307
Blackflies
  larvae, 31
  list of species, 308
Bloodworm, 31
Boraria carolina, 233
Boreoheptagyia lurida, 152, 183
Brachyiulus pusillus, 298
Brachyspirobolus, 298

Caddis worms, 306
Caecidotea laticauda, 206
Californiobolus uncigerus, 56, 232
Callianassa
  brachyophthalma, 247
  californiensis, 30, 247, 252
  gigas, 247, 252
  uncinata, 246
Callibaetis pacificus, 171
Cambarus affinis, 302
Cancer oregonensis, 252
Capillus, 270
  intestinalis, 270, 298
  iuli, 270, 298
Capnia bifrons, 158
Capniomyces, 156
  stellatus, 156, 310
Carcinus maenas, 252
Carouxella, 140
  scalaris, 140–142
Centroptilum luteolum, 161
Cercyon obsoletus, 306
Cestodella, 270
  allopori, 270, 299
  attenuata, 270, 299
  glandulosa, 270, 299
  operculata, 270, 299, 300
  pachyboli, 270, 298
  parva, 270, 299
  pseudonannolenis, 270, 299
  rhinocrici, 270, 299
  straeleni, 270, 299

Chaetarthria siminulum, 306
Chaetophiloscia elongata, 201
Cherokia georgiana, 234
Chironomus, 31, 147, 149, 180, 182, 193, 264, 312
  alternans, 311
  hawaiiensis, 149
  plumosus, 194, 307
  plumosus complex, 147, 193
  zealandicus, 149
Chloroperla, 306
Cingalobolus carli, 234
Cladostreptus castaneus, 299
Cloeon dipterum, 171
Cnephia, 143
Coelostoma orbiculare, 306
Coemansia sporocladia, 276
Cordyloporus ornatus, 286
Corophium volutator, 219
Crabs
  fiddler, 30
  list of species, 303–304
Crayfish, list of species, 302
Cricotopus, 147, 184, 185, 196
Cryptops
  anomolans lusitanicus, 286
  hortensis, 286
Culex, 181, 311
  halifaxii, 193
  pervigilans, 311
Culiseta, 181
  impatiens, 181, 311
  incidens, 181, 311
Cultures of fungi
  abnormalities in, 130–131
  axenic methods, 21–25
  conditions for growth, 107–110
  experimental uses, 105–131
  list of axenic isolates, 310–312
  media, 22
  storage, 25
Cylindroiulus
  friseus, 298
  londinensis, 235
  nitidus, 298

Daloala, 270
  mardonii, 270, 299

Daphnia
  magna, 265, 266
  pulex, 265
Dasyhelea, 140, 142, 182, 311
  lithotelmatica, 140, 142, 182
Diamesa, 150, 152, 176, 189, 312
  nivoriunda, 184
  valkanovi, 196, 197
Dictyosomes, 51
Dicyrtocercyon ustulatus, 306
Dissection techniques, 16–19
Distribution
  geographic, 35–37
  mechanisms, 101–104
Dixa, 308
Dixidesmus tallulanus, 235
Dixidium, 270
  dixae, 270, 308
Dragonflies, list of species, 305
Drunella
  coloradensis, 31
  spinifera, 165

Eccrina, 270
  brevis, 270, 298
  flexilis, 223
  gigantea, 270, 299
  longa, 270, 299
  longipes, 270, 298
  moniliformis, 242
  montana, 270
  pseudoramosa, 270, 297
  uncigeri, 270, 298
Eccrinaceae, 210
  generic characters, 212
  key to genera, 210–211
  life cycle, 45
Eccrinales, 209
  asexual reproduction, 78–86
  key to families, 138–139
  life cycles, 44–46
  list of species, 294–296
Eccrinella, 270
  corophii, 219
  gammari, 219
Eccrinidus, 222
  flexilis, 64, 223–224
Eccrinoides, 224
  broelemanni, 224–225

Eccrinoides (cont.)
  helleriae, 225
  henneguyi, 225–226
  key to species, 224
  monticolae, 226
  stammeri, 270, 297
Eccrinopsis, 270
  attenuatus, 230
  helleriae, 225
  hydrophilorum, 240
  leidyi, 270, 306
  ligidii, 270, 301
  mercieri, 261
  monticolae, 226
  sciarae, 270, 308
Ecdyonurus, 305
Ecdysis, effects on fungi, 98–101
Echinogammarus berilloni, 219
Electron microscopy techniques, 20–21
Elliptra astigmatica, 193
Emerita
  analoga, 239
  talpoida, 239
Enterobryus, 56, 80, 226–227
  adjanohouni, 228–229
  ahlesi, 229
  allobrogicus, 271, 298
  apheloriae, 229–330
  attenuatus, 64, 230–232
  bifurcatus, 232
  borariae, 59, 61, 136, 232–233
  brachyspiroboli, 271, 298
  bresiliensis, 271, 297
  broelemanni, 271
  cherokiae, 233–234
  cingaloboli, 234
  compressus, 234–235
  cylindroiuli, 235
  dixidesmi, 235–236
  doscari, 271, 298
  duboscqui, 271, 298
  elegans, 64, 67, 230, 236–238
  eurydesmi, 271, 297
  euryuri, 53, 78, 230, 238–239
  fertilis, 271, 297
  flavus, 243, 271, 298
  gracilis, 271, 299
  halophilus, 239–240
  helocharei, 271, 306
  hyalinus, 271, 299
  hydrophilorum, 240
  inflatus, 271, 298
  isoporostrepti, 240
  iuli-terrestris, 271, 298
  key to species, 227–228
  leptodesmi, 271, 297
  leptoiuli, 241–242
  moniliformis, 78, 242
  nudatus, 271, 298
  onychostrepti, 242–243
  orthomorphae, 271, 298
  oxidi, 243
  oxydesmi, 244
  pennatus, 271, 298
  pentodoni, 271, 307
  peridontopygei, 244
  philhydri, 271
  pseudoeurydesmi, 271, 297
  rectus, 271, 298
  robini, 271, 298
  sao-pauloi, 271, 297
  schubarti, 271, 297
  seychelloboli, 271, 299
  spiralis, 271, 298
  spirostrepti, 271, 300
  strongylosomae, 271, 298
  tumidus, 271, 299
  tuzetae, 244–246
  vulgaris, 271, 299
Enteromyces, 246
  callianassae, 246–248
Enteropogon, 248
  sexuale, 248–249
Epeorus longimanus, 198
Ephemerella, 198
  inermis, 198
  infrequens, 165
Eukiefferiella, 191
Eupagurus
  cuanensis, 214
  excavatus, 252
  prideuxi, 303
Eurydesmus acutus, 297
Euryurus erythropygus, 238
Evolutionary relationships
  among trichomycetes, 279–284
  with other fungi, 275–279
Excluded and doubtful taxa, 269–273

Exosphaeroma
   amplicauda, 211, 218
   oregonensis, 257

Fairy shrimp, 300
Flies, list of species, 307–309
Fungi, list of species, 291–296

Gaetice depressus, 252
Galathea strigosa, 215, 252
Gammarus
   duebeni, 222
   locusta, 221, 222
   pulex, 219, 221
   roeselii, 219, 220
Gauthieromyces, 157
   microsporus, 157
Genistella, 164
   chironomi, 178
   choanifera, 271, 307
   mailleti, 272, 304
   microspora, 157
   ramosa, 165
   rhitrogenae, 272, 305
Genistellaceae, 152
Genistelloides, 157
   hibernus, 158–159, 310
Genistellospora, 159
   homothallica, 50, 64, 73, 75, 159–160
Geographic distribution, 35–37
Ghost shrimps, 30
   list of species, 303
Globotherium actaeon, 299
Glomeris
   annulata, 223
   connexa, 223, 297
   conspersa, 223
   hexasticha, 223, 297
   marginata, 223
Glotzia, 161
   centroptili, 161
   ephemeridarum, 56, 161–162
   key to species, 161
Glycogen, 52
Graminella, 162–163
   bulbosa, 163
   key to species, 163
   microspora, 163–164

Gribbles, list of species, 302
Growth of fungi *in vitro*, 107–110
   growth curves, 107–109
   pH values, 109–110
   temperature effects, 108–109

Habitats of arthropods, 29–34
Habrophlebia fusca, 305
Halophiloscia couchii, 256
Haplogaster mertensii, 216
Harpella, 142
   key to species, 142
   leptosa, 64, 142–143
   melusinae, 54, 56, 64, 74, 143–144
      var. *eyziesi*, 144
Harpellaceae, 139
   key to genera, 139–140
   trichospores, 141
   zygospores, 141
Harpellales, 139
   appendage types, 71
   asexual reproduction, 69–77
   key to families, 138
   life cycle, 43, 44
   list of species, 291–294
   zygospore types, 89–90
Harpellomyces, 145
   eccentricus, 145–146
Harpochytrium, 278, 285
Helleria brevicornis, 225
Helochares, 306
   lividus, 306
Hemigrapsus
   nudus, 252
   oregonensis, 252
   penicillatus, 247
Historical résumé, 6–10
Holdfasts, 55–67
   attachment to non-cuticle substrates, 60–62
   definition of, 55
   mucilaginous secretion, 57
   multiple system, 55, 57
   penetration of gut lining, 58–59
   ultrastructure, 63–67
   variations in, 56
Hosts, *see also* Arthropods
   experimental infestation, 38–41, 124–126

Hosts (*cont.*)
  fungal effects on development, 112–116
  fungal specificity in, 38–42
  killed by fungal growth, 97–98, 114, 115
Hydrobius fuscipes, 306
Hydrochus, 306
Hydroisotoma schaefferi, 209
Hydrous
  flavipes, 240
  piceus, 240
  pistaceus, 240
Hyloniscus riparius, 302

Ischiotrichus fossulatus, 299
Isoporostreptus bouixi, 241
Isotoma, 208
Isotomurus palustris, 208

Julus
  albipes, 298
  marginatus, 236
  scandinavius, 298
  terrestris, 298

Kickxellales, 275–279

Labyrinthiform organelle, 277, 278
Laccobius nigriceps, 306
Lactella, 272
  aphodii, 272, 306
  cercyonis, 272, 306
  chaetarthriae, 272, 306
  coelostomatis, 272, 306
  helocharidis, 272, 306
  laccobii, 272, 306
  philhydri, 272, 306
  platystethi, 272, 307
Lajasiella, 258
  aphodii, 259
Lectin, 62
Legeriomyces, 164
  aenigmaticus, 164–165
  key to species, 164
  ramosus, 165–166

Legeriomycetaceae, 152
  key to genera, 152–156
  trichospores, 154
  zygospores, 155
Leptaulax dentatus, 231
Leptodesmus
  gasparae, 297
  juncudus, 297
  paulistus, 297
  ramosus, 297
  vagans, 297
Leptoiulus belgicus, 242
Ligia, 204, 205
  exotica, 30, 204
  italica, 204
Ligidium hypnorum, 301
Limnoria
  lignorum, 213
  tripunctata, 213
Lipids, 52
  production of, 112
Lirceus hoppinae, 206
Loboglomeris
  pyrenaica, 226
  rugifera, 226

Mardonius piceus, 299
Maruina, 149
Mayflies
  larva, 31
  list of species, 304–305
Media
  for culturing fungi, 22
  for culturing mosquito larvae, 113
    vitamin-deficient, 113–116
    sterol-deficient, 113–116
Mesocapnia, 158
Microasellaria, 272
  funicularia, 272, 299
Microdeutopus
  anomalus, 250
  gryllotalpa, 251
Microeccrina, 272, 285
  fertilis, 272
  glomeri, 272
  iuli, 272
  leptophylli, 272
  ligidii, 272
  orthomorphae, 272

parva, 272
siccophila, 272
Microscopy, preparation of specimens, 19–21
Microtrichella, 272
hydrophilorum, 272, 285
Midges
larva, 31
list of species, 307–309
Millipedes, 30
list of species, 297–300
Mononema, 286–287
Mosquitoes, list of species, 308
Mud shrimps, list of species, 303

Narceus, 236
americanus, 30, 236
annularis, 236
Nematodes, attachment of thalli, 60–62, 66
Nemura
cinerae, 167
variegata, 167, 269
Nephelopteryx nebulosa, 306
Nodocrinella, 272
hylonisci, 272, 302
Nuclei, 52–54
Nutrition of fungi *in vitro*, 106–107

Oedignathus inermis, 252
Oedogoniomyces, 285–286
Oesophagomyces lithobii, 287
Omphalocystis plateaui, 287
Oniscus asellus, 261
Onychostreptus
aoutii, 243
assiniensis, 243
Opuntiella, 272
digitata, 272, 307
Orchesella villosa, 207
Orchesellaria, 206
key to species, 206–207
lattesi, 207
mauguioi, 207–208
pelta, 207, 208–209
podurae, 207, 209
Orchestia
bottae, 254

mediterranea, 255
traskiana, 222
Orconectis nais, 216
Orphella, 166
coronata, 166–167
culici, 179
Orthocladius, 189, 196, 197, 312
rubicundus, 190
Orthomorpha
coarctata, 286, 298
gracilis, 243
Oxidus gracilis, 243
Oxydesmus granulosus, 244

Pachybolus, 298
ligulatus, 245
Pachyiulus communis, 298
Pagastia, 184
Paguristes turgidus, 217, 252
Pagurus
berlinganus, 252
ganosimanus, 252
kennerlyi, 252
maculatus, 214
spinimanus, 215
Palavascia, 255
beauforti, 256, 257
key to species, 256
philosciae, 256
sphaeromae, 56, 64, 83, 256–258
Palavasciaceae, 255
life cycle, 45
Palloiulus distinctus, 298
Panopeus herbstii, 240
Paraenterobryus multiformis, 248
Paraleptophlebia, 305
Paramacrinella, 250
microdeutopi, 250
Paramoebidium, 266–267
arcuatum, 272, 304
chattoni, 272, 308
curvum, 64, 93, 267–268
dispersum, 272, 305
eccriniformis, 272, 305
fuscum, 272, 306
geniculatum, 272, 305
giganteum, 272, 306
inflexum, 268–269
key to species, 267

Paramoebidium (cont.)
  pavillardi, 273, 304
  procloeoni, 273, 305
  simulii, 273, 308
  thrauli, 273, 305
Parataeniella, 78, 259
  armadillidii, 260
  binucleata, 261
  dilatata, 261
  intermedia, 261
  key to species, 259–260
  mercieri, 261–262
  scotonisci, 262
Parataeniellaceae, 258
  key to genera, 258
  life cycle, 45
Paratendipes, 147
Paratrichella, 273
  pentodona, 273
Passalus cornutus, 231
Pathogenicity, 97–98
Pennella, 167
  angustispora, 64, 168
  arctica, 56, 168–169
  grassei, 169–170
  hovassei, 170
  key to species, 167
  simulii, 170–171
Pentodon punctatus, 307
Perapolydesmus progressus, 286
Peridontopyge gasci, 244
Philhydrus, 306
  testaceus, 306
Phylogeny, 275–284
Pill bugs, 30
  list of species, 301
Pilumnus hirtellus, 252
Pistillaria, 273
  bresiliensis, 273
  flavus, 273
  plagiodesmi, 273, 298
Plagiodesmus, 298
  aotypus, 298
  occidentalis, 298
Platystethus cornutus, 307
Plethocrossus acutiformis, 229
Podura aquatica, 209
Polydesmus
  complanatus, 286, 299
  denticulatus, 299

  granulatus, 242
  rubromarginatus, 299
  virginiensis, 299
Polypedilum, 147, 151
Popilius disjunctus, 31, 231
Porcellio
  laevis, 201, 261
  lamellatus, 201, 261
  monticola, 226
Portunus puber, 252
Procladius, 192
Procloeon bifidum, 305
Prosimulium, 143, 159, 170, 310
  exigens, 168, 191
  ferrugineus, 168
  onychodactylum, 191
Protoglomeris vasconica, 225
Protonemura humeralis, 167
Protracheoniscus occidentalis, 201
Psectrocladius sordidellus, 187, 312
Psectrotanypus, 147
  varius, 177
Pseudochironomus, 151
Pseudoeurydesmus baguassuensis, 297
Pseudokiefferiella, 312
Pseudonannolene tricolor, 299
Pteromaktron, 171, 282
  protrudens, 171–172, 276

*R*amacrinella, 250
  raibauti, 251
Recticharella, 203
  aselli, 202–203
Recticoma, 273
  cambari, 273, 302
Relationships
  host-fungus, 95–104
  nutritional, 96
Rhabdomyces lobjoyi, 287
Rhinocricus, 299
  divaricatus, 299
Rhitrogena alpestris, 305
Ribosomal RNA, 129–130, 277
Rock louse, 30
  list of species, 301
Rubetella, 273
  culicis, 179
  inopinata, 181, 182
  simulii, 192, 193

Sand fleas, list of species, 301
Scaphiostreptus sjoestedti, 299
Schizophyllum, 298
　sabulosum, 298
Sciara, 308
Scotoniscus marcomelos, 262
Scuds, list of species, 301
Scytonotus granulatus, 242
Sechellobolus dictyonotus, 299
Septum
　in Asellariales, 50–21
　in Coemansia aciculifera, 275
　in Eccrinales, 51, 283
　in Harpellales, 50–51, 283
　perforate, 50–51, 275, 283
Serology, 128–129, 276
Sesarma hematocheir, 252
Sexual reproduction, 87–93
　in Eccrinales, 92–93
　in Harpellales, 87–92
Simuliomyces, 172
　key to species, 172
　microsporus, 61, 172–183
　spica, 173
Simulium, 143, 159, 168, 170, 180, 191, 193, 267, 311, 312
　arcticum, 168
　argus, 168, 193, 312
　bezzii, 195
　canonicolum, 142
　defoliarti, 191
　equinum, 169, 195, 308
　monticola, 170
　ornatum, 195
　variegatum, 195
　venustum, 170
　venustum complex, 142
　virgatum, 168
　vittatum, 31, 168, 170, 180, 182, 312
Smittia, 177
Smittium, 174, 312
　arcticum, 176
　alpinum, 176
　arvernense, 176–177
　bisporum, 177
　cellaspora, 177–178
　chironomi, 64, 178
　culicis, 179–181, 310–311
　culisetae, 70, 115, 121, 179, 181–183, 311
　dimorphum, 183–184
　elongatum, 184
　external attachment, 115
　gigasporus, 184
　incrassatum, 184–185
　inopinatum, 181, 182
　key to species, 174–176
　longisporum, 185
　macrosporum, 185–186
　megazygosporum, 186
　morbosum, 186–187, 188–189, 312
　mucronatum, 187, 189, 312
　orthocladii, 189–190
　ouseli, 190–191
　pennelli, 191–192
　pusillum, 70, 192
　simulii, 24, 76, 116, 192–193, 312
　typhellum, 193–194
Sow bugs, list of species, 302
Spartiella, 194
　barbata, 194–195
Sphaeroma
　quadridentatum, 257
　serratum, 257
Spirobolus, 236
　americanus, 236
Spirostreptus, 300
　fossulatus, 299
　ibanda, 299
　virgator, 299
Sporangiospore
　development, 78–86
　primary infestation, 78–79, 81–84
　secondary infestation, 79–81
Spore mother-cell, 80, 81
Springtails, list of species, 304
Stachylina, 146
　chironomidarum, 147
　euthena, 147
　grandispora, 64, 147–149
　intermedia, 150
　key to species, 146–147
　longa, 149
　lotica, 148, 149
　macrospora, 150
　manicata, 148, 150–151
　minuta, 151
　nana, 151
　pedifer, 59, 151
　penetralis, 148, 152

Sterol
  deficiency in larvae, 113–114
  production by fungi, 110–112
Stipella, 195
  vigilans, 195–196
Stoneflies
  larva, 31
  list of species, 305–306
Strongylosoma guerini, 298
Stypella, 195
Survival mechanisms, 101–104
Sympotthastia, 178
Syncrotopus rufiventris, 150, 186
Syndiamesa macronyx, 150

Taeniella, 251–252
  carcini, 252–253
  galatheae, 252
  grandis, 252
  longa, 252
Taeniellopsis, 253
  flexilis, 254
  key to species, 253
  orchestiae, 254
  susplugasi, 255
Tanytarsus, 147, 149, 178
Taxonomic problems, 135–137
Tecticeps japonicus, 257
Tendipes, 147
Thallus
  attachment to non-cuticle substrates, 60–62
  development, 48–53
  dimorphism, 48
  penetration of gut lining, 58–59
Thaumalea, 145
Thraulus bellus, 305
Tracheoniscus rathkii, 261
Trichella, 273
  attenuatus, 230
  coelostomatis, 273, 306
  compressus, 234–235
  helocharei, 273
  hydrobii, 273, 306
  hydrophilorum, 240
  leidyi, 273
  longa, 273, 306
  pentodoni, 273
  philhydri, 273, 306

  sciarae, 273
  stammeri, 273, 306
Trichellopsis, 273
  polydesmi, 273, 299
  schizophylli, 273, 298
Trichocera, 200, 309
Trichoceridium, 200, 273
  ramosum, 200, 273, 309
Trichomycetes
  description, 3–5, 139
  keys to orders and families, 138–139
  list of species, 291–296
  summary of orders, 5
Trichoniscus roseus, 261
Trichospore, see also Appendage
  attachment to algae, 76
  development, 69–77
  germination, 116–117, 120–125
  longevity, 124–125
  osmotolerance, 33
  production in vitro, 111, 116–120
    effect of glucose, 117
    effect of tryptone, 117–118
    pH values, 120
    temperature, 118–119
Trichozygospora, 196
  chironomidarum, 196–197
Trissocladius, 307
Typhella, 273
  chironomi, 178
  choanifera, 273

Uca
  crenulata, 240
  pugilator, 30, 240, 247
  pugnax, 240
  minax, 240
Unciger foetidus, 298
Upogebia
  affinis, 247
  pugettensis, 247, 249, 252

Virus-like particles, 93–94

Wall composition, 48–51, 127
Water fleas, list of species, 300

Xantho pilipes, 252

Zygomycetes, 128–129, 275, 277
Zygopolaris, 197
  borealis, 198
  ephemeridarum, 198–199
  key to species, 197
Zygospore
  development, 87–92
  of *Harpella melusinae,* 53, 54
  phylogeny, 277
  types, 89–90

DATE DUE			
AUG 1 1 87			
INTERLIBRARY LOAN			

DEMCO 38-297